Improving Product Reliability and Software Quality

Wiley Series in Quality & Reliability Engineering

Dr Andre Kleyner
Series Editor

The Wiley series in Quality & Reliability Engineering aims to provide a solid educational foundation for both practitioners and researchers in Q&R field and to expand the reader's knowledge base to include the latest developments in this field. The series will provide a lasting and positive contribution to the teaching and practice of engineering.

The series coverage will contain, but is not exclusive to,

- statistical methods;
- physics of failure;
- reliability modeling;
- functional safety;
- six-sigma methods;
- lead-free electronics;
- warranty analysis/management; and
- risk and safety analysis.

Wiley Series in Quality & Reliability Engineering

Improving Product Reliability and Software Quality
by Mark A. Levin, Ted T. Kalal, Jonathan Rodin
April 2019

Design for Safety
By Louis J Gullo, Jack Dixon
February 2018

Next Generation HALT and HASS: Robust Design of Electronics and Systems
by Kirk A. Gray, John J. Paschkewitz
May 2016

Reliability and Risk Models: Setting Reliability Requirements, 2nd Edition
by Michael Todinov
September 2015

Applied Reliability Engineering and Risk Analysis: Probabilistic Models and Statistical Inference
by Ilia B. Frenkel, Alex Karagrigoriou, Anatoly Lisnianski, Andre V. Kleyner
September 2013

Design for Reliability
by Dev G. Raheja (Editor), Louis J. Gullo (Editor)
July 2012

Effective FMEAs: Achieving Safe, Reliable, and Economical Products and Processes using Failure Mode and Effects Analysis
by Carl Carlson
April 2012

Failure Analyis: A Practical Guide for Manufacturers of Electronic Components and Systems
by Marius Bazu, Titu Bajenescu
April 2011

Reliability Technology: Principles and Practice of Failure Prevention in Electronic Systems
by Norman Pascoe
April 2011

Improving Product Reliability: Strategies and Implementation
by Mark A. Levin, Ted T. Kalal
March 2003

Test Engineering: A Concise Guide to Cost-effective Deign, Development and Manufacture
by Patrick O'Connor
April 2001

Integrated Circuit Failure Analysis: A Guide to Preparation Techniques
by Friedrich Beck
January 1998

Measurement and Calibration Requirements for Quality Assurance to ISO 9000
by Alan S. Morris
October 1997

Electronic Component Reliability: Fundamentals, Modelling, Evaluation, and Assurance
by Finn Jensen
November 1995

Improving Product Reliability and Software Quality

Strategies, Tools, Process and Implementation

Second Edition

Mark A. Levin
Teradyne, Inc.
California, USA

Ted T. Kalal
Retired
Texas, USA

Jonathan Rodin
Teradyne, Inc.
California, USA

This edition first published 2019
© 2019 John Wiley & Sons Ltd

Edition History
John Wiley & Sons, Ltd (1e, 2003)

Registered Offices
John Wiley & Sons, Inc., 111 River Street, Hoboken, NJ 07030, USA
John Wiley & Sons Ltd, The Atrium, Southern Gate, Chichester, West Sussex, PO19 8SQ, UK

Editorial Office
The Atrium, Southern Gate, Chichester, West Sussex, PO19 8SQ, UK

For details of our global editorial offices, customer services, and more information about Wiley products visit us at www.wiley.com.

Wiley also publishes its books in a variety of electronic formats and by print-on-demand. Some content that appears in standard print versions of this book may not be available in other formats.

Library of Congress Cataloging-in-Publication Data

Names: Levin, Mark A., 1959- author. | Kalal, Ted T., author. | Rodin, Jonathan,
    1957- author.
Title: Improving product reliability and software quality : strategies,
    tools, process and implementation / Mark A. Levin, Teradyne, Inc.,
    California, USA, Ted T. Kalal (Retired), Texas, USA, Jonathan Rodin,
    Teradyne, Inc., California, USA.
Other titles: Improving product reliability
Description: 2nd edition. | Hoboken, NJ : John Wiley & Sons, Inc., [2019] |
    Revised edition of: Improving product reliability : strategies and
    implementation / Mark A. Levin and Ted T. Kalal. c2003. | Includes
    bibliographical references and index. |
Identifiers: LCCN 2018061430 (print) | LCCN 2019000421 (ebook) | ISBN
    9781119179412 (Adobe PDF) | ISBN 9781119179436 (ePub) | ISBN 9781119179399
    (hardcover)
Subjects: LCSH: Reliability (Engineering) | Manufacturing processes–Data
    processing. | Computer software–Evaluation.
Classification: LCC TS173 (ebook) | LCC TS173 .L47 2019 (print) | DDC
    620/.00452–dc23
LC record available at https://lccn.loc.gov/2018061430

Cover Design: Wiley
Cover Images: (top to bottom): © teekid/Getty Images, © ez_thug/Getty Images, © AK2/Getty Images,
Courtesy of Universal Robots/Teradyne Inc.

Set in 10/12pt WarnockPro by SPi Global, Chennai, India

Printed in Great Britain by TJ International Ltd, Padstow, Cornwall

10 9 8 7 6 5 4 3 2 1

Cary and Darren Kalal

To my beautiful wife, Dana Mischel Levin, for her endless love, support, and patience, and to our sons, Spencer Nathan Levin and Andrew Dylan Levin.

To Brigid, Sam, and Molly Rodin for their support and encouragement.

Contents

About the Authors

Mark A. Levin is the reliability manager at Teradyne, Inc. and is based in Agoura Hills, California. He received his bachelor of science degree in Electrical Engineering (1982) from the University of Arizona, a master of science degree in Technology Management (1999) from Pepperdine University, a master of science in Reliability Engineering (2009) from the University of Maryland, and all but dissertation for a PhD in Reliability Engineering from the University of Maryland. He has more than 36 years of electronics experience spanning the aerospace, defense, consumer, and medical electronics industries. He has held several management and research positions at Hughes Aircraft Missiles Systems Group, Hughes Aircraft Microwave Products Division, General Medical Company, and Medical Data Electronics. His experience is diverse, having worked in manufacturing, design, and research and development. He has developed manufacturing and reliability design guidelines, reliability training classes, workmanship standards, quality programs, JIT manufacturing, and ESD safe work environments, and has established a surface mount production facility. (Mark.levin@Teradyne.com)

Ted T. Kalal is a reliability engineer (now retired) who has gained much of his understanding of reliability from hands-on experience and from many great mentors. He is a graduate of the University of Wisconsin (1981) in Business Administration after completing much preliminary study in mathematics, physics, and electronics. He has held many positions as a contract engineer and as a consultant, where he was able to focus on design, quality, and reliability tasks. He has authored several papers on electronic circuitry and holds a patent in the field of power electronics. With two partners, he started a small manufacturing company that makes high-tech power supplies and other scientific apparatus for the bioresearch community.

Jonathan Rodin is a software engineering manager at Teradyne, Inc. A graduate of Columbia University (1981), Jon has 39 years of experience developing software, both working as a programmer and managing software development projects. His experience spans companies of many sizes, ranging from early stage startups to companies of greater than 100 000 employees. Prior to joining Teradyne, Jon held executive engineering management positions at FTP Software, NaviSite, and Percussion Software. He has led software process reengineering projects numerous times, most recently driving the effort to bring Teradyne's Semiconductor Test Division to CMMI Level 3.

List of Figures

List of Tables

Series Editor's Foreword

Engineering systems are becoming more and more complex, with added functions, capabilities and increasing complexity of the systems architecture. Systems modeling, performance assessment, risk analysis and reliability prediction present increasingly challenging tasks. Continuously growing computing power relegates more and more functions to the software, placing more pressure on delivering faultless hardware-software interaction. Rapid development of autonomous vehicles and growing attention to functional safety brings quality and reliability to the forefront of the product development cycle.

The book you are about to read presents a comprehensive and practical approach to reliability engineering as an integral part of the product design process. Various pieces of the puzzle, such as hardware reliability, physics of failure, FMEA, product validation and test planning, reliability growth, software quality, lifecycle engineering approach, supplier management and others fit nicely into a comprehensive picture of a successful reliability program.

Despite its obvious importance, quality and reliability education is paradoxically lacking in today's engineering curriculum. Few engineering schools offer degree programs or even a sufficient variety of courses in quality or reliability methods. Therefore, a majority of the quality and reliability practitioners receive their professional training from colleagues, engineering seminars, publications and technical books. The lack of formal education opportunities in this field greatly emphasizes the importance of technical publications, such as this one, for professional development.

We are confident that this book, as well as the whole series, will continue Wiley's tradition of excellence in technical publishing and provide a lasting and positive contribution to the teaching and practice of engineering.

Dr. Andre Kleyner
Editor of the Wiley Series in Quality & Reliability Engineering

Series Foreword Second Edition

There is a popular saying, "If you fail to plan, you are planning to fail." I don't know if there is another discipline in complex product development where this is more true than designing for product reliability. When products are simple, it is possible to achieve high reliability by observing good design practices, but as products become more complex, and include thousands of components and hundreds of thousands of lines of software, a systematic approach is required.

This has played itself out inside of Teradyne over the last decade through two product lines in our Semiconductor Test Division. One product line, the UltraFLEX Test System, was designed internally. Another, the ETS-800 Test System, was designed in a company that Teradyne acquired in 2008.

The UltraFLEX platform was designed using Teradyne's internal Design for Reliability standards. The principles embodied in those standards are described by the authors. We religiously used an approved parts list of qualified components and suppliers, we analyzed the electrical stress on every circuit, and we calculated predicted reliability for every instrument and the whole system. Once the system was fielded, we tracked MTBF and executed our failure response, analysis, and corrective action system (FRACAS) on repeat failure modes. The result is that the UltraFLEX platform, our most complex product, has a field reliability about three times higher than prior-generation products. What makes this more remarkable is that the UltraFLEX has the capability to test two or even four more semiconductor devices in parallel compared to prior testers.

During the development of the UltraFLEX and over the past decade, we also began to deploy and came to rely upon more formal methods to improve software reliability. To be frank, our organizational maturity in software reliability lagged behind our hardware best practices. But through the application of tools like defect models, and especially tracking the reliability of deployed software through automated quality monitors, we were able to both improve the quality of the deployed product and also improve our development methods. A key tool we use to evaluate software reliability is a metric we call *clean sessions*. A clean session is a session where an operator starts up the tester, loads a program, executes a task like developing tests, debugging, or just testing devices, finishes the task, and then unloads the program, without encountering any anomalous behavior. When we started tracking this metric at the launch of the UltraFLEX, only about half of the sessions were clean. It took us nearly five years to get to 95% clean sessions, and this has set a benchmark that our competitors struggle to reach. Through the learning achieved in this long struggle, we have been able to achieve 95% clean sessions within three months of the release of our next-generation product.

The ETS-800 is the next generation version of the successful tester for mixed signal and power devices. When Teradyne acquired the business in 2008, there was no formal reliability program in place, but their products were well regarded in the marketplace and reasonably reliable. The ETS-800 was a big step up in terms of capability from the prior generation. The instruments were two to four times as dense, and the system could support almost twice as many instruments. Further, the tester included a promising new feature that would greatly simplify customer test programs by providing the switching needed to share tester resources between different device pins.

From a functional and performance perspective, the ETS-800 was a fantastic success. A single ETS-800 could replace up to eight prior generation testers. But we found out the hard way that the informal approach to reliability that worked for simple products did not work for more complex ones. When we initially fielded the ETS-800, it was not a reliable tester. The weak link in the design was the inclusion of thousands of mechanical relays. These relays provided superior electrical performance, but are challenging to use from a reliability perspective. Mechanical relays are highly reliable if they are not hot switched, or switched while a current is flowing through the contacts. A hot-switching event causes an arc across the contacts surface that causes a rapid degradation to the contact surface and the life of the relay. If the relays were designed for reliability, the hot-switching event could have been avoided. The ETS 800 reliability was an order of magnitude below the much more complex UltraFLEX platform, and this put a blemish on the reputation we worked hard to develop for delivering highly reliable products.

We worked for a long time to try to improve the robustness of the relays, and reduce the occurrence of hot switching without making much progress. Ultimately we decided to redesign all of the instrumentation using guidelines from the Teradyne reliability system. We are just beginning the deployment of the redesigned instruments, but in side-by-side testing, they are demonstrating about 100 times higher reliability than the ones that they replace. It was a hard but effective lesson that a systematic approach to hardware reliability and software quality as the authors have described is the best way to achieve both high customer satisfaction and good profits.

Gregory S. Smith
President, Semiconductor Test Division
Teradyne, Inc.

Series Foreword First Edition

Modern engineering products, from individual components to large systems, must be designed and manufactured to be reliable. The manufacturing processes must be performed correctly and with the minimum of variation. All of these aspects impact upon the costs of design, development, manufacture, and use, or, as they are often called, the product's life cycle costs. The challenge of modern competitive engineering is to ensure that life cycle costs are minimized whilst achieving requirements for performance and time to market. If the market for the product is competitive, improved quality and reliability can generate very strong competitive advantages. We have seen the results of this in the way that many products, particularly Japanese cars, machine tools, earthmoving equipment, electronic components, and consumer electronic products have won dominant positions in world markets in the last 30–40 years. Their success has been largely the result of the teaching of the late W. E. Deming, who taught the fundamental connections between quality, productivity, and competitiveness. Today this message is well understood by nearly all the engineering companies that face the new competition, and those that do not understand lose position or fail.

The customers for major systems, particularly the US military, drove the quality and reliability methods that were developed in the West. They reacted to a perceived low achievement by imposing standards and procedures, whilst their suppliers saw little motivation to improve, since they were paid for spares and repairs. The methods included formal systems for quality and reliability management (MIL-Q-9858 and MIL-STD-758) and methods for predicting and measuring reliability (MIL-STD-721, MIL-HDBK-217, MILSTD781). MIL-Q-9858 was the model for the international standard on quality systems (ISO9000); the methods for quantifying reliability have been similarly developed and applied to other types of products and have been incorporated into other standards such as ISO60300. These approaches have not proved to be effective and their application has been controversial.

By contrast, the Japanese quality movement was led by an industry that learned how quality provided the key to greatly increased productivity and competitiveness, principally in commercial and consumer markets. The methods that they applied were based on an understanding of the causes of variation and failures, and continuous improvements through the application of process controls and the motivation and management of people at work. It is one of history's ironies that the foremost teachers of these ideas were Americans, notably P. Drucker, W.A. Shewhart, W.E. Deming, and J.R Juran.

These two streams of development epitomize the difference between the deductive mentality applied by the Japanese to industry in general, and to engineering in particular, in contrast to the more inductive approach that is typically applied in the West. The deductive approach seeks to generate continuous improvements across a broad front and new ideas are subjected to careful evaluation. The inductive approach leads to inventions and "break-throughs," and to greater reliance on "systems" for control of people and processes. The deductive approach allows a clearer view, particularly in discriminating between sense and nonsense. However, it is not as conducive to the development of radical new ideas. Obviously these traits are not exclusive, and most engineering work involves elements of both. However, the overall tendency of Japanese thinking shows in their enthusiasm and success in industrial teamwork and in the way that they have adopted the philosophies of western teachers such as Drucker and Deming, whilst their western competitors have found it more difficult to break away from the mold of "scientific" management, with its reliance on systems and more rigid organizations and procedures.

Unfortunately, the development of quality and reliability engineering has been afflicted with more nonsense than any other branch of engineering. This has been the result of the development of methods and systems for analysis and control that contravene the deductive logic that quality and reliability are achieved by knowledge, attention to detail, and continuous improvement on the part of the people involved. Therefore, it can be difficult for students, teachers, engineers, and managers to discriminate effectively, and many have been led down wrong paths.

In this series we will attempt to provide a balanced and practical source covering all aspects of quality and reliability engineering and management, related to present and future conditions, and to the range of new scientific and engineering developments that will shape future products. The goal of this series is to present practical, cost-efficient and effective quality and reliability engineering methods and systems.

I hope that the series will make a positive contribution to the teaching and the practice of engineering.

Patrick D.T. O'Connor
February 2003

Foreword First Edition

In my 26 years at Teradyne, I have seen the automated test industry emerge from its infancy and grow into a multibillion-dollar industry. During that period, Teradyne evolved into the world's leading supplier of automated test equipment (ATE) for testing semiconductors, circuit boards, modules, voice, and broadband telephone networks. As our business grew, the technology necessary to design ATE became increasingly complex, often requiring leading-edge electronics to meet customer performance needs. Our designs have pushed the envelope, demanding advancements in nearly every technological area including process capability, component density, cooling technology, ASIC complexity, and analog/digital signal accuracy.

Our customers, too, insist on the highest performance systems possible to test their products. But performance alone does not provide the product differentiation that wins sales. Customers also demand incomparable reliability. Revenue lost when an ATE system goes down can be staggering, often in the area of tens of thousands of dollars per hour. Furthermore, because of design complexity and system cost, the warranty cost to maintain these systems is increasing. Low reliability severely impacts the bottom line and impedes the ability to gain and hold market share.

To improve product reliability, changes had to be made to the reliability process. We learned that the process needed to be proactive. It had to start early in the product concept stage and include all phases of the product development cycle. In researching solutions for improving product reliability, we found the wealth of information available to be too theoretical and mathematically based. Clearly, we didn't want a solution that could only be implemented by reliability engineers and statisticians. If the training were overly statistical, the message would be lost. If the process required training everyone to become a reliability engineer, it would be useless. The process had to reduce technical reliability theory into practical processes easily understood by the product development team.

For the reliability program to be successful, we needed a way to provide both management and engineering with practical tools that are easily applied to the product development process. The reliability processes presented in this book achieve this goal.

The authors logically present the reliability processes and deliverables for each phase of the product development cycle. The reliability theory is thoughtful, easily grasped, and does not include a complex mathematical basis. Instead, concepts are described using simple analogies and practical processes that a competent product development team can understand and apply. Thus, the reliability process described can be implemented into any electronic or other business, regardless of its size or type, and ultimately helps give customers products with superior performance and superior reliability.

Edward Rogas Jr.
Senior Vice President
Teradyne, Inc.

Preface Second Edition

When this book was first published, the primary focus was on improving product reliability, why reliability improvement efforts fail, and how poor reliability negatively impacts current and future business. We discussed the ease with which consumers can research a product to determine consumer satisfaction and discover issues related to product reliability. To improve product reliability, we presented a comprehensive process for product development and an implementation strategy that any business can start. We also discussed ways to change the corporate culture so that it strives to design reliable products.

The importance of designing reliable products has not changed since the book was first published. However, much has changed in regard to the types of products being developed today compared to when the book was first published. The most significant change is the amount of software and firmware required for new products. The other significant change is the number of new products being developed that connect to the internet (IOT) to provide ease of use, communicate with other devices, aid in customer support and update software remotely. The internet provides the consumer with greater ease of use and a better user experience, but brings with it a new set of risks regarding security and privacy.

We changed the book title to *Improving Product Reliability and Software Quality* to convey the importance of software in product development. There are many books written about hardware reliability and likewise about software quality. The hardware reliability books do not cover software quality and the software quality books do not address hardware reliability. However, successful product development is dependent on the synergy of these two functional groups working well together. Hardware engineers and software engineers are very different and communicate in different languages; therefore, they do not effectively integrate each other's requirements and dependencies. Assumptions are often made regarding what other functional groups are delivering, which later turn out to be wrong.

Hardware and software engineers look at bugs very differently. The hardware development team strives to release products without any reliability issues and assumes last-minute discoveries will be fixed with software. Hardware requirements can be fully defined and validated to ensure the release of a reliable product. The software development team does not set a requirement for a 100% bug-free product before product release. In fact, for most products, the software requirements and validation cannot define every use condition and possible state. When the software is released, the team is already working on the next update, tier release, or patch.

In addition to software quality, there is also the issue of software security. Many new products access the Internet as a way to quickly and efficiently send out software bug fixes and as a way to improve customer use experience through user apps. A good example is the Nest™ programmable learning thermostat. This connectivity raises software security concerns and new challenges that are often overlooked or underestimated. Some products can communicate via Bluetooth and Global Positioning Services (GPS), which also have the potential to be compromised.

Each new generation of electronic products incorporates significantly more software and firmware than the previous generation. This goes for simple products like a home thermostat to complex ones. Even the mix of development engineers required for product development is shifting. The goal of the book is to provide insight, process, and tools to help meet these changing demands.

Preface First Edition

Nearly every day, we learn of another company that has failed. In the new millennium, this rate of failure will increase. Competitors are rapidly entering the marketplace using technology, innovation, and reliability as their weapons to gain market share. Profit margins are shrinking. Internet shopping challenges the conventional business model. The information highway is changing the way consumers make buying decisions. Consumers have more resources available for product information, bringing them new awareness about product reliability.

These changes have made it easier for consumers to choose the best product for their individual needs. As better-informed shoppers, consumers can now determine their product needs at any place, any time, and for the best price. The information age allows today's consumer to research an entire market efficiently at any time and with little effort. Conventional shopping is being replaced by "smart" shopping. And a big part of smart shopping is getting the best product for the best price.

As the sources for product information continue to increase, the information available about the quality of the product increases as well. In the past, information on product quality was available through consumer magazines, newspapers, and television. The information was not always current and often did not cover the full breadth of the market. Today's consumer is using global information sources and internet chat to help in their product-selection process. An important part of the consumer's selection process is information regarding a product's quality and reliability. Does it really do what the manufacturer claims? Is it easy to use? Is it safe? Will it meet customer expectations of trouble-free use? The list can be very long and very specific to the individual consumer.

Quality versus Reliability

From automobiles to consumer electronics, the list of manufacturers who make high-quality products is continuously evolving. Manufacturers who did not participate in the quality revolution of the last two decades were replaced by those that did. They went out of business because the companies with high-quality systems were producing products at a lower cost. Today, consumers demand products that not only meet their individual needs, but also continue to meet these needs over time. Quality design and manufacturing was the benchmark in the 1980s and 1990s; quality over time (reliability) is becoming the requirement in the twenty-first century. In today's marketplace, product quality is necessary in order to stay in business. In tomorrow's marketplace, reliability will be the norm.

Quality and reliability are terms that are often used interchangeably. While strongly connected, they are not the same. In the simplest terms:

- Quality is conformance to specifications.
- Reliability is conformance to specification *over time.*

As an example, consider the quality and reliability in the color of a shirt. In solid-color men's shirts, the color of the sleeves must match the color of the cuffs. They must match so closely that it appears that the material came from the same bolt of cloth. In today's manufacturing processes, several operations occur simultaneously. One bolt of cloth cannot serve several machines. The colors of several bolts of cloth must be the same, or the end product will be of poor quality. Every bolt of cloth has to match to a specified color standard, or the newest manufacturing technologies cannot be applied to the process. Quality in the material that goes into the product is as important as the quality that comes out. In fact, the quality that goes in becomes a part of the quality that comes out. After numerous washings, the shirt's color fades out. The shirt conformed to the consumer's expectations at the time of first use (quality), but failed to live up to the consumers' expectations later (reliability).

Reliability is the continuation of quality over time. It is simply the time period over which a product meets the standards of quality for the period of expected use. Quality is now the standard for doing business. In today's marketplace and beyond, reliability will be the standard for doing business. The quality revolution is not over; it has just evolved into the reliability revolution.

This book is an effort to guide the user on how to implement and improve product reliability with a product life cycle process. It is written to appeal to most types of businesses regardless of size. To achieve this, the beginning of each chapter discusses issues and principles that are common to all businesses, independent of size. We also segregate business into three categories based on size: small, medium, and large. Definitions are summarized in Table I.1.

The finance department can, more precisely, quantify the lost revenue due to warranty claims and poor quality. This loss represents the potential dollars that are recoverable "after" the reliability process improvements have been implemented and have begun to bear fruit.

Table I.1 Business size definition.

Metric	Company size		
	Small	Medium	Large
Employee count	<100	>100 and <1000	≥1000
Gross sales dollar	<$10 M	>$10 M and <$100 M	≥$100 M
Dollars available from the warranty budget (approx.)	<$1 M	$1 M to <$10 M	≥$10 M

Gaining Competitive Advantage

Manufacturers who have no reliability engineering in place typically have warranty costs as high as 10–12% of their gross sales dollar. A company that implements reliability into their processes can see warranty costs diminish to below 1% of the gross sales dollar. The total amount that can be recovered from the warranty budget represents the dollars that could be reinvested (from the warranty budget) or added to earnings. If research and development is 10% of the gross sales dollars, then the annual warranty dollar savings from reliability can cover the costs to develop future products. Of course, this only addresses the tangible benefits from a reliability program. There are many intangible benefits that are gained by improving product reliability. Examples include better product image, reduced time to market, lower risk of product recall and engineering changes, and more efficient utilization of employee resources. These intangible assets are addressed later in the book.

Acknowledgments

Special thanks to Dana Levin, Molly Rodin, Larry Steinhardt, Anto Peter, Ken Turner, Steve Hlotyak, Chris Behling, Thomas Mayberry, Joel Justin, Jim McLinn, Pat O'Connor, Steve King, Joel Justin, Kevin Giebel, and Debra Levin for their technical edits, and proofreading of the second edition. Thanks to Glenn Hemanes for his patience and help with some of the artwork for the second edition. Finally, we would like to thank Gregory Smith for the second edition Foreword and for supporting our work.

We would like to recognize and thank Harding Ounanian for his significant contribution in the first edit of the first edition of the book and to Ed Rogas for the first edition Foreword.

We would like to thank the following people who have brought a better awareness about reliability and continue to influence our way of thinking; Benton Au, Joe Denny, Dave Evans, Jim Galuska, Ray Hansen, Dr. Greg Hobbs, Jim McLinn, Pat O'Connor, Roy Porter, Dr. David Steinberg, and Michael Pecht.

Glossary

ALT	Accelerated life testing is a test designed to identify unintended early life wearout mechanisms.
AGC	Automatic gain control is a closed feedback loop that is used to regulate or control a signal level.
API	Application program interface is code that allows two software programs to communicate with each other.
AQL	Acceptable quality level is used to define what the lowest acceptable quality level is for a process.
ARG	Accelerated reliability growth is a process to accelerate product reliability growth rate.
ASIC	Application-specific integrated circuit is a custom-designed integrated circuit for an application specific use.
ATE	Automated test equipment is an automated test designed to perform device testing and evaluate/diagnose results to identify defective material.
Bathtub curve	A set of three curves that combine to illustrate the likely product failure rate, namely: "infant mortality (having a decreasing failure rate)," "random or constant failure rate," and "wearout (having an increasing failure rate)."
Benchmarking	The process of comparing the performance of something (e.g. process, technology, etc.) against a standard or against other equivalent objects.
BOM	Bill of materials is a list of material required for to build a product.
CDU	Coolant distribution unit is a cooling unit designed with a separate cooling loop that uses water to remove heat from a heat exchanger.

CFF	Conductive filament formation is an electrochemical process resulting in a conductive filament that grows in a printed circuit board and can result in an electrical short.
CMMI	The Capability Maturity Model Integration is a software development process framework, developed by the Software Engineering Institute (SEI), which describes elements and attributes of a comprehensive software development process.
COTS	Commercial off-the-shelf products (hardware or software) are readily available for sale to the general public.
CTE	Coefficient of thermal expansion is a measure of the amount matter changes in shape (x, y, and z) due to a change in temperature.
DFM	Design for manufacturing is the engineering practice of designing a product for ease of manufacturing from raw materials to a finished product.
DFR	Design for reliability is a proactive engineering process to design a product to meet defined reliability requirements for the useful life of the product.
DFS	Design for service (and maintainability) is the engineering practice of designing a product for ease of service to ensure that maintenance can be performed within the quality and time requirements.
DFT	Design for test is the engineering practice of designing a product for ease of testability during manufacturing to ensure finished products are within the product specifications and is defect free.
DTS	A defect tracking system, also known as a bug tracking system, is an application that records and monitors defect (bug) status.
DOE	Design of experiments is an applied statistical method involving planning, conducting, analyzing, and interpreting a set of controlled tests to evaluate the factors that govern the value of a parameter or parameters.
DVT	Design verification test is the process of verifying that a design meets specification.
DUT	Device under test refers to any electronic assembly, component, or device that is undergoing testing.
ECO	Engineering change order is the process of documenting a change at the component, module, or assembly level with detailed instructions to implement the change order.
ED	Escape or escaped defect. Any defect discovered by an end user after the software has been released.

ELT   Early life testing is a process during product development to identify unintended early life wearout mechanisms.

EOS   Electrical overstress is potential damage that occurs when electrical signals applied to a circuit or a device exceed the specification limits.

ESD   Electrostatic discharge is potential damage caused by a sudden flow of electrical current between two electrically charged objects that create an electrical short or dielectric breakdown.

ESS   Environmental stress screening is a process of exposing a component, device, or product to stresses such like thermal cycling and vibration in order to precipitate latent defects to manifest themselves by permanent or catastrophic failure during the stress screening process.

FA   Failure analysis is the process of collecting and analyzing data to determine the root cause of a failure.

FBD   A functional block diagram is a pictorial view that describes the functions and interrelationships of a system or process.

FIFO   First in, first out is a process to manage the inventory of goods, raw materials, parts, and components in a way that uses the oldest inventory first.

FIT   Failures in time/billion hours is a way to describe the failure rate or frequency with which an system, device, or component fails.

FMEA   Failure modes and effects analysis is a tool to identify potential failures that could occur and determine its severity, probability of occurrence, and likelihood of detection in a way that identifies the most significant and urgent risks issues to investigate.

FMMEA   Failure modes, mechanisms, and effects analysis is a tool to identify potential failures and associated mechanisms that could occur and determine its severity, probability of occurrence and likelihood of detection in way that identifies the most significant and urgent risks issues to investigate.

FRACAS   Failure reporting, analysis, and corrective action system is a process for continuous improvement involving collecting failure data, prioritizing and investigating the failures according to risk with the objective of determining root cause and implementing corrective actions to eliminate the failures from occurring in the future.

FRU   Field-replacement unit is any assembly, component or printed circuit board that can be quickly and easily repaired or replaced by the user or service technician without the need to send it to a repair facility.

FTA	Fault tree analysis is a graphical logic diagram used to identify potential causes of device or system failures.
GOBI	Get out of burn-in is a procedure to stop burn-in when it meets an accepted yield level.
HAL	A hardware abstraction layer (HAL) is a thin layer of software that exposes a functional interface for the hardware to hide the hardware details from the calling software.
HALT	Highly Accelerated Life Test is a step stress technique designed to discover weak links in a product.
HASA	Highly Accelerated Stress Audit is a method to switch from stress testing all production material to a sampling test to ensure product quality.
HASS	Highly Accelerated Stress Screen is an accelerated reliability screen that can reveal latent product defects that are process or a manufacturing related.
HAST	Highly Accelerated Stress Test is a highly accelerated stress test involving temperature and humidity.
HTOL	High-Temperature Operating Life is an accelerated reliability stress test applied to integrated circuits to determine their reliability.
ICM	Identify, communicate, and mitigate is a process to manage and mitigate technology risk.
IDE	Integrated development environment is a software development tool that combines editing, compiling, and debugging features. Example IDEs are Microsoft Visual Studio and Eclipse.
IOT	The Internet of Things is a general term used to refer to the plethora of devices that are connected to a network and controlled or monitored remotely via that network.
JIT	Just in time is way to manage inventory whereby materials, processes, and labor are scheduled to arrive when it is needed in the production process.
LOC	Line of code represents a single programming language statement of compilable code.
MRB	Material review board is a group of individuals that meet to determine the disposition of non-conforming material.
MTBF	Mean time between failures is the estimated time between failures for a repairable system that is calculated from the arithmetic mean time between all failures.
MTBI	Mean time between incident is a less commonly used term to describe the frequency of an event that may require user intervention to fix.

MTBM	Mean time between maintenance is used to describe the reliability of a repairable system. It is a measure of the average time between maintenance (preventive maintenance and repair) for all the systems in the field.
MTTF	Meat time to failure is a commonly used term to describe the reliability of a nonrepairable system.
MTTR	Mean time to repair is the sum of the time required to fix all failures in the fleet divided by the total number of failures in the fleet.
MTTRS	Mean time to restore system is similar to MTTR but includes the additional time associated with obtaining parts to fix the problem.
Next-generation product	A newer release of a product containing improved technology or features.
NIST	National Institute of Standards and Technology is a measurements standards laboratory.
NRE	Nonrecurring engineering are the one-time charges associated with research, design, and development for a new or derivative product.
Pareto chart	A type of chart showing a sorted ranking of occurrences with the most frequent on the left of the x-axis diminishing to the least frequent to the right.
PDCA	Plan-do-check-act is a four-step method used for continuous improvement of processes and products.
PHM	Prognostics and health management provides a process and methodology to compare real-time data to an expected good set of data so a determination can be made regarding its health.
POS	Proof of screen is a repetition of a HASS profile to determine if the precipitation stress is too severe or ineffective.
PPM	Parts per million.
QFD	Quality function deployment is the process of identifying all factors that might affect the ability of the product to satisfy customer needs and requirement.
Quality	Quality is conformance to specifications.
RAC	Reliability Analysis Center is funded by the US Department of Defense to provide technical expertise in the engineering disciplines of reliability, maintainability, supportability and quality.
RCA	Root cause analysis is a process by which weaknesses in the overall development process can be identified with the intent to improve them over time.

RDT	A reliability demonstration test is a process that statistically ensures that the reliability goal set early in the program is met.
Reliability	Reliability is the probability that a product will meet it requirements for the intended period of time and under specified operational conditions – i.e. quality over time.
RPN	Risk priority number is a numerical assessment of risk determined by taking the product of occurrence, severity, and detection ranking.
SDK	Software development kits are libraries and/or sample code that facilitates development of an application using a specific software platform or product.
SDLC	A software development life cycle is a set of processes and procedures used to develop, test, and release software.
SFTA	The software fault tree analysis, like an FMEA, is a tool to anticipate failures and their underlying causes.
Sprint	A sprint is a short development iteration used in Agile SDLCs.
SQA	Software quality assurance is the process of verifying that software works as intended without error. SQA can also be used to refer to the people who perform software quality practices.
SPC	Statistical process control is a method of quality control that employs statistical methods to monitor and control a process.
TQM	Total quality management is a quality-control system in which every department in the organization sets goals and processes to attain them.
UML	Unified Modeling Language is a general-purpose modeling language that is used to describe various aspects of a software system
Validation	The process of making sure a product specification meets the customer's need, i.e. building the right thing.
Verification	The process of making sure the product complies with the specification, i.e. building the thing right.
VOC	Voice of the customer is the process of capturing customer input regarding expectation, experiences, preferences and future requirements.
Wearout	Wearout is when a product can no longer be used because it has deteriorated beyond an acceptable state due to chemical, electrical, mechanical, physical, radiation, or thermal degradation.

Part I

Reliability and Software Quality – It's a Matter of Survival

1

The Need for a New Paradigm for Hardware Reliability and Software Quality

1.1 Rapidly Shifting Challenges for Hardware Reliability and Software Quality

Hardware reliability and software quality, why do you need it? The major US car manufacturers saw their dominance eroded by the Japanese automobile manufacturers during the 1970s because the vehicles produced by the big three had significantly more problems. The slow downward market slide of the US automobile industry was predictable when the defect rate of US automobiles was compared with the Japanese automobile industry. In 1981, a Japanese-manufactured automobile averaged 240 defects per 100 cars. The US automobile manufacturers, during the same time period, were manufacturing vehicles with 280–360% more defects per 100 vehicles. General Motors averaged 670 defects per 100 cars, Ford averaged 740 defects per 100 cars, and Chrysler was the highest, with 870 defects per 100 cars.

Much has been written about how this came about and how the US manufacturers began implementing total quality management (TQM), quality circles, continuous improvement, and concurrent engineering to improve their products. Now the US automobile industry produces quality vehicles, and the perception that Japanese vehicles are better has eroded significantly. J.D. Powers and Associates reported in its 1997 model year report that cars and trucks averaged about 100 defects per 100 vehicles. This represented a 22% increase from 1996 and a 100% decrease from 1987. Vehicles such as the GM Saturn and Ford Taurus are a tribute to that success, both in financial terms and in the improved perception that automobile manufacturers in the United States can produce reliable, quality automobiles. Quality programs like TQM have dramatically improved American manufacturing quality. The automotive industry has also benefited from the quality of the components going into automobiles, which is also at a very high quality level. Counterfeit components and counterfeit material is still a major concern for the electronics and automotive industry that requires constant diligence and an effective program to minimize the risk of counterfeit material entering into the production stream.

In the 1970s, the typical automobile warranty was for 12 months or 12 000 miles. In 1997, automobile manufactures were offering 3-year/36 000-mile bumper-to-bumper warranties. Three years later, these same automobile manufacturers were offering 7-year/100 000-mile warranties. Jaguar is now advertising a 7-year/100 000-mile warranty on its used vehicles! BMW has responded with a similar type of program. The reason these manufacturers can offer longer warranty periods is because they

Improving Product Reliability and Software Quality: Strategies, Tools, Process and Implementation, Second Edition. Mark A. Levin, Ted T. Kalal and Jonathan Rodin.
© 2019 John Wiley & Sons Ltd. Published 2019 by John Wiley & Sons Ltd.

understand why and how their vehicles are failing and can therefore produce more reliable vehicles.

A 1997 consumer reports survey of 604,000 automobile owners showed a dramatic improvement in the perception of the reliability of US-manufactured automobiles. The improvement by the big three automobile manufacturers did not occur overnight. It was the result of a commitment to provide the necessary resources along with a credible plan for producing reliable vehicles. It was a paradigm change that took years and evolved through many steps.

The process to improve hardware reliability has made significant progress over the past 20 years. If you follow the process outlined, there can be significant improvements to your product reliability. The hardware reliability errors are often the result of either not following or poorly executing the hardware reliability process rather than being a weakness in the reliability process.

Reliability research and development continues in some areas, such as prognostics and health management (PHM). PHM can improve product maintainability, reduce unscheduled downtime, and lower the cost of ownership over the lifetime of a product. PHM uses real-time sensors to monitor the health of a system. The sensor data is then compared to a good set of data to determine if the system is degrading and to estimate the time to failure. PHM strategies are being use in the automotive, aerospace, and other industries.

Even though hardware reliability has improved significantly for many companies, software quality and software security have become a bigger issue. This is partly due to the fact that many of the new products being developed require significantly more software and firmware. For example, McKinsey & Company estimates that over 10% of automobile vehicle content today (2018) is software and that software will reach 30% of vehicle content by 2030 [1]. Many of the hardware products being developed increasingly need software and firmware to function properly. Increasingly, software is also being deployed to manage critical safety and health operations such as robotic surgery and self-driving cars. It is not uncommon for the software development team to be inadequately sized for the staff and skill level needed to support software development. Automation, the Internet of things (IOT), advances in Wi-Fi and Bluetooth, and greater use of the internet to improve customer experience all drive the need for increased software code development and an improved software quality system. The Internet can be used to push out software updates effortlessly to the end user, but a poor software quality process results in injecting more software bugs than it fixes.

The number of software-to-hardware bugs that need to be fixed during product development can be in the order of 50 to 100 : 1. The complexity of software continues to increase along with new software languages and new drivers that create compatibility issues. About 40% of software development cost is to support testing for software verification. Software quality bugs are design faults that need to be identified, prioritized, and fixed. The traditional software quality systems addressed software bugs downstream as part of the final production testing. For many of the products being developed, this is too late in the development process. This is driving the need to improve software quality and the software development process from the perspective of project management, software requirements, and performance. This includes improvements to the software development process and staffing the team with domain experts.

1.2 Gaining Competitive Advantage

Companies successfully competing in the twenty-first century share a common thread. They all produce quality products that meet or exceed customer expectations over time. This may not seem like a revelation, but the process and tools that these companies will use to achieve this success must be new. In some industries, technology moves so fast that customers tend to trade up to the next-generation product before the first model stops performing to specification. This may seem like the ideal environment for a manufacturer because the product life expectations of the consumer are shorter. However, in reality, achieving product reliability with decreasing product development times requires a change in the way we develop products. Platform product development times have shortened to 18 months and their derivatives (product offshoots) have shrunk to 12 months or less. Of course, this is highly dependent on the product complexity and regulatory and safety requirements, but the trend cannot be ignored. Companies pay a heavy price for releasing a product that is "buggy" or unreliable. Satisfied customers are repeat customers. It is a well-known fact that it costs 5–10 times more to acquire new customers than it does to retain existing ones. It doesn't matter whether you are competing on cost or product differentiation; reliable products result in repeat customers and product growth through word of mouth. A faulty product usually results in the customer communicating dissatisfaction to anyone who will listen until the product or service is replaced with a more reliable one.

1.3 Competing in the Next Decade – Winners Will Compete on Reliability

The business practices of the past few decades will not be sufficient to ensure success in the twenty-first century. Through the years, we've learned to master the skill of building quality products. Higher-quality products have resulted in improved profit margins. In fact, consumers make buying decisions based on their perception of which products have better quality when the competing products were of the same approximate price. In the past few decades, reliability was not a deciding factor for most consumers. This is mostly the result of the consumer's lack of knowledge about product quality. However, the average consumer in the twenty-first century will make buying decisions based not only on price and quality but also on the perceived reliability of the product. Consumers make buying decisions based on which product offers the best value. We can define product value as

$$\text{Product value} = \frac{\text{Customer-perceived value}}{\text{Price}}$$

Here the customer-perceived value is related to the quality and reliability of the product. One of the key advantages of implementing reliability throughout the organization and at every phase of the product life is that the product value increases because of an improved customer perception of the value of the product and a lower cost of production. There is a common misperception that implementing reliability delays the product development time and increases the cost of the product (both in material and production costs). The reality is the exact opposite. Products that are more reliable generally have lower production costs. The reason for this is the result of many factors that contribute to reducing product costs and the product development cycle. For example, products that are reliable generally have

- Higher first-pass yield in test,
- Less material scrap,
- Less product rework (which helps to lower product cost and improve product reliability),
- Fewer field failures,
- Reduced warranty costs (this saving can be passed onto the consumer to provide a competitive price advantage),
- Lower risk of recall,
- Superior designs that are easier to manufacture.

Looking back at the definition of what the consumer considers to be of value, it becomes clear that product reliability will increase the perceived product value and lower the cost of production. This is an important fact about product reliability that is often misunderstood.

1.4 Concurrent Engineering

An important ingredient for successful design and implementation of new technologies into manufacturing involves the establishment of concurrent engineering practices. Concurrent engineering is a process used from design concept through product development and into manufacturing in which cross-functional representatives from all relevant departments provide input on key decisions. These decisions have a direct impact on the price, performance, quality, and development time required for the product. The concept was discussed extensively in the 1990s and has resulted in better products, shorter product development times, and greater profits for those who use it. However, the cross-functional teams consisted of marketing, design, test, and manufacturing. The teams did not include a separate representative for reliability, since this was considered part of the design and test engineers' responsibility. (We will see in Part II of this book that the tools used to improve product reliability are unknown to most design and test engineers.)

This convention needs to be changed and a more encompassing version of concurrent engineering must be developed that takes into account the entire product life cycle. The product life cycle approach includes reliability, serviceability, and maintainability inputs that begin in the design concept phase and continue through product development and product life. This cradle-to-grave approach ensures that the lessons learned along the way are captured and incorporated into the next development cycle.

Previous approaches to product development relied heavily on early design for manufacturing (DFM) effort and prototype testing to catch design flaws prior to product release. The problem with this approach is that DFM engineers (being highly skilled in the manufacturing process) primarily ensure that a product is manufacturable and can be rapidly ramped in production to meet market forecasts. Put another way, DFM ensures that the products designed can be ramped in production with ease (high-quality products), but the effort contributes little to product reliability.

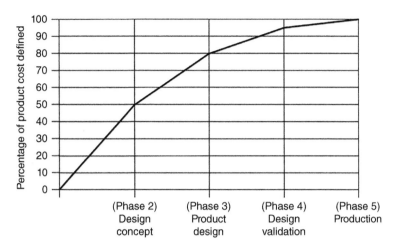

Figure 1.1 Product cost is determined early in development.

Testing performed at the prototype stage will validate product performance to specification prior to engineering release. This does not, however, consider the ability of every product produced to meet specifications in manufacturing. The problem with this approach is that decisions are continuously made in product development that have significant impact on the product performance, reliability, and the ease with which the product can be serviced and maintained. At this stage, decisions need to be made fast, and they should include inputs from everyone affected: marketing, design, test, manufacturing, field service, and reliability.

As stated earlier, it is important to involve all the relevant organizations and support groups early in the product development cycle in order to ensure the lowest product cost and highest product reliability. Design programs including DFM, as well as design for test (DFT), design for reliability (DFR), design for service (DFS), and maintainability must be considered early in the product concept phase. Representatives of each of these functions provide inputs based on guidelines developed from industry standards, lessons learned, intellectual property, and internal process development. These decisions must be made on the basis of facts, not perceptions.

Applying these design guidelines concurrently to product development will continuously reduce cost and cycle time, while also optimizing reliability. Figure 1.1 illustrates how a product life cycle approach to product development will have a direct, positive impact on the cost of the product. Typically, 80% of product cost is committed by the time it goes into prototyping. *Consequently, the greatest opportunity to reduce the cost of a product is in the design phase.* The product life cycle approach addresses all issues that affect the cost, service, reliability, and maintainability of the product for the entire life cycle of the product. These activities include involvement of the entire team on decisions that affect new technologies, packages, processes, and designs, and are based on a cost–benefit analysis, which includes market research, risk, and reliability.

Another driving reason for incorporating reliability as early as possible into product development is the cost of a change based on manpower and capital, when it is made after the design concept phase. Figure 1.1 illustrates how dramatic this impact can be on product cost. The greatest opportunity to reduce cost is in the development and

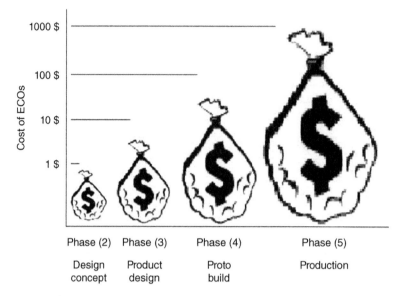

Figure 1.2 Cost to fix a design increases an order of magnitude with each subsequent phase. Source: Courtesy of Teradyne, Inc.

design concept phase, when risk issues relating to technology, components, and processes determine the majority of the product cost. By applying these practices early in the design phase, the cost and labor resources required for implementing engineering changes can be greatly reduced.

This point is illustrated further in Figure 1.2 in which the total cost of an engineering change can increase by several orders of magnitude when it is made late in the product development cycle.

1.5 Reducing the Number of Engineering Change Orders at Product Release

In Part IV, "Reliability Process for Product Development," we will show how using tools such as Highly Accelerated Life Test (HALT™), Highly Accelerated Stress Screening (HASS™), failure modes and effects analysis (FMEA), and risk mitigation early in the product development cycle will reveal hidden problems that are usually not caught until the product has been in production for some time. The product life cycle approach will also reduce the number of engineering change orders (ECOs) at product introduction and increase long-term product reliability. This idea is best illustrated in Figure 1.3 in which a product life cycle approach ensures that the majority of the engineering design changes occur early in the development cycle. This is the best way to reduce the risk of field returns after product release. The graph illustrates how the number of field returns and engineering changes is significantly reduced through early implementation of reliability in the product development cycle.

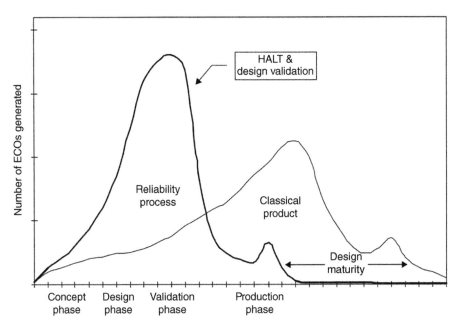

Figure 1.3 The reliability process reduces the number of ECOs required after product release.

1.6   Time-to-Market Advantage

One of the driving forces affecting product reliability is the greatly reduced product development cycles that organizations are facing. Coincidentally, this is also the biggest reason cited for why a product cannot undergo the additional activities required for reliability. But this argument is contrary to what actually happens when reliability is included early in the design concept phase. A major advantage to the implementation of a product life cycle approach to reliability is the reduced development time for product introduction. When time-to-market goals are achieved, the benefits include product name recognition, the ability to set industry standards, recognition as a leader, expansion of the customer base, and the maximization of profits. Using a product life cycle approach, product development time will be significantly reduced, as is shown in Figure 1.4.

Finally, Figure 1.5 illustrates how the timing of product introduction can affect product profitability. Introducing a new product at the same time as the competition will lead to average profits over the life of the product. By releasing a product ahead of the competition, the opportunity for profits increases. Conversely, when releasing a product after the competition, the opportunity for profits is much lower. It is important to point out that getting too far ahead of the market can be undesirable. This point is illustrated in the article *"A Survey of Major Approaches for Accelerating New Product Development"* [2], in which a late entrant in the memory chip market may not receive any profit or even recoup its investment.

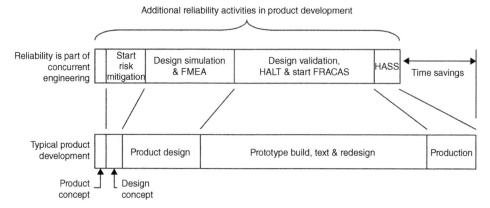

Figure 1.4 Including reliability in concurrent engineering reduces time to market.

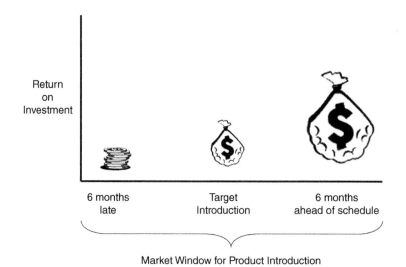

Figure 1.5 Product introduction relative to competitors.

1.7 Accelerating Product Development

There are many ways to accelerate the development of a new product. Product development can be accelerated by simplifying the product design process, improving communications between the cross-functional organizations, implementing an escalation process to resolve conflicts, eliminating unnecessary steps, maintaining development workload at no higher than 85% of capacity, parallel processing as much as possible, and most importantly, eliminating delays.

In today's competitive technology environment, companies can no longer afford to be late with a new product release, especially in a market in which product life continues to decrease. It is important for a company to eliminate the "not invented here" attitude that can often lead to overlooking a simpler solution or a new opportunity.

Another way to simplify manufacturing and to shorten the product development cycle is by standardizing common designs with a modular structure. Hewlett-Packard has been very successful in creating products with modular design. Modularity renders itself to easy upgrades and new design features. In addition, by standardizing certain common features inherent to all products, the development time is greatly reduced. This also eliminates the problem of having many different versions of a common feature. For example, a common circuit like an amplifier with 10 dB of gain could be designed differently by each engineer, but would perform the same function. If the 10-dB amplifier were standardized, then engineering time would not be wasted on "reinventing the wheel" and there would be a high assurance that the new circuit would work.

1.8   Identifying and Managing Risks

The key to any reliability program is the identification of risk. This concept is addressed in great detail in Part IV in which the tools and process for implementing reliability into the product life cycle process are presented. A credible reliability program must focus on the high-risk issues in the project. There will be risk issues at every stage of the product life cycle. Early in the concept phase, decisions are made relating to the features and specifications needed to capture the target market. Marketing uses extensive voice of the customer (VOC) to identify the next high-growth opportunity. These growth opportunities usually involve new technologies. For businesses that compete on the cutting edge of technology, new technologies represent a significant portion of the risk to the program and long-term reliability.

To develop the new platform product, a list of challenges must be devised. Each of these challenges represents risk to the program. To manage the risk, each item should be ranked on the severity of the risk and those items representing the highest risk should be tracked through the program. The role of reliability in the concept phase is to ensure that the risk issues are properly identified. They should be ranked by severity with corrective actions listed so that, when completed, the risk is mitigated. The risk plan needs to include all the functions that are affected in the life cycle of the product. Risk issues that relate to maintenance, manufacturability, design, safety, and environment are included. Unfortunately, these activities are reactive and thus add to the program development time. However, it will be shown later that the net result of these activities is the reduction in product development cost.

There are also proactive reliability activities that occur in advance of product development that help reduce product development time and improve product reliability. An example of a proactive reliability activity to reduce technology risk is the technology roadmap. Early VOC will identify future market needs that require new technologies. Mitigating technology risk issues in advance of need provides the required time to fully mitigate all risk issues. By mitigating technology risk in advance of need, you can gain competitive advantage by being the first to market, capturing a greater market share than would otherwise have been possible.

1.9   ICM, a Process to Mitigate Risk

One of the biggest challenges in any development program is to identify and mitigate the risk issues early in the program. This can be best achieved through a technique called identify, communicate, and mitigate (ICM), which is illustrated in Figure 1.6. The ICM approach is a three-step process to identify significant risk to the program,

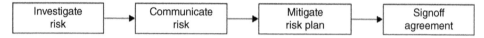

Figure 1.6 The ICM process.

to communicate its impact, and to develop an agreed-upon strategy to mitigate the risk. The ICM approach is an effective way to allocate and utilize limited resources in a way that will most benefit the program goals.

To identify the risk, each functional group reviews the product concepts, designs, and processes. Each group then identifies issues in which the present technology, methods, materials, processes, and tools cannot ensure success. These are the risks. Early risk identification makes the entire team focus on the concept of product reliability at the earliest possible time. All risk issues, no matter how small, need to be identified during this phase of the process. The low-risk issues will not have the same visibility and priority as the high-risk issues. In order to ensure that the major risk issues get completed prior to first customer ship, all risk issues are captured at the earliest possible point, the severity is ranked, and a risk mitigation plan is put in place based on the severity of each issue. Risk identification is a process that must be formalized and documented with the following captured information:

1. Description of the risk
2. Brief description of the activity needed to mitigate risk
3. Impact of risk to program and other functional groups
4. Severity of the risk defined on a scale from insignificant to catastrophic
5. Alternate solutions.

1.10   Software Quality Overview

Delivering high-quality software is critical to the success of a product and company. The lack of quality processes adds a lot of cost to software development organizations. The National Institute of Standards and Technology (NIST) determined in a 2002 study [3] that the cost to software development organizations in the United States due to a lack of adequate quality assurance infrastructure and process was estimated between $22 billion and $59 billion. With the growth of the software industry and inflation, that economic impact is now roughly double that 2002 estimate. This is a heavy productivity burden for a software development group to carry in a competitive marketplace and is a major drain on profitability. In addition to the overhead costs that comes from inadequate quality processes, a product's success often depends on its quality. If the product does not work or is too buggy, it will not succeed.

There is a trend in the industry to deliver more frequent software releases. This can be due either to a need to stay ahead of the competitors in consumer device markets or to the insatiable customer appetite for new features on top of existing hardware. Trying to support low-quality and buggy software slows down the ability to deliver timely software releases.

Fortunately, the process to create high-quality, reliable software is not very complicated. But, that does not mean it is easy. Developing quality software requires discipline in the application of techniques, processes, metrics, and controls. The quality of the software cannot be known unless it is measured. The production of quality software cannot occur unless procedures and tools are used to prevent, discover, and fix defects in the software during the development process. This book provides an introduction to basic metrics, tools, and procedures that can be used to produce high-quality software and to improve software quality over time.

The software described is a data-driven style of software development management. Quantifiable goals are set. Techniques and procedures are deployed throughout the software development process to prevent and find the software defects. Measures are taken and updated throughout the development process to track the current quality level and determine readiness for software release.

There are two overlapping sets of processes described here. The first consists of a set of techniques and metrics for delivering a high-quality software release. The second set of processes consists of methods to improve the software quality capabilities over time, resulting in continuous improvement from release to release.

The general software quality process cycle consists of four steps:

1. Set a quality goal for each software release.
2. Track the quality of the software as early as possible during the development process.
3. Employ techniques and procedures to prevent, detect, and fix defects during the development of a software release.
4. Analyze the quality misses after the release and evolve the processes to improve quality for subsequent software releases.

This book covers many aspects of the software life cycle, processes, methodologies, tools, and metrics that apply to creating and releasing high-quality software. However, this is not a book about software life cycles and processes; there are many excellent books that cover those topics in depth. This book only covers those topics insofar as they apply to producing quality software.

This book contains examples of pseudocode, test plans, and metrics. It should be noted that these examples are for illustrative purposes only. They are not intended to be complete, compilable, or production worthy.

References

1 Burkacky, O., Deichmann, J., Doll, G. et al. (2018). *Rethinking Car Software and Electronics Architecture*. McKinsey & Company.
2 Millson, M., Raj, S.P., and Wilemon, D. (1992). A survey of major approaches for accelerating new product development. *The Journal of Product Innovation Management* 9: 55.
3 Tassey, G. (2002). *The Economic Impact of Inadequate Infrastructure for Software Testing*. National Institute of Standards and Technology.

Further Reading

Hoffman, D.R. (1998). *An Overview of Concurrent Engineering*, 1997 Proceedings Annual Reliability and Maintainability Symposium (1998).

2

Barriers to Implementing Hardware Reliability and Software Quality

2.1 Lack of Understanding

Probably the greatest barrier to improving reliability is the lack of understanding of what *reliability* actually means. Much of the resistance you will experience in implementing reliability will come from individuals who believe that quality and reliability are the same thing. But as we have shown before, they are not. Quality is conformance to specification, while reliability is *quality over time.* Reliability from a customer perspective is that "the product works the way it was supposed to work for its desired period of use." Before you start improving reliability, you must be aware of what you are improving.

As a practical matter, most companies have quality improvement processes in place throughout their manufacturing process. Understanding the difference between quality and reliability is the key, because if you don't understand the difference, you may just be improving your product quality while not impacting your product's overall reliability at all. How can you separate the two?

First, focus on what's been going wrong in the manufacturing processes. Then take a look at the data that describe what has been going wrong. Review the data to see how the problem has been corrected. If it has been corrected by anything other than a design or process change, all you did was manage the process to maintain conformance to a specified parameter. You maintained conformance to specifications. If the improvement was made by an added inspection step, measurement step, or any other added human intervention, this is a clue that all you did was improve the quality. If, on the other hand, a change was made which improved the ability to maintain conformance to specification, in all likelihood you made a reliability improvement. If that is actually what occurred, current data will show that there was little or no subsequent corrective action required to maintain this quality specification. The data then will show that the problem area has been greatly improved or completely corrected.

Study the information that you already have. Gather data from the manufacturing process and from field failure data; this can have immediate impact on lowering warranty costs. Problems that can be corrected by design take longer. Very often, the decision has to be made to apply your resources on new designs rather than on old or current designs. If it is believed that the current design will be in production for a long time, resources should be expended on making changes to this design. This is also the time to begin using in- and out-of-warranty repair data. Compare this data to the changes put in place that were intended to make these problems disappear. If some problems did disappear,

Improving Product Reliability and Software Quality: Strategies, Tools, Process and Implementation, Second Edition. Mark A. Levin, Ted T. Kalal and Jonathan Rodin.

Figure 2.1 Overcoming reliability hurdles bring significant rewards. Source: Courtesy of Teradyne, Inc.

because the data show that they stopped reoccurring, then there probably has been a reliability improvement. If some problems did indeed disappear, while others emerged, that information indicates that the processes are not in control.

After gathering the data and analyzing the information of failures, corrections, and the related processes, there will be a better understanding of where to start the reliability improvement process. In most cases, it will be apparent that there are both design and process problems. Select from the data the low hanging fruit and take corrective action to make these easy problems disappear. These easy problems will usually return corrections that yield quality and reliability improvements with small utilization of resources. The harder problems will take more resources and a longer time to rectify, but they oftentimes return greater improvements. The gains from improving product reliability are multifaceted, as shown in Figure 2.1.

2.2 Internal Barriers

Installing reliability into a company can be a very difficult and trying task. *Expect to encounter barriers all along the way.* The barriers you encounter will be both internal and external. The internal barriers will be the most difficult to overcome, and they are real. The external barriers are less significant. We will show that many of them are based on invalid perceptions. We begin by first discussing the internal barriers and suggest ways to break them down. The internal barriers will seem insurmountable at the beginning. Many companies find that a couple of years after they implemented their reliability program, the improvements they have made in product reliability are not significant.

Obviously, the smaller the organization, the easier it will be to implement reliability into the organization. Perseverance is necessary.

You will be amazed at what happens within the organization once internal barriers begin to break down. This phenomenon is similar to that of a long-term investment in a retirement account. In the beginning, the retirement investment is small and the amount routinely contributed does not appear to bring you any closer to that final goal. However, over time this amount becomes significant as it begins to grow exponentially. The same effect will be seen within the company once the organization begins to see the benefits of reliability.

The internal barriers are the most difficult to overcome in implementing an effective reliability program. We have found that companies that are successful make it part of their core competencies. Selling reliable products will distinguish your company from that of your competitors. We begin by looking at these internal barriers. Here are 13 of the most common internal barriers:

1. Resistance to change
2. Lack of knowledge about reliability in management
3. Lack of knowledge about reliability in engineering
4. Inadequate training
5. Lack of management support for the process
6. Capital resources lacking to support the process
7. No adequate staff to work on the issues
8. Goals arbitrarily set or not well defined
9. Adequate time not put in the schedule for the process
10. Adequate time not put in the schedule to fix the problems found
11. Process for implementation not well defined
12. Engineers wanting to move on to the next design rather than returning to make improvements and fixes on older products
13. The attitude that "It won't work for us"

2.3 Implementing Change and Change Agents

Not surprisingly, the greatest barrier to successful implementation will be the resistance to change. During the early stage of implementation, the resistance to change will be across all cross-functional organizations within the company. The greatest resistance to change will be experienced in the engineering organization. Engineers, in general, are very set in their ways. It won't take long before you start hearing these phrases from engineering: "We have to do it this way to make it work," "The system wasn't designed to work that way," and "We have been designing quality products for years; how is this going to make things better?"

One of the problems that we commonly see, again and again, is that engineers do not understand the difference between quality and reliability. Sure, you may be designing and manufacturing quality products that meet customer's expectations. This was the goal for most companies during the 1990s, but more is needed to stay competitive now. Quality addresses the ability to meet the customer's expectations at the time of purchase. Reliability addresses the ability of a product to meet those expectations over time. If you are building quality products that do not meet the test of time, you have a reliability problem. This is the message that we need to drive into the organization; the way we are doing business today will no longer work in tomorrow's competitive environment.

A common and universal reason for the internal resistance toward implementing reliability is the lack of knowledge about the process for product reliability. Many employees prefer to work in their comfort zone. It is human nature to fear what you don't know. It is natural to oppose something where success can't be guaranteed. The only way to resolve the fear and discomfort about the reliability process is to remove doubts through education. As the organization becomes more knowledgeable about the reliability process, the resistance toward implementation slowly diminishes.

There are two necessary requirements that are needed to deliver the knowledge of the process. The first requirement is to have a charismatic leader or champion, who is highly knowledgeable about the process and can communicate clearly what needs to be done. Salesmanship here can be of great value. The second requirement is that management needs to support the necessary changes required to implement the new process. Both these requirements are necessary ingredients for success. Management support particularly is an absolute necessity.

The champion should be the reliability manager who is responsible for implementing the reliability process. Most companies lack this individual. Companies must go outside the organization to find this person. Selecting the right champion will be the difference between success and failure. Not only will this individual need to be a good communicator and motivator but he/she must also blend well into the culture of the organization. One of first steps the champion needs to take is to establish credibility in the organization. The resistance level will drop dramatically once the champion establishes credibility. One common mistake is to select someone inside the organization, who has been successful in a different area to be the champion for product reliability. Often, the quality manager is given this new responsibility. Unfortunately, most quality managers lack the required skills and experience needed to establish credibility. This is not to say that quality managers can't make excellent reliability managers. However, during the implementation phase when the new process is being developed, problems will arise that are best resolved by someone who already has experience in similar situations.

Selecting the right candidate is vital, but equally important is the way the individual is introduced to the organization. It is a guarantee that there will be resistance to the implementation of reliability; the way you introduce the reliability manager to the organization will either initiate the breaking down of these barriers or will increase resentment and resistance.

At a small job shop where I once worked, the boss called a meeting on the manufacturing floor and introduced me to everyone as the new reliability "consultant." All 50 employees listened as the boss made it clear what was about to happen, and that the faster it happened, the better. He went on to say that everyone's cooperation and teamwork with the consultant would speed his (my) departure. "In fact, your next pay raises are on hold and being paid to the consultant as of this day; so the sooner we get this 'reliability thing' going, the sooner your next raises will be forthcoming." While I certainly had support from management, this introduction made the hill I had to climb much steeper.

By contrast, in a very large company where a high-cost item was only part of the whole corporation's output, the management was considerably more diplomatic: "Let's see how this reliability thing works, and in four or five months we'll take a closer look at the progress." But, that's not management commitment.

The second necessary ingredient in the implementation process is commitment. The commitment needs to be in manpower, capital resources, schedule allotment, and in management. The management commitment must come from the highest level. In addition, the commitment should be part of the five-year planning, since the first couple of years the payback may not seem apparent. Once senior management has made the decision to implement reliability into the organization, a meeting should be planned with middle management and outside consultants in which shared goals for implementation can be set and the foundation for implementation established. Some companies use weekend retreats for this meeting. The implementation of reliability should not be viewed as an experiment.

If the need to implement reliability is real, so too must be the commitment. Typically, when a reliability manager is hired, senior management is high on the possibilities and low on the belief that it will be successfully implemented. This disbelief is even greater with the support staff. This disbelief can be diminished if the method of implementation includes the best practices in reliability. Many individuals in the organization, both management and staff, will usually have a negative attitude toward the whole idea. While others in the organization may have prior experience with reliability implementation, there is a 50–50 chance that they too may have a negative attitude toward the whole process. Simply put, disbelief in success is high.

In general, the design engineers' feelings toward the process will be unified in a group consensus. The reliability manager will be looked upon as an outsider. The need for senior and middle management commitment in the process is, therefore, paramount to the success of implementation. On the surface, the staff will consider this to be the latest management fad that doesn't seem to last more than a year. The boss has done things like this in the past. It usually doesn't pan out, so we'll just humor him and let this latest fad die on its own. This view can quickly be addressed by sharing with the organization the five-year implementation plan and the commitment in resources to achieve this goal.

2.4   Building Credibility

The third necessary ingredient in the implementation process is establishing the internal knowledge of what product reliability is and how it can be achieved. Knowledge can best be achieved through routine training sessions. The training should proceed in a logical fashion. First, you need to establish a common understanding of why product reliability is of concern to the company. This is an excellent opening to discuss the missed opportunities (i.e. problems) in previous products and products that were late to enter the market. Next, there should be training on the reliability process. What changes from the way you have developed products in the past? How does it impact different organizations? How will it benefit them and what resources will be available to achieve the goal? It is recommended that training meetings or mini seminars take place on a routine basis.

The training should not be limited strictly to reliability. Other product development organizations like the mechanical group, circuit board designers, field service, marketing, component engineering, and so on should hold classes to communicate issues and guidelines that affect product reliability. One topic that could easily be started is a training session on things that have failed on the manufacturing floor and in the field. There will probably be a lot of data to help with the class preparation. The class could point out what the problems were and what was done to correct them and then identify if the problems were design- or manufacturing-related. The staff could learn a lot from this session and it would help jump start the attitude change that this reliability effort is real.

The classes are especially important to new employees and will help build a stronger and more effective team. For example, you can select from training sessions on manufacturing guidelines, mechanical guidelines, maintainability and availability, serviceability, testability, thermal management, product life cycle, accelerated testing, mean time between failure (MTBF), failure modes and affects analysis (FMEA), design of experiments (DOEs), physics of failure, component reliability, mechanical reliability, and system reliability. Periodic training sessions (one every two weeks or so) will develop the required knowledge base for achieving product reliability. Use internal experts to teach the sessions. This training process communicates who the resident experts in these particular areas are. The staff will learn the names and faces of the experts and will seek their help when needed. Something serendipitous that the authors found was that these classes introduced personnel with cross-functional skills to one another. The meetings created an improved working relationship that didn't exist before.

2.5   Perceived External Barriers

There are five perceived external barriers facing companies that want to implement reliability:

1. Time to market
2. Product development cost
3. Competitors not doing it
4. No local test facilities
5. No local experts

The first two barriers, the time-to-market and product development costs, are the primary arguments used by those opposing implementation of reliability. This is a very shortsighted view; the perception is quite the opposite of reality. As we discussed in Chapter 1, when reliability is implemented in a concurrent manner, total product development time and cost are reduced. The idea may seem apparent, but the perception will be quite different early in the implementation process. One reason is that one output from reliability drives a need for design changes. These design changes are interpreted by the design team as being unnecessary and will cause further delays to the project. However, if the reliability activities are performed concurrently with the design process, then the design changes will be implemented to the program at the lowest cost. For example, doing an FMEA with a cross-functional, multidiscipline team prior to design layout will identify design issues that would not have been caught until the prototype build or sometime after product release. The net result is a reduction in the number of revisions required prior to the engineering release. Likewise, performing a Highly Accelerated Life Test (HALT) prior to releasing the product to manufacturing will remove design problems before first customer ship. This, in turn, will greatly reduce your warranty costs and can help to reduce product recalls.

The third barrier, and another common misperception, is that your competitors are not implementing reliability processes as part of their product development. This perception becomes more and more inaccurate every day as more companies embrace product reliability. If you dawdle, your competition gains a competitive edge. Besides these advantages, faster product development times, lower product development cost, and lower warranty costs, all come with improved product reliability. Making reliability a strong component in your product has several very desirable results that actually dovetail into one another. You get greater brand equity through product reliability, which commands a higher premium for your product, which generates increased profit margins overall.

The companies that do have product reliability programs consider this as part of their core competency. From FMEA, HALT, and from previous programs, the lessons learned can be applied to future products. The lessons learned can be captured in a database and made available to everyone through a computer-based retrieval system. Because these databases can get quite large, there should be search engines capable of finding studies based on key words or subjects. Second, lessons learned should be summarized into a design for reliability guideline that is updated and communicated to the design community.

One method that companies use to learn of these best practices in product reliability (prior to implementing the reliability program) is to benchmark their competitors. While this is an excellent idea in concept, it may be quite difficult to implement in practicality. One reason discussed earlier is that companies don't discuss their reliability programs. Second, companies rarely publish information discussing how they achieve product reliability. Here, much insight can be gleaned from interviewing and hiring reliability expertise to help you get started. Some companies use outside consultants and test facilities for part of their reliability activities. These companies have strict nondisclosure agreements in place that prohibit outside sources from discussing these issues. Some companies may even consider buying the competitor's product and performing a tear down to learn how they achieve product reliability. Fortunately, or unfortunately, competitors cannot tear down your product to learn how you achieve product reliability – not yet anyway.

2.6 Time to Gain Acceptance

After a short time, everyone could see that the reliability manager wasn't going to be deterred, and day by day some swung over and got on board. In a few months, they could see small successes and more people were persuaded to join the reliability improvement process. Shortly after that, there seemed to be an avalanche of enthusiasm for the new "reliability thing." The reliability manager was swamped with requests to implement his reliability processes on their specific assemblies. They saw the light and wanted to get the benefit. Some diehards remained skeptical and indeed were relative speed bumps to the growth of the process. These doubters have to be discovered and persuaded to help make the reliability process happen.

In every case, the new reliability process is on a critical path. Resistance to it is easy to mount, because early market entry is extremely important for profits. Any added activity delays market entry. Sometimes, as an alternative, management decides to place the reliability process on a sidetrack so that the existing product development is not slowed down by anything, especially the untried reliability processes. Placing the reliability process in parallel to the regular product development flow can work, but the results will be very small and the savings in development time will not be realized. Later, the evaluation of the new reliability process will have earned little support, because the data to support continuing the reliability effort will be almost nonexistent.

Even when the reliability tools reveal designs that have to be corrected, the time and resources needed to implement the improvements can cause delays that management may not want to tolerate. If the time to implement reliability fixes are not made part of the product development schedule, the reliability process is doomed from the start.

Engineers are intelligent people. They have a technical understanding of how things work. They often have explanations for how things work (or will not work) on the basis of an instant analysis of a situation. When presented with the concept of reliability engineering, their instant analysis often finds reasons why it just won't work in their environment. They believe they know all about how to design and mount an electrical component, cool a system or select a fastener, and so on. What most engineers have little knowledge of is the feedback from the field failure data that has resulted from their latest design. As a result, they believe that everything they ever designed is fantastic, when, in many cases, there are parts of their designs that could have been made more reliable if the reliability tools had been applied to their designs before final approval. They need to be approached slowly with the reliability tools of the trade. Today, many engineers move from project to project or even company to company and have little opportunity to learn of their design oversights. By the time the feedback is available from field failures, they have moved on to new things. It is understandable that educating the engineer is critical to the success of the reliability effort.

2.7 External Barrier

Logistics can be a barrier to applying HALT and other reliability tests to the product development process. New product developers can rent these services and spend a week (typically) to uncover product weaknesses. There are many of these sophisticated test laboratories throughout the country and the world. Some have HALT capabilities that can help the new user of HALT to design stress test regimens. Qualmark is a HALT chamber manufacturer. They have several HALT facilities that are located in major cities that are available for product developers (Qualmark at www.qualmark.com). A list of companies that provide HALT services and HALT equipment can be found in Appendix A. By contacting HALT chamber manufacturers, you can find the most local test houses where you can perform HALT. This list grows and changes often. (We recommend that you use the list of HALT chamber manufacturers provided in the appendix.) Unfortunately, the test houses are few and scattered around the country. In most cases, the facilities will not be close by.

The cost and logistics issues to perform the HALT testing at an outside facility can mount up fast. Design engineers may have to travel to get to the HALT facility, often requiring air, hotel, car rental, and other travel expenses. This, added to the facility charges for HALT testing, can make the cost a significant barrier to improving product reliability. Distance to the test house is thus very important to consider (the closer the better). If you are fortunate to have a test facility nearby, then travel cost is not a significant factor. The HALT service providers are in high demand, so you will need to schedule and book the testing time needed – typically four to six weeks in advance.

An engineer's time, like everyone else's, is valuable. Engineers do not like to spend a lot of time traveling to off-site facilities to do their work. Minimizing this factor can make or break the reliability test implementation planning.

When the HALT process is not well understood and HALT testing is outsourced to a distant test facility, there needs to be a knowledgeable person who ensures that all the preparation work is complete before traveling to the test facility. In some cases, the new product developers may have to hire a consultant or contract engineer who specializes in this testing to join the HALT process. These skills can either be bought or learned. Often, both are needed at the start. This adds to the cost and is an additional barrier to improved reliability.

There are consultants and HALT machine manufacturers who provide HALT training. These services can be made available either on-site or at training seminars. A list of consultants for HALT testing can be found in Appendix A.

If a company has determined that the costs for implementing these reliability tests are prohibitive, they should take another look at the possibilities. If the manufacturer's end product does not give the customer the reliability that is expected, then the costs of not doing HALT will be much higher. The cost may be a loss of the customer base that could eventually bankrupt the business. This could mean that looking at purchasing the HALT machine is a better, long-term solution.

The upfront HALT testing costs appear, at first, to be prohibitive. When compared to the impact on the business (increased warranty spending, product recalls, and lost customers), they are small. The returns from long-term reliability improvements that become part of the product reliability will more than offset the dollars spent on implementing HALT testing. The manufacturer will benefit by retaining customers and by developing new customers who insist on nothing but the best reliability.

2.8   Barriers to Software Process Improvement

Improving software quality requires changes to the software development process and the software development culture. Expect resistance to adding or changing the software development process because of claims that the new processes will increase development time and reduce responsiveness. However, the predicted increased development time and cost is a red herring. Increasing software quality and decreasing defect escape rate results in fewer resources and time spent on sustaining engineering. When there is no separate sustaining organization to support software quality issues and there is a list of new features planned for next release, the focus quickly will shift to fixing escaped defects. The development team can be quickly overwhelmed by fixing software defects, delaying or canceling the release of planned new feature enhancements. A high escaped defect rate is a distraction to the development team, as well as being costly. The time to market takes two hits, first for having to support a product with a high escaped defect rate and a second for having to delay the development of new features because developers have to spend more of their time just fixing the escaped defects. Additionally, productivity takes a hit when engineers have to context switch between working on new features and fixing escaped defects, so even the time spent by developers on task is not as productive.

Improved software quality, as measured by the escaped defect rate, is a competitive advantage. Delivering software releases on the promised schedule is also a competitive advantage. Customers prefer vendors they can trust to deliver software on time that works well. Better schedule predictability results from understanding the defect model and adequately planning for the effort to find and remove defects. Improving software release predictability results in improved customer confidence in your ability to meet software delivery commitments. This, in turn, leads to a better customer experience, increased customer satisfaction, and ultimately a competitive advantage.

Another line of argument against using analytical techniques to manage software projects is that the models and metrics are inherently inaccurate and thus cannot really be relied on. It is true that models are estimates and therefore cannot be exactly predictive. However, not using any quantifiable measurements to plan and track projects is often disastrous. There is simply no way to know the status of any given project without using some quantifiable metrics. The fact that the metrics are not 100% accurate does not mean they are not extremely useful.

Finally, some people will assert that software development is a creative endeavor, an art, and applying processes and metrics to it will stifle the creativity of its developers. In fact, software development is creative, but it cannot be treated as an art. Art is in the eye of the beholder; it is interpreted individually by the end user. Many artists may be delighted that different viewers or readers find different meanings in their art. However, if you believe your software has a specific purpose and you want to ensure it achieves that purpose, then you must use rigorous processes to guarantee the software is successful.

3

Understanding Why Products Fail

3.1   Why Things Fail

When we talk about product reliability, we are describing the trouble-free time period before a product fails. A failure is anytime the product does not function to specification when the product or service is needed. There are degrees of failure; for example, a color television that only displays black and white images, or a remote control that can change channels by using the number keypad, but not with the channel up and down control. A new shirt that quickly loses a button and hangs unworn in the closet constantly reminds the consumer of their dissatisfaction. These are not complete failures, but the effects they have on subsequent purchases are the same. An automobile that stops on the way to a job interview, a computer that crashes in the middle of tax preparation, or a parachute that doesn't open are much more severe failures. These failures are often communicated to others and have a more devastating effect on profit and future market share. The list is endless; the degree of failure can be varied, but the negative effect on your business is the same. A dissatisfied consumer results in the loss of repeat business.

What causes these failures? They can be due to inadequate design, improper use, poor manufacturing, improper storage, inadequate protection during shipping, insufficient test coverage, or poor maintenance, to name just a few. A product can be designed to fail, although unintentionally.

For example, a large and expensive industrial product requires a high amount of airflow to cool the machine. When the fan stops working, the machine fails. In this example, a \$20 fan caused a multimillion-dollar system to stop producing products. The failure results in your customer having a complete production shutdown with significant unrecoverable dollar losses to the business.

Design engineers should know this and expect it to happen because fan manufacturers specify fan-life expectancies. To achieve product reliability, we must ask the questions, "What will wear out before the end of the customer-expected product life, and why?" By identifying the things that will fail in the field, design changes can be made to improve product performance or a maintenance program can be established.

Materials expand and contract with temperature variation. Larger temperature variations cause greater material expansion and contraction. The amount of material

Improving Product Reliability and Software Quality: Strategies, Tools, Process and Implementation,
Second Edition. Mark A. Levin, Ted T. Kalal and Jonathan Rodin.
© 2019 John Wiley & Sons Ltd. Published 2019 by John Wiley & Sons Ltd.

expansion and contraction can vary for different materials. The greater the amount of temperature variation or the bigger the difference in material expansion rates between materials, the greater the stresses will be. Stresses due to temperature variations can cause component or solder connection fractures that eventually lead to failure. Designers can mitigate these failures with better environmental control, improving attachment structures, or by proper selection of mating materials themselves.

Manufactured assemblies are a commingling of many subparts. These parts are attached by a variety of fasteners. Often, the fastener is a screw and nut. In shipping or through normal use, the associated vibration(s) eventually accumulate, which can loosen the screw. Over time, failures in fasteners can cause larger failures. These breakdowns can be avoided by selecting a fastener that will not come apart in the expected environment. Perhaps the proper torque with a split-ring lock washer would be a solution in some applications; sometimes a press fit pin will do the job. Through design changes, the poorly chosen fastener that slowly leads to an eventual failure can be removed from the possibility of causing a failure.

Materials are stored to failure. When we think of hard manufactured goods, we don't think of the parts getting stale while they wait to be placed in the manufacturing process. The meat and produce industries must move their products to the end customer rapidly; otherwise they will suffer losses through spoilage and pests. Other industries have similar concerns, but usually to a lesser degree.

The grocery store places its products on shelves. When a new shipment arrives, the old product is rotated to the front of the shelf and the new product is placed at the rear. This is commonly known as rotating or facing the shelves. This ensures that some items do not rest on the shelf too long to spoil. This is done with dairy products every day. The term used in industry is *FIFO*, First-In, First-Out.

Many electronics components actually start to wear out right after they are produced. How soon after they arrive at the manufacturing location they are installed in the product and shipped to the customer can be important. These parts also have to be used on a FIFO basis to ensure that the decaying process does not accumulate to lower the part's life expectancy.

The tin-plated copper leads on electronic component parts will corrode if left on the shelf at room temperatures for several months. These corroded parts do not solder to the circuit board well and tend to exhibit solder-connection failures sooner than they would have if the corrosion was not allowed to occur.

Adhesives have short shelf lives. If not used for several months, many adhesives are susceptible to early failure. Sticky-backed labels are often purchased in large quantities to get good pricing. Often these labels are in storage for several years before the last ones are applied to the product. In the field, these old labels will usually fall off in a few months, and as such their value is lost.

Products are transported to fail. Assemblies arrive at the customer's destination only to be found inoperative because something broke during shipment. This is often a function of shipping cartons or crates that were not designed to stand the stress of shipping shock and vibration. The shipping carton can be a major cost item in the whole cost picture of a product. Sometimes, manufacturers save on these costs only to suffer even greater losses from returned goods from the customer. Manufacturers should require that the shipping carton be part of the total design effort. This should include appropriate testing to ensure safe delivery.

Products can be tested or operated (and the list goes on) to failure. When a product has failed, the failure mechanism must be learned to determine the root cause of the failure. The design of the product or the process must be updated to remove the failure possibility from happening.

At a meeting in a very large medical diagnosis manufacturer's facility where there were 20 design engineers and managers assembled, a new chief of new product development was heading the meeting. The purpose of the meeting was to begin the planning of a new model of a large medical diagnosis system that is found in nearly every hospital in the country, and in fact, the world. After some initial discussion, the new chief asked how much more reliable this system would be relative to last year's model.

There was some laughter; then a respected designer explained that the new system would, of course, be less reliable than the older model because there were many new features that raised the component count and therefore lowered the overall reliability. The room full of engineers agreed. To them, this was obvious. The chief remained quiet and scanned the men at the table. Then, he asked one question.

"How many of you here have a VCR to record television programs?" Everyone said that they did. Then the chief asked if they remembered the older, reel-to-reel tape recorders, even the simple audio ones. They all remembered them. He went on to discuss the problems that the older units had. Many chimed in with their own horror stories of when the tapes got wound up all over the place and one channel was out of commission and so on. Then the chief asked another question.

"How many of you have had problems with the newer cassette VCRs?" There was silence. It appeared that the room full of engineers and managers had experienced no failures in their units ever. This came as no surprise. The Japanese manufactured them. Reliability was expected. Then the chief made a startling remark.

"These new VCRs are more complex and do many more things. They record video, audio in stereo and hi-fi stereo, they edit, and so on. Then, he asked a question and the room fell quiet.

"So why do the new VCRs seem to run forever?"

There was no answer from the table.

The chief claimed that the improved reliability of the current technology was due to sound design and the removal of faults that are inherent in the design itself. This can be accomplished by identifying and removing faulty manufacturing processes with rapid corrective actions on field failures. New tools need to be found that put sound designs into bulletproof manufacturing processes so that the customer sees no reliability problems.

That meeting was held in the early 1980s. Since then, the reliability of components (in general) has improved two to four orders of magnitude. Parts were often specified in failures per million hours of operation (the term used is *lambda*). Today, parts are specified in failures per billion hours of operation, which are referred to as *FITs* (failures in time). If parts were the main contributor to failures, then, with the vastly improved complexity of new devices, they would be failing constantly. We can all attest that they are not.

Televisions, radios, and automobiles all have more parts and last longer. This is due to the inherent design and the manufacturing processes, not the parts count. What is needed to improve the reliability of a manufactured assembly is to improve the design and the manufacturing process.

3.2 Parts Have Improved, Everyone Can Build Quality Products

Many things contribute to good quality and reliability. Nothing has been more important than the quality of the components that go into a product. Much work has been done over the past few decades to improve the quality and reliability of components. This effort has, for the most part, been very successful. In fact, the measurement used to describe the quality of components has been changed three orders of magnitude as well (from lambda to FIT).

3.3 Hardware Reliability and Software Quality – The New Paradigm

Today we hear from friends and colleagues, and we know from personal experience, that the rules are different. With company takeovers, buyouts, mergers, and downsizing it is becoming clear that the old rules no longer hold up. The world is changing. Companies have to change to stay in business. Today's managers have to adapt their companies to these new paradigms.

What is a *paradigm?* The word has grown popular, but can be vague. It means a model or set of rules; if you do something a certain way, things will come out as expected. Do this and you get that. Tip a glass of water and the contents pour out. Do it in space and it might not, since there is no gravity. The force on Earth that makes the water flow down is not present in space, so the water does something new, something unexpected. In outer space, the rule that gravity pulls things down doesn't apply to the water in the glass. There is a new outcome when you tip the glass. Change the rule and you change the outcome. What we see in the marketplace is that the same old rules don't work anymore. The paradigm has changed.

Paradigms don't change rapidly. Rules do one at a time. Paradigms are what we believe to be true, not necessarily what really is true. Paradigms are made up of an assortment of rules. With more rules, the paradigm is more entrenched. With more established rules, our belief in the paradigm is stronger.

Five hundred years ago, the world was flat. It wasn't really flat, but almost everyone believed and behaved as if it were. We think of the discovery that the world is round as an event in time, almost as if this transformation happened all at once. Then, after the new realization, we all behaved as if the world had always been round, but this is not accurate. It took many years before the world was round (in everyone's mind). It took time before this new paradigm was well established. Now, we all agree that the world is round.

When many rules support the old paradigm, more obstacles have to be overcome to move into the new paradigm. More rules mean the paradigm changes slowly. At first, one rule doesn't work anymore. It hardly gets noticed. Then a second rule changes, and another, and so on. They begin to mount up. When many of the rules fall away, it becomes apparent that what used to work doesn't work anymore. In space, what goes up goes up. For people in space, this is a new paradigm; all the things that they did with water on Earth have to be changed because the behavior of the world in outer space is different. The water will do its own thing even if you don't adapt. To make water work for you, you have to adapt. Your life will be better if you do.

Rules change on a continuous basis. Today, they change even faster and we must adapt faster just to keep from falling behind. When the rules of the market change, you had better take notice. The sooner you recognize that the world (the rules) is changing, the better your life will be. If these rules have been continuously changing, why are so many people aware of paradigms now? What has happened that has made paradigms become more important?

Information is more abundant now than ever before; there is television, 24-hour network news, newspapers and magazines, tweets, phone apps, and the internet. Some are just databases or repositories of information. When taken as a whole, the result is knowledge.

If you know what your competitor is doing that allows it to sell at lower prices and still remain competitive, that's knowledge. How that company manages to do this is better knowledge. Seeing what you are doing (or not doing), in comparison, can make or break your company. Adapting to meet the competition will keep you in the marketplace. Investing in new methods can take you past your competitor. This is what you need for your business to stay alive.

Do you ask yourself, "What is it that my competition is doing that allows it to sell below me and still maintain the margins needed to remain in business?" It may be as simple as the fact that your competitor keeps nearly all the money received from sales. A significant portion of the sales dollars don't have to go back to customers by way of warranty returns because products deliver the promise of quality, day after day. The customers are satisfied and happy. The products are reliable. Customers reorder more of your competitor's product because they have experienced good quality for a long time. That's reliability.

We often overlook how our suppliers help us to stay ahead of our competitors; instead we focus on our major cost burdens and how to incrementally squeeze cost from our suppliers. And then, the suppliers must think about how to reduce costs to stay competitive. They work to improve or eliminate whatever it is that is hampering business. They reorder whichever components that continue to deliver the quality and reliability they need to stay in business. In effect, suppliers look for products that generate repeat sales with no added sales cost.

Another new paradigm is reliability. When your designs are mature and your processes are in control, the reliability of your product will be high. The return is in dollars not lost to warranty claims and upset customers. You, as a manager, have to make the changes that ensure quality and reliability. Otherwise, the market will look to those who have learned these new rules earlier.

3.4   Reliability vs. Quality Escapes

When this book was originally published, the focus was on the hardware reliability process. Hardware reliability represented the major challenge to designing reliable products. This is no longer true; there has been a paradigm shift where software quality escapes often significantly outnumbers the number hardware reliability issues. The shift occurred because products today have a significant amount of embedded firmware and software that is released with significant quality and design issues. These quality escapes represent the majority of the product fixes that occur after product launch. In fact, if you compare the number of software bug fixes to the number of hardware reliability issues after product launch, you will find that they typically differ by orders of magnitude. It is common practice for software updates to be released on a routine basis and the consumer has come to expect software updates. The challenge is not to design a process that will produce perfect software, but to design a process that ensures software bugs are of a minor nature when released and do not pose significant risk.

3.5   Why Software Quality Improvement Programs Are Unsuccessful

There are several software process improvement models an organization can select from to improve software quality. The models help an organization evaluate their current software quality process, infrastructure, and capabilities and they also provide a framework on how to make incremental improvements. It is not uncommon for this process to take years before reaching the desired software quality level. The most common software quality improvement models are the Capacity Maturity Model and ISO 9001: 2015:

- *The Capability Maturity Model (CMM).* Discussed in Chapter 11, this model was developed at Carnegie Mellon University and provides a framework to improve software process quality into one of five maturity levels. All software development efforts fall into one of the five categories. The model provides the guidance and key processes to improve software quality to the next Capability Maturity Model Integrated (CMMI) level. The CMMI level is not a certification; instead, it is an appraisal of the organizations capability to produce quality software.
- *ISO9001: 2015.* This is an international standard detailing the requirements for a quality management system (QMS). An organization can be certified to ISO9001 by demonstrating that it follows the guidelines in the ISO9001 standard. The ISO standard details the requirements for quality assurance from design, development, production, installation, and maintenance.

However, there are many reasons why a software quality improvement program might turn out to be unsuccessful. Improving software quality is an incremental process that cannot be rushed through the organization. If the expectations and enthusiasm set by senior management are unrealistic, it can set the software quality improvement process up for failure. Improving software quality involves changing the software development culture, which takes time and requires incremental success to reinforce the importance and benefit from developing higher-quality software. The software improvement process requires a commitment from senior management down through the organization for the resources, risk, and potential impact to the development schedule as the process improvement changes are implemented. If the management commitment is not there, it will be very difficult to make any significant improvement. The commitment needs to also include adjustments to the development schedule that accounts for training and process improvement activities. The process improvement work needs to be viewed with the same priority and importance as the rest of the development process and the effectiveness of these activities evaluated with equally high importance.

Improving software quality requires rolling out education and training material to the software development team and management. If the training is inadequate or nonexistent, it can lead to a failure to properly implement process improvement, false starts, and potentially having the software quality get worse instead of better. Training and education are important steps for everyone involved in the software development process.

It is important that an organization doesn't rush to achieve a desired software quality level, and once achieved, lose management support and focus phasing out training and education. Maintaining the improved software quality requires institutionalizing the process and training and making sure all new employees are fully trained.

Further Reading

Steinberg, D.S. (2000). *Vibration Analysis for Electronic Equipment*, 3e, 8–9. New York: Wiley.

Stewart, W.S. (1955). *Determining Bolt Tension*. Machine Design Magazine.

Dicely, R.W. and Long, H.J. (1957). *Torque Tension Charts for Selection and Application of Socket Head Screws*. Machine Design Magazine.

4

Alternative Approaches to Implementing Reliability

4.1 Hiring Consultants for HALT Testing

The decision has been made to implement reliability processes on the new product. At the start, there is no one with skills in the company to begin making reliability improvements. Where to begin? Who to call? The first thing to know is what not to do. Do not locate a Highly Accelerated Life Test (HALT) facility, bring the product to be tested, and expect optimum results. A little planning goes a long way.

Contact several contract-engineering agencies. They will often have resumes of engineers who have experience in HALT and other reliability processes. This will help locate local experts. Interview these specialists and determine if their experience is a good match with your product. During the interviewing process, you will learn if there are local HALT facilities. While looking for a HALT facility, you should inquire about: availability, flexibility, cost, staffing, and so on. There is a list of HALT testing facilities in Appendix A. Do these facilities have additional reliability-testing capabilities? Can they do failure analysis (FA)? FA will help you with in-process and field failures that have been difficult to correct.

When you contact these test houses, learn as much as you can about their capabilities and the costs of their services. They will usually have a fee for the test equipment, the test engineer, support materials, and so on. You will be surprised at how this fee will differ in one community. Often, if the test house is aware that there is a considerable amount of business coming from more new products, then the potential for a long-term relationship might net lower pricing. Visit the test houses. Meet the individuals who will interface with your personnel. Bring your engineers on these site visits. Essentially, start a business relationship. If you have decided to hire one of the contract engineers you interviewed, bring him or her too.

4.2 Outsourcing Reliability Testing

Take the time needed to identify the scope of the long-term reliability goals and objectives. Then, seek advice from consultants and experts at test houses. Tell the test house what you are planning to do. Get their input. They often can offer support and guidance to your reliability improvement goals.

Seeking advice from consultants can be very beneficial, especially if the magnitude of the reliability improvement plan is large. This will require the guidance of a specialist.

Improving Product Reliability and Software Quality: Strategies, Tools, Process and Implementation, Second Edition. Mark A. Levin, Ted T. Kalal and Jonathan Rodin.

The specialist will usually be able to provide immediate training at your facility. Working with a specialist will also tend to speed up the start of the new reliability program.

Using consultants to evaluate a new technology or process, device package, custom-design, or new component for reliability is an effective way to manage quality and reliability risks that are outside the core competency or comfort zone of the organization. In some cases the consultant may have already studied the problem and can provide the needed information or test plan to mitigate the risk. If not, the consultant can submit a proposal for the evaluation and test costs. It's best to get at least two quotes from different consultants to compare the methodology, cost, and turnaround time.

4.3 Using Consultants to Develop and Implement a Reliability Program

When implementing a reliability program into an organization that lacks a formal reliability process, the challenge can be overwhelming. The first step is to define and document the reliability process for each phase of product development. There are numerous companies and educational organizations that can help with this vital first step. Appendix A has a list of reliability consultants, educational and professional organizations, and professional reliability societies that can help with this process. Before a consultant can develop a reliability program for the organization, they first need to understand your process for developing new products and any quality and reliability processes that are in place. The consultant needs to understand the types of quality and reliability problems the organization is working to solve and the resources the organization is willing to commit long-term. The consultant can then tailor a program that will meet the organization's needs and develop process documents and training tools that can be rolled out to the organization.

This training should, at first, be at the top levels of the organization. This ensures understanding, commitment, and direction. Trained managers will then carry the message to the rest of the staff. This is important for continuity. Managers can select the reliability person(s), hire contract engineers and/or consultants, set the schedules, and track the performance of the reliability effort.

This is a general description of the early steps needed to get the reliability process started. There is more to getting reliability implemented than what is stated here, but as the readers continue, they will see that there are different strategies for different companies, depending on their special needs.

4.4 Hiring Reliability Engineers

Hiring a reliability engineer can be a difficult position to fill. The reason is twofold. First, there are not many universities teaching reliability engineering, so the pool of talent is small to begin with. This makes it especially challenging if the goal is to hire a college recruit with a graduate degree in reliability engineering. The appendix includes a list of universities with master's and doctorate programs in reliability engineering. Most of the schools offer a master's degree in reliability engineering and the University of Maryland has a PhD program in reliability engineering. Most of the master's degree programs in reliability engineering are offered in combination with another field of engineering. It could be in combination with statistics, availability, maintainability, safety, risk assessment, or asset management. Depending on the type of reliability engineer you're looking for, one program may be better suited for your needs.

This is a hard position to fill because the talent pool with the right skill set is small, and it can be hard finding these people through a talent search. You can use a job recruiter, but it's very likely that the recruiter will not have experience recruiting reliability engineers. Search engines such as LinkedIn.com, Monster.com, and Indeed.com can help you search for reliability engineers. You can narrow the search by selecting engineers from the universities with programs in reliability engineering that you want to target. However, not all reliability engineers have degrees in reliability engineering. In fact, it's not uncommon to find senior reliability engineers who developed the skills and talent over time. You can also narrow the search by looking for reliability engineers with work experience from companies in fields similar to yours. It's not unusual to find self-taught reliability engineers with certificates in reliability engineering. There are many places an engineer can get a certificate in reliability engineering; Appendix A lists some of these programs.

If your strategy is to internally develop a reliability engineer, then having the engineer complete an online reliability certification program is a great way to get a broad foundation in reliability engineering. To become a Certified Reliability Engineer (CRE) you need eight years of on-the-job related work experience. The amount of on the job work experience is reduced if you have completed a degree from a college, university, or technical trade school. The amount of work experience waived, depends on the type of degree you have. It can range from one year credit for a technical diploma, two years for an associate's degree, four years for a bachelor's degree, and five for a master's degree. To become a CRE a candidate needs to develop an understanding of a broad array of reliability principles that can range from reliability systems and reliability block diagrams (RBDs), maintainability and availability, statistical theory and mathematical modeling, failure modes and effects analysis (FMEA), reliability testing, failure reporting, analysis, and corrective action systems (FRACASs), reliability performance analysis, and reliability program management.

Part II

Unraveling the Mystery

5

The Product Life Cycle

Many choices must be made before a reliability program can be put in place. These decisions will have a significant impact on the organization and the time required to implement the reliability program. A small company will have different barriers and decision criteria than a large one. A large company that has made the commitment to improve its products through reliability may be willing to spend whatever it takes to be successful. Of course, this alone will not guarantee success, and if implemented poorly, leads to lower profits with marginal returns on product reliability. A small company will have different constraints driving the need for more reliable products and will implement change in a different way from a large company. Before tailoring a reliability program that is best suited for you, an understanding of the reliability process, concepts, and tools is needed. In Part II, we will unravel the mystery behind a successful reliability program.

5.1 Six Phases of the Product Life Cycle

We begin this chapter with a brief overview of the reliability process. The process is the same for all companies implementing a reliability program. The degree to which this process is formalized will depend on the size of the company and the time-to-market constraints. (A more detailed description of the reliability process is presented in Part IV, "The Reliability Process for Product Development.") The reliability process needs to include the entire product life cycle. The product life cycle is a cradle-to-grave approach, where decisions made in any phase of the product life cycle will have an impact on product reliability, customer satisfaction, profit, and product image. In addition, these decisions must be made by considering the impact they will have on the life cycle of the product. The product life cycle consists of six phases:

1. Product concept phase
2. Design concept phase
3. Product design phase
4. Validate design phase
5. Production phase
6. End-of-life phase, as shown in Figure 5.1

The reliability process is a multidiscipline effort that is conducted concurrently in each of the six phases of the product life cycle. The multidiscipline team consists of

Improving Product Reliability and Software Quality: Strategies, Tools, Process and Implementation, Second Edition. Mark A. Levin, Ted T. Kalal and Jonathan Rodin.

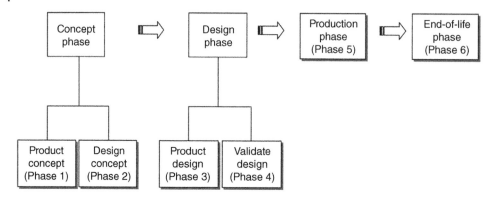

Figure 5.1 The six phases of the product life cycle.

representatives who possess the knowledge, lessons learned, and expertise from their particular functional activity. Traditionally, concurrent engineering has been a multidiscipline approach that is implemented in the design phase and production phase of the product life cycle. Concurrent engineering, quality circles, continuous improvement, and other such programs have brought about a significant increase in the quality of products being manufactured in the United States. These programs are successful because they utilize the knowledge and expertise in the organization from all functional groups early in the product design phase, where decisions are made impacting product quality.

As a result of these activities, manufacturers in the United States are producing some of the highest quality products in the world. These quality programs are more than 30 years old and are now practiced globally by competitors. Unfortunately, a strategy of competing on quality alone no longer provides a competitive advantage. Today's consumer is knowledgeable about product quality and can explain which products are of higher quality and why. These same consumers who have developed this understanding will soon be knowledgeable about product reliability. The time is not too far off when these same consumers will be capable of explaining which products are more reliable and why.

Easy and ready access to the internet has helped further accelerate consumers' knowledge about product quality and reliability. Now, there are many search engines available where consumers can do comparison shopping for a particular product. These same search engines offer chat rooms where consumers can discuss questions and concerns that they have about a product. Some of these search engines have consumer product reviews in which you can read about a particular product of interest. Although the full impact that the internet and comparison shopping search engines have on the consumer decision-making process is unknown, they should be viewed as powerful communicators of product quality and reliability. The internet will provide a competitive advantage in the next decade for those who produce quality products that are reliable for the expected time of use.

The best way to design and manufacture quality products that continue to meet consumer expectations is to take a multidisciplinary, concurrent engineering approach to the product life cycle. It is also necessary to establish a reliability program that is integrated into the concept of the product life cycle. The reliability program considers any issues that relate to product quality and reliability, along with any significant

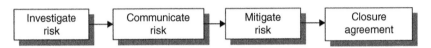

Figure 5.2 The ICM process.

technology risk, which can impact a program. In the next section, we present an overview of the reliability program centered on the six phases of the product life cycle. Table 5.1 contains a summary of the functional activities that take place in the product life cycle. Some concepts will be further explained in Chapter 7.

5.2   Risk Mitigation

Risk mitigation is a three-step process, resulting in a closure agreement, as shown in Figure 5.2. First, investigate to identify risk issues. Next, communicate the risk issues to all involved for acceptance of risk. Finally, develop a plan to mitigate the risk. As part of the process, meet periodically to update status and review risk issues, close resolved issues, and add new ones when they surface. Each of these three parts is next described in further detail.

5.2.1   Investigate the Risk

To identify the technology risk, each functional group reviews the product concept and identifies issues where the present technology, methods, processes, and tools will not ensure success. Early risk identification focuses the entire team on the concept of product reliability at the earliest. All risk issues, no matter how small, need to be identified during this phase of the process. Low-risk issues will not have the same visibility and priority as high-risk issues. Therefore, to ensure that everything gets completed prior to first customer ship, all risk issues get captured at concept phase and are recorded as shown in Figure 5.3. Risk mitigation can be planned on the basis of each of these issues.

Risk identification is a process that must be documented and must address the following issues:

1. Description of the risk
2. Brief description of the activity needed to mitigate the risks
3. Impact of risk to program and other functional groups
4. Severity of the risk defined on a scale from insignificant to catastrophic
5. Alternate solutions

5.2.2   Communicate the Risk

Once the risk issues have been identified, they need to be communicated to the key shareholders of the program. They are the representatives of the functional groups that make up the concurrent engineering team. The shareholders should also include the senior management of the organization. Since the risk is shared, shareholders need to agree on each identified risk of the program and determine if it is necessary. The process of communicating the risk is best achieved in a formal meeting with all shareholders. The

Table 5.1 Functional activities for cross-functional integration of reliability.

Functional activities	Concept phase		Design phase		Production phase		
	Product concept	Design concept	Product design	Validate design	Production ramp	Volume production	End-of-life phase
Marketing	Risk Mitigation – external VOC, reliability goals, lessons learned	Risk mitigation – internal VOC	Implement risk mitigation plan				
Electrical design	Risk mitigation – external VOC, reliability goals	Risk mitigation – internal VOC, apply design guidelines	HALT, Implement risk mitigation plan, design FMEA	Risk mitigation closure, design and performance validation, operate FRACAS	ECO verification		
Mechanical design		Risk mitigation – internal VOC, apply design guidelines	Implement risk mitigation plan, design FMEA	Risk mitigation closure, design and performance validation, operate FRACAS	ECO verification		
Software design		Risk mitigation – internal VOC	Implement risk mitigation plan, design FMEA	Risk mitigation closure, design and performance validation	ECO verification		
PCB design		Risk mitigation – internal VOC, apply design guidelines	Implement risk mitigation plan, design FMEA	Risk mitigation closure	ECO verification		
Reliability	Risk mitigation – external VOC, technology risk, reliability goal, reliability process, lessons learned	Risk mitigation – internal VOC, lower level reliability goals, define reliability design guidelines, technology risk, reliability capital budget	Implement risk mitigation plan, design FMEA, reliability estimates, root cause failure analysis, apply design guidelines, install FRACAS, HALT planning	HALT, risk mitigation closure, proof of design, proof of screen, root cause failure analysis, document findings into lessons learned, operate FRACAS	HASS, ECO verification, HASS, FRACAS, reliability growth, Burn-in, ARG, early life testing (ELT)	HASA, FRACAS, reliability growth	HASA, FRACAS, Obsolescence

Function					
Software quality and reliability	Risk mitigation – internal VOC	Implement risk mitigation plan, design FMEA	Risk mitigation closure		
Quality	Risk mitigation – internal VOC	Implement risk mitigation plan, design FMEA	HALT, risk mitigation closure, operate FRACAS	HASS, FRACAS, SPC, 6-sigma	HASA, FRACAS, SPC, 6-sigma	HASA, FRACAS, SPC, 6-sigma
Test engineering	Risk mitigation – internal VOC	Implement risk mitigation plan, design FMEA	Risk mitigation closure, operate FRACAS	FRACAS, SPC, 6-sigma	HASA, FRACAS, SPC, 6-sigma	HASA, FRACAS, SPC, 6-sigma
DFM	Risk mitigation – internal VOC	Implement risk mitigation plan, design FMEA, apply design guidelines	Risk mitigation closure, operate FRACAS, Process FMEA	FRACAS, SPC, 6-sigma	FRACAS, SPC, 6-sigma	FRACAS, SPC, 6-sigma
DFT	Risk mitigation – internal VOC	Implement risk mitigation plan, design FMEA, apply design guidelines	Risk mitigation closure, process FMEA, operate FRACAS	FRACAS, SPC, 6-sigma	FRACAS, SPC, 6-sigma	FRACAS, SPC, 6-sigma
Manufacturing	Risk mitigation – internal VOC	Risk mitigation, design FMEA	Risk mitigation closure, process FMEA, operate FRACAS	FRACAS, SPC, 6-sigma	HASA, FRACAS, SPC, 6-sigma	HASA, FRACAS, SPC, 6-sigma
Customer support	Risk mitigation – internal VOC	Implement risk mitigation plan, design FMEA	Risk mitigation closure	FRACAS	FRACAS	FRACAS
Material management	Risk mitigation – internal VOC	Implement risk mitigation plan, design FMEA	Risk mitigation closure	FRACAS	FRACAS	FRACAS
Component engineering	Risk mitigation	Implement risk mitigation plan, design FMEA	Risk mitigation closure	FRACAS	FRACAS	FRACAS Obsolesce
Safety and regulation	Risk mitigation – external VOC	Implement risk mitigation plan, design FMEA	Risk mitigation closure	FRACAS	FRACAS	FRACAS

Note: DFM, Design for Manufacturing; DFT, Design for Test; ECO, Engineering Change Order; FMEA, Failure Modes and Effects Analysis; FRACAS, Failure Reporting, Analysis, and Corrective Action System; HALT, Highly Accelerated Life Test; HASA, Highly Accelerated Stress Audit; HASS, Highly Accelerated Stress Screens; PCB, Printed Circuit Board; SPC, Statistical Process Control; VOC, Voice Of the Customer; ARG, Accelerated reliability Growth: ELT, Accelerated life testing.

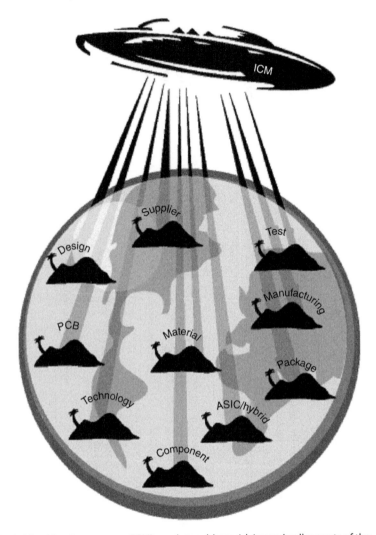

Figure 5.3 A risk mitigation program (ICM) needs to address risk issues in all aspects of the development program. Source: Courtesy of Teradyne, Inc.

sole purpose of the meeting is to present the risk issues, agree on which risks are necessary, and sign a formal agreement to resolve the key risk issues prior to first customer shipment. This agreement is a commitment to spend the necessary resources to resolve key risk issues and communicate to the team the risks that cannot be resolved in the scheduled time frame.

5.2.3 Mitigate the Risk

The last step in the identify, communicate, and mitigate (ICM) approach is the risk mitigation plan. The risk mitigation plan captures at a high level the activities that will take place to ensure that the product is reliable and can be manufactured at the volumes necessary to meet the marketing projections. The risk mitigation plan should not contain the steps necessary to achieve the deliverables, because it will be read by all the shareholders on the team and they may not be interested in all this detail. Included in the risk mitigation plan should be a list of all the experiments, environmental stress tests, tooling, and so on that will take place, along with the desired outcome.

The plan should also include a time frame during which each activity will start and close. The final necessary ingredient in the risk mitigation plan is the deliverable required for each activity. This is the one area in which many companies falter. The deliverable, as to what is required to have closure for each risk issue, needs to be clearly stated. For example, will a formal report, which contains all the necessary information for someone to repeat the activity and achieve the same desired outcome, be produced? Extremely high-risk issues will require an alternative path that is performed parallel to the primary effort but at a lower resource level. Finally, some of the risk activities will merge with other groups because the risk is common to both groups. In these situations, a team effort is necessary so that there is no duplication of effort and the results are agreed upon by the groups affected. The concurrent effort will ensure that there are no missed activities because it was mistakenly assumed that another group was doing that effort.

The ICM process requires an official signoff, which allows the program to proceed to the next stage. This procedure is a gate that requires agreement that the risk is manageable in order to proceed to the next development stage. The signoff can be either from the senior management team member or by the functional team members themselves. Often included in this signoff is a definition of mandatory deliverables that must be resolved with an associated time frame usually aligned with a future phase of the product development process.

5.3   The ICM Process for a Small Company

The reliability process for a small company may be different from that for a large or a medium-sized one. The small company often lacks the numerous functional support groups that are present in a medium and large-size company. The small company does not have the communication barriers that are present in their larger counterparts. As a result, a small-size company is often less process-restrained; it can solve problems informally and spend less time documenting and generating reports.

The small company may face the same reliability issues and risks in developing technology-driven products that a medium- and large size company may face. However, because small companies are more flexible, they are able to respond fast to change and take on added risk. The process for managing risk and ensuring reliability in product development is the same no matter how small the company is. Early in the product concept phase, an assessment needs to be performed to identify all significant risk(s). The small company mitigates risk differently due to its limited financial resources, technical expertise, number of qualified technical staff and lab capabilities.

In today's global environment, it is possible for a small company to compete with a large or a medium-sized company. The company that will win the market share will be the one to market first, with an effective business plan, and with the most reliable product. Since small companies still face many of the same risks and technology challenges as their large- to medium-sized counterparts, it is necessary to use the ICM approach to identify, communicate, and mitigate all significant risks. The small company needs to document the entire ICM process in order to ensure that no issues slip through without resolution.

The most significant difference for the small company is the implementation strategy for mitigating risk. Because the small company, in many cases, will be unable to adequately mitigate the risk internally, it should consider looking outside for assistance. Some of the alternative ways to mitigate the risk are as follows:

1. Hire a consultant with expertise in that area.
2. Outsource parts of the work to a university.
3. Hire temporary contract help.
4. License the technology.
5. Outsource to another company (not recommended that you outsource anything that is a core competency).

5.4 Design Guidelines

An important ingredient for successful implementation of reliability involves implementation of design guidelines. These guidelines include design for manufacturing (DFM), design for test (DFT), and design for reliability (DFR). These three guidelines, when implemented in product development, using concurrent engineering, will ensure that the product will meet the minimum standard for manufacturing, test, and product reliability. An integral part of successful implementation of any "design for" guideline in product design and development involves concurrent engineering. Concurrent engineering is a way of ensuring that things get done right the first time and that there is timely communication with all the groups involved in the design decisions. Concurrent engineering causes the developers of the product to consider all elements of the product life cycle, including manufacturability, serviceability, test, cost, schedule, user requirements, quality, and reliability.

5.5 Warranty

Every sales figure has buried in it a small dollar amount that is set aside to ensure that the seller has some capital available in the future to cover the costs of warranty claims made by their customers. The higher the sales, the higher this dollar figure will be. More importantly, the lower the reliability, the greater the amount of revenue that needs to be set aside to honor warranty claims. One way of looking at the contingency for the warranty claims figure is that this is the expected reduction in profit due to reliability escapes in the design and manufacturing process.

If the design calls for more nuts and bolts than needed, the cost of goods to manufacture the product will be higher. Cost-reduction efforts should be initiated to remove the unneeded fasteners to save some money on the cost of materials. After the unneeded screws have been removed, the cost of goods to manufacture the product will decrease and the profit will increase. The manufacturer then has the option of receiving a slight increase in profits or reducing the product cost with the opportunity of greater market share. Having options like these help ensure the company's future.

A warranty is like unnecessary screws. The more unreliable the product, the greater the amount of monetary reserves that have to be set aside to ensure that funds are available to cover warranty costs. It's like paying a little extra for the materials that went into the shipped product. Unfortunately, the manufacturer will pay in terms of warranty costs and in terms of lost sales never made because the customer takes his/her business elsewhere.

Companies with little or no reliability as part of the new product development product can have warranty expenses that reach 10–12% of the annual sales dollar (author's experience). Companies that have some reliability imparted during the development process can lower this figure to 6–8%. Only those companies that have implemented a cross-functional reliability process in their new product development process ever get this figure below 1%.

For companies with $10 million in annual sales and poor reliability, in essence, a million dollars is being handed over to the cost of doing business. To recoup this million dollars, how many salespeople would have to be hired to return this figure to the bottom line? How hard will the purchasing department have to negotiate to keep the cost of the product competitive? How many manufacturing people will have to be eliminated to maintain profitability? If this revenue was available for staffing more designers for product development, it would reduce time to market and increase the profit.

Every time you send a service person to your customer with replacement parts, you are paying for poor reliability. All those extra parts in the stockroom to support field service are really dollars set aside as a contingency for warranty claims and poor reliability. All those materials parked in the stockroom are costing you money that can otherwise be actively making money, finding more customers, and hiring more employees.

Reliable products can help you prevent lost warranty dollars. To reverse this loss, the reliability of your products must be improved. Every improvement returns a portion of these lost warranty dollars. Not one, not two, but many reliability improvements will accumulate to return a significant portion of the lost sales dollar for better use.

Further Reading

Reliability Process

H. Caruso, *An Overview of Environmental Reliability Testing*, 1996 Proceedings Annual Reliability and Maintainability Symposium, pp. 102–109, IEEE (1996).

U. Daya Perara, *Reliability of Mobile Phones*, 1995 Proceedings Annual Reliability and Maintainability Symposium, pp. 33–38, IEEE (1995).

W. F. Ellis, H. L. Kalter, C. H. Stapper, *Design For Reliability, Testability and Manufacturability of Memory Chips*, 1993 Proceedings Annual Reliability and Maintainability Symposium, pp. 311–319, IEEE (1993).

Evans, J.W., Evanss, J.Y., and Kil Yu, B. (1997). Designing and building-in reliability in advanced microelectronic assemblies and structures. *IEEE Transactions on Components, Packaging, and Manufacturing Technology: Part A* 20 (1): 38–45. IEEE (1997) & Fifth IPFA '95 Singapore.

S. W. Foo, W. L. Lien, M. Xie, et al. Reliability By Design A Tool To Reduce Time-To-Market, Engineering Management Conference, pp. 251–256, IEEE (1995).

W. Gegen, *Design For Reliability – Methodology and Cost Benefits in Design and Manufacture*, The Reliability of Transportation and Distribution Equipment, pp. 29–31 (March, 1995).

Golomski, W.A. (1995). *Reliability & Quality in Design*, 216–219, IEEE. Chicago: W. A. Golomski & Associates.

R. Green, *An Overview of the British Aerospace Airbus Ltd.*, Reliability Process, Safety and Reliability Engineering, British Aerospace Airbus Ltd., Savoy Place, London WC2R OBL, UK, IEEE (1999).

D. R. Hoffman, M. Roush, *Risk Mitigation of Reliability-Critical Items*, 1999 Proceedings Annual Reliability and Maintainability Symposium, pp. 283–287, IEEE (1999).

J. Kitchin, *Design for Reliability in the Alpha 21164 Microprocessor*, Reliability Symposium 1996. Reliability – Investing in the Future. IEEE 34th Annual Spring. 18 April 1996.

I. Knowles, Reliability Prediction or Reliability Assessment, IEEE (1999).

Leech, D.J. (1995). Proof of designed reliability. *Engineering Management Journal* 5 (4): 169–174.

S. M. Nassar, R. Barnett, Applications and Results of Reliability and Quality Programs, 2000 Proceedings Annual Reliability and Maintainability Symposium, IEEE (2000).

Novacek, G. (2001). Designing for reliability, maintainability and safety. *Circuit Cellar*, January 126: 28.

DFM

D. Baumgartner, The Designer's View, *Printed Circuit Design* (January, 1997).

R. Prasad, Designing for High-Speed High-Yield SMT, *Surface Mount Technology* (January, 1994).

C. Parmer, S. Laney, DFM & T Guidelines for Complex PCBs, *Surface Mount Technology* (July, 1993).

6

Reliability Concepts

The information in this book is intended primarily for people in the design community, managers, CEOs, company presidents, associate reliability engineers – just about everyone except the reliability engineer. The reliability engineer understands the mathematics behind these reliability concepts that is of little importance to everyone else. Fortunately, the mathematics behind the reliability concepts is beyond the scope of this book. We, instead, will focus on the reliability concepts to provide understanding of what they are, how they get applied, and how to interpret the results. You do not need to be a reliability engineer to understand and discuss these concepts. By understanding the concepts and tools, you will have a heightened awareness about product reliability and how it is achieved.

One of the more difficult challenges for someone wanting to implement a reliability program is developing an understanding of all the terms, definitions, and concepts used to describe product reliability. Many of these concepts are mathematically based and can be highly theoretical when pursued to minute detail. This has prevented all but a statistician or a reliability engineer from fully grasping these concepts. Obtaining this great depth of understanding is outside the focus of the book. However, it is vital to have a working understanding of the reliability terms, definitions, and concepts. One reason for developing such an understanding is that it will enable you to hire the right people when you begin to develop a reliability process. Once the reliability process has been established, these terms and concepts will be in common use when discussing product reliability. In some cases, we have oversimplified the explanation in order to avoid messy and confusing mathematics. In our opinion, it is important to understand these concepts because reliability is everyone's responsibility. As the organization becomes more knowledgeable about reliability, the products designed will be more robust and profitable to the company. Therefore, we present these concepts to develop this understanding without laboring over the mathematics behind them. Some fundamental reliability concepts commonly used in product development include the following:

1. The bathtub curve
2. Mean time between failures (MTBF)
3. Warranty costs
4. Availability
5. Reliability growth
6. Reliability demonstration testing (RDT)
7. Maintenance and availability

Improving Product Reliability and Software Quality: Strategies, Tools, Process and Implementation,
Second Edition. Mark A. Levin, Ted T. Kalal and Jonathan Rodin.
© 2019 John Wiley & Sons Ltd. Published 2019 by John Wiley & Sons Ltd.

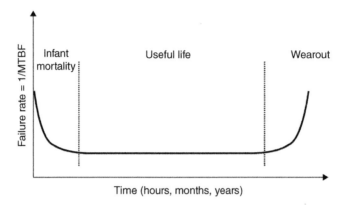

Figure 6.1 The bathtub curve (timescale is logarithmic).

6.1 The Bathtub Curve

The most fundamental concept commonly discussed in product reliability is the *bathtub curve*. Shown in Figure 6.1, the bathtub curve is another way of looking at the cumulative number of failures for a product population operated over time. The bathtub curve is derived from the *cumulative failure curve* shown in Figure 6.2, which is a plot of the running cumulative number of failures over time. For example, suppose you shipped 1000 nonrepairable widgets and then kept track of the total number of widgets that failed through the life of a product. The plot will look similar to Figure 6.2. Often, in the first year, you will see a higher "rate of failure" for the product. Some of the more common causes are variations in the manufacturing process, using parts that have marginal tolerance, insufficient design margin, or an inadequate test process. The failures that occur in this region are referred to as *infant mortality failures*. The failures due to "infant mortality" are considered quality related and are also called early life failures. The next part of the failure curve is called the *useful life*. The failure rate in the "useful life" region has stabilized and may be characterized by a relatively constant failure rate. The failure rate

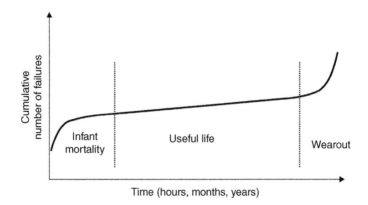

Figure 6.2 Cumulative failure curve.

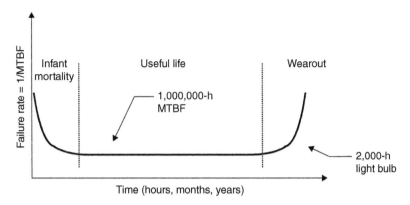

Figure 6.3 Light bulb theoretical example.

in this region is sometimes said to be *randomly occurring events*. After a long time, the product may exhibit a greatly increasing failure rate. This region of the bathtub is called *product wearout*. Here, the time has been reached when the product has consumed its useful life. It may be time for it to be replaced or upgraded.

The failure rate of a light bulb illustrates well the difference between useful life and wearout. Suppose a light bulb has a life expectancy of 2000 hours. The light bulb packaging shows this as the rated life (or median life). The 2000 hours represent the knee of the curve in wearout. The light bulb is not expected to operate after 2000 hours. However, incandescent light bulbs are extremely reliable; for this example, we will assume that it has a one million-hour MTBF. The MTBF represents the mean or average failure rate of the light bulb during its useful life. Figure 6.3 illustrates this difference.

Suppose now, you build a light panel to accommodate one million of these light bulbs. In addition, all the lights are wired together so that they operate at the same time, much the same way Christmas lights operate. During the useful life (the first 2000 hours of operation), one of the one million light bulbs will fail (stop illuminating) every hour. In reality, the light bulbs fail at a random rate. However, during the first 2000 hours, 2000 of the one million light bulbs will have failed. This, by definition, is how the one million-hour MTBF is calculated. Now, if we continue to operate this light panel past the 2000 hours, the rate at which the light bulbs fail increases quickly. It may only take another 400 hours before the majority of the million light bulbs no longer operate. This illustrates how the rate at which failures occur during the useful life is low, but once the product reaches its rated life, the rate at which the light bulbs begin to fail dramatically increases. In essence, failure is expected after 2000 hours, so we can plan for this event. But, failures during the useful life are random and unexpected (one out of a million every hour).

6.2 Mean Time between Failure

The most common term used to describe product reliability is the MTBF. This term measures the failure rate of the product during its normal life. There are other ways to describe the failure rate of a product, which are explained in the following sections.

6.2.1   Mean Time between Repair

Mean time between repair (MTBR) is another way of describing the basic measure of reliability for a repairable system. It is a measure of the average time between all repairs for the systems in the field.

6.2.2   Mean Time between Maintenance (MTBM)

Mean time between maintenance (MTBM) is a commonly used term to describe the reliability of a repairable system. It is a measure of the average time between maintenance (preventive maintenance and repair) for all the systems in the field.

6.2.3   Mean Time between Incidents (MTBI)

Mean time between incidents (MTBI) is a less commonly used term to describe the frequency of an event that may require user intervention to fix. It could be used in reference to an automated system (i.e. a system transport or a tote on a conveyor) that can get stuck and require a user intervention to remedy. MTBI can also be used to describe the mean time between interrupt. Using MTBI requires crisp definition for what constitutes an incident.

6.2.4   Mean Time to Failure (MTTF)

Mean time to failure (MTTF) is a commonly used term to describe the reliability of a nonrepairable system. MTTF describes the average time a collection of systems runs until the next system failure. This term is usually used in cases where the product will not be repaired. Because it is not repaired, it cannot have time "in between" failures in the normal operating sense.

6.2.5   Mean Time to Repair (MTTR)

Mean time to repair (MTTR) is the most commonly used term to describe the maintainability of a system. It is the sum of the time required to fix all failures divided by the total number of failures. The time required to fix the failure typically includes troubleshooting, fault isolation, repair, and any testing that is required to verify that the problem has been fixed. Simply stated, it is the time from when the customer could not use the product to the time the customer could use it again.

6.2.6   Mean Time to Restore System (MTTRS)

Mean time to restore system (MTTRS) is similar to MTTR but includes the additional time associated with obtaining parts to fix the problem.

These definitions are used to describe the frequency of events for a predefined environmental and operating condition. The frequency of these events can vary significantly for different environmental and operational conditions. There are many different ways to discuss a product's reliability. A company producing products globally will be more concerned about MTTRS because it includes the effectiveness of its field service and spare parts logistics. A company producing a disposable electronic product or a product with a short useful life will measure its MTTF. The method you choose to measure your product's reliability may be determined on the basis of your customer's needs, the market forces, and your ability to collect valid data.

The MTBF is defined as the reciprocal of the failure rate in the "constant failure rate" portion of the bathtub curve. MTBF is usually described in terms of hours between failures. The MTBF does not include the infant mortality failure rate and product wear out. To illustrate this point, we consider the reliability of a group of printers. This group may have an MTBF of 10 000 hours (this MTBF number was made up and does not represent the actual MTBF of a printer). This statement implies that the average failure rate of these printers is about 1 per 10 000 hours of operation. However, some printers last much longer. There will be other customers whose experience is that their printers last much less than the average life. In fact, customer experience may vary such that the observed printer time to failure ranges between 6000 and 30 000 hours of use. The reason a printer that has a 10 000 hours MTBF rating and actually operates for only a few hundred to a few thousand hours relates to the bathtub curve. The operational life includes the infant mortality, constant life, and the wearout phase of a printer. Along this timeline, eventually all printers can be expected to fail. Wearout is not usually considered part of the useful life of the product, even though it may be hard to know when the wearout phase began. MTBF should really address only the constant failure rate portion of the bathtub curve, so care must be taken when observing failures. Observed field MTBFs can show a wide range of times-to-failure, while the average (MTBF) is typically determined from testing and field experience. Some printers last much longer than the average and some do not. MTBF is this runtime average that is often used to judge conformance.

6.3 Warranty Costs

When manufacturers use the term MTBF with customers, they must be careful. Typically, the customer sees this number as a sort of guaranteed number. So, when a single purchased unit fails in less than the "specified MTBF time," some form of compensation is often sought from the manufacturer. In general, for a given product with a given MTBF, there will be some products that fail before the MTBF (average) failure rate and some that will operate beyond the expected MTBF. The occurrences of the failure events are random and cannot be predicted; they can only be estimated. The ones that fail, at times less than the MTBF specification, usually impact warranty budgeting as well as customer opinion.

Manufacturers can (and should) use the MTBF number to budget their warranty costs. If a product has an MTBF of 10 000 hours and the product is typically operated continuously, then a one-year warranty (8760 hours) may be appropriate. The manufacturer may absorb any warranty costs when there are early failures (i.e. under warranty). Knowing the MTBF will give the manufacturer a basis to budget the warranty reserve figure for a given product. Using this example, we will next look at a simple situation to get a better understanding of how MTBF impacts warranty costs.

First, MTBF is a failure rate average of many units in operation that were not all manufactured at the same time. It is the fleet average. Keeping it simple, we can look at 100 units that were manufactured in the same short time period manufacturing cycle, say one week. They were probably placed in operation by their many users at about the same time, so their individual accumulated runtimes will often be similar.

In this example, a group of units will exhibit a fleet MTBF that is close to the specified MTBF figure, say 10 000 hours. So, this minifleet of 100 units will exhibit failures (randomly), where the average failure rate will be 10 000 hours. Some customers will see no failures; while others will experience one or more failures. Without going into the detailed math (see Appendix B), there will be approximately 37 units that will not fail during the specified 10 000 hours. These customers are the lucky ones selected by the randomness of failure events. Some users will be unlucky and may have two, three, or even four failures in the same 10 000 hour time period! A simple example follows.

Suppose you have a product with a 10 000 hours MTBF and you just sold the first 100 units. During the first 10 000 hours of use, the following can be estimated:

1. A failure about every 100 hours. This is the average (mean).
2. A total of 100 failures in the 10 000-hour time period.
3. The occurrence of failures is randomly distributed.
4. Some units (26) will have more than one failure.
5. Some units (37) will exhibit no failures.

The mathematics behind these estimations is presented in Appendix B.

How can you avoid much of this grief? Use reliability techniques so that your MTBF is very high. Even a product with a one-million-hour MTBF may exhibit a first failure (in a fleet of 100) in about 10 000 hours. But there will be 99 very happy customers because the next failure isn't (statistically) expected for another 10 000 cumulative hours.

A warranty requires a set-aside of sales dollars, intended to absorb warranty costs, during the warranty portion of the sold product. A product unit will experience one failure, on average, for every accumulated MTBF time period. Manufacturers must take this expected failure accumulation into consideration so that there are funds available to absorb this expected cost.

Manufacturers who sell products that have very high reliability and therefore high MTBF figures typically have very low warranty costs. They have happier customers who reorder products from this same manufacturer time and again. When we consider the number of failures in this example, all the failed units will be quickly repaired and placed back into service. Table 6.1 illustrates how MTBF relates to in-warranty failures for several warranty periods for repairable units.

See Appendix B for a more detailed discussion on MTBF and warranty budgeting.

Table 6.1 Failures in the warranty period w/different MTBFs.

Warranty period (yr)	MTBF = 8760 h	MTBF = 87 600 h	MTBF = 876 000 h
1	100 failures	10 failures	1 failure
2	200 failures	20 failures	2 failures
5	500 failures	50 failures	5 failures

6.4 Availability

When a system fails and becomes dysfunctional to the customer, the time it takes to rectify the problem is critical (MTTR). Long repair times take a lot of productivity out of the usefulness of a continuously operating product, so having short MTTRs is a critical availability issue. Simply stated, static availability may be expressed as

Availability = MTBF/(MTBF + MTTR)

Example: With an MTBF of 10 000 hours and a 10-hour MTTR:

Availability = 10,000/(10,000 + 10) = 10,000/10,010 = 0.999 or 99.9%

The reader can see that bigger MTBFs can drive availability to nearly 100%. Small MTBFs drive availability down, even with the same MTTR figure. A customer who experiences frequent outages needs to correct the problem quickly. If the MTTR is very short, less than one hour, the impact is relatively small. But when the MTTR is long, such as several days, the availability can become very small. This customer will be unhappy. It is in the manufacturer's best interest to have high MTBFs, but they must also drive toward short MTTRs. Because parts availability is a major contributor to bringing a product back into operation, addressing this issue early in the reliability improvement planning is recommended.

Short availability is not a savior when the MTBF is low. Some manufacturers produce complex equipment that requires skilled, on-site, service personnel. If they can fix everything in a few hours on average, but there are many units operating, they will be fighting an uphill battle. For illustration, we will consider a customer who has 100 units in operation. With an MTBF of 1000 hours and a 4-hour MTTR, there will be a new failure every 10 hours on average. Service will have to fix these failures that are expected to accumulate at about 2.4 failures per day if the equipment runs 24/7 hour. The repair person can be expected to fix an average of two units per 8-hour shift and have another 0.4 units to repair before going home. The repair service may require more service personnel. Logistical constraints often drive MTTR requirements even higher. Long procurement from a few hours to a few days or even weeks may be observed. With an average of 2.4 accumulating failures per day, it is easy to see that the repair team will be quickly overwhelmed.

As failures appear, customers may have reduced output from their production. The customers will consider other manufacturers the next time they select equipment for the production line.

Shorter MTTRs translate into service capabilities that can correct field failures quickly. This means trained service personnel and readily available spare parts. Service/repair personnel have to be available to the customer almost immediately. This can be accomplished in eight ways:

1. On-site manufacturer service personnel
2. Customer trained service personnel
3. Manufacturer training for customer service personnel
4. Easy-to-use service manuals

5. Rapid diagnosis capability
6. Repair and spare parts availability
7. Rapid response to customer requests for service
8. Failure data tracking

6.4.1   On-site Manufacturer Service Personnel

Customers, particularly new customers, often do not have employees who are knowledgeable in troubleshooting, diagnosis, corrective action, and spare parts selection for their newly acquired devices. Providing factory service personnel to the customer at the customer's facility will shorten long MTTRs.

6.4.2   Trained Customer Service Personnel

Knowing that the customer will need trained personnel, the customer's service team training should be completed before delivery of the product.

6.4.3   Manufacturer Training for Customer Service Personnel

The manufacturer should provide training to the customer's service personnel. If the customer prefers "before delivery training," it can be done at the manufacturer's facilities. To save the customer money, some training might be provided at the customer's facilities with prototype systems. Customers often desire this added service, but it may be at the expense of not having a physical unit to work on if their own unit has not yet been delivered.

6.4.4   Easy-to-Use Service Manuals

Part of the new product development process for increased reliability is the failure modes and effects analysis (FMEA) process. One of the deliverables (outputs) of an FMEA is a detailed understanding of how the product can fail. This information is invaluable to the service manual writers. They can use the FMEA to create a service manual that considers all the issues discussed by the design team. Service manuals will vary greatly, depending on the end product, but the FMEA process will always be an abundant source of information to help in the development of the service manual.

Manuals should be readily available to the customer. Many companies now place the manuals online. Providing a place on your web page to link to these manuals will shorten the MTTR of the end product. Placing manuals on CDs is another option.

6.4.5   Rapid Diagnosis Capability

Factory training, along with the service manuals, can be very useful, but some special designs can be difficult to diagnose. This may mean that special tools, test equipment, and software are required to help speed the diagnosis. Part of the service manual should have a recommended list of tools available for rapid diagnosis. Expensive tools or special tools that can be obtained only from the manufacturer often hinder rather than help in driving down the MTTR. Even a common tool that can be obtained almost anywhere in the United States may not be very easy to find in other parts of the world.

6.4.6 Repair and Spare Parts Availability

Wear items may need to be on hand at the customer's facility because of frequent failures. These include filters, fluids, lubricants, computer disks and tapes, for example. A list should be available to the customer with a kit at the customer's facility, so that the delays caused by these frequently used items are eliminated.

Other parts and subassemblies should always be in stock with the manufacturer for rapid delivery to the customer. A complete system of parts ordering, inventorying, and shipping should be in place to address the needs of the business. Globally dispersed customers require several parts locations to help speed their delivery to the point where needed.

6.4.7 Rapid Response to Customer Requests for Service

The parts and service department at the manufacturer's facility should be designed to make it easy for the customer to identify what is needed and to obtain the necessary pieces rapidly. Special software and internet capability may well be the solution. Here is where the quality of the naming and numbering of parts is very important. The customer may not have the same detailed knowledge as the manufacturer with the parts system(s) and may order the wrong part. This adds to the repair delay and greatly increases the MTTR. Make the parts ID system easy to use.

You must learn from the salesforce if the customer feels that the system is useful. Find out what needs improvement and make changes accordingly.

6.4.8 Failure Data Tracking

Gather the data from the field to learn about troublesome areas. Use this information to eliminate problems. Often, this information leads to design changes that can improve new products. These changes may save a lot of money and time during subsequent product development. Tracking the failure data everywhere can lead to rapid discovery and corrective action. A comprehensive failure reporting, analysis, and corrective action system (FRACAS) will pay large dividends by lowering warranty costs and increasing the MTBF, thus lessening the impact of existing MTTRs.

Figure 6.4 illustrates the relationship between, MTBF, MTTR, and the resultant availability.

6.5 Reliability Growth

Reliability growth is the process of measuring the product reliability improvement when failure mechanisms are permanently removed. This was first described by J.T. Duane, who derived that as time passed the interval between failures increased [1]. Simply stated, the MTBF increased after each failure mechanism was designed out of the product. For every defect that is removed from the system (meaning the failure cannot ever happen again in any system), the resultant MTBF must increase.

There are essentially two ways in which companies deal with reliability problems. The first choice is the one many manufacturers take, that is, to accumulate field failure information. When field data overwhelmingly indicates there is a problem, they investigate

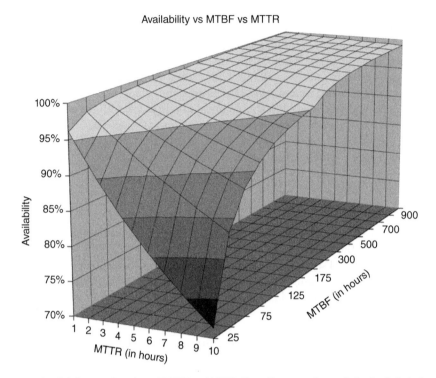

Figure 6.4 Availability as a function of MTBF and MTTR. *Note:* The curve has a slight ripple in it due to change in the MTBF axis. For the range between 0 and 200, it is marked in 25-hour increments and in 100-hour increments thereafter. This was done for resolution purposes to illustrate the impact of both low MTBFs and long MTTRs.

the failure to find the root cause. The problem then gets fixed and is closed with an engineering change. Hopefully, the design change becomes a recommendation for future designs. The total process from identifying a problem to implementing a fix may take considerable time, occasionally years. By then, the cost impact of the problem may have become significant. Because the resolution of the problem can take years, there is a good chance that new designs would be completed without the benefit of this knowledge. The same old mistakes get designed into new products and cause similar problems. This is a form of reactive field failure analysis and corrective action. The process of finding failures and fixing them usually delays the product from reaching design maturity by a couple of years. Design maturity should ideally be achieved before the first units are shipped. Product recalls are the result of not identifying what will fail in the field before shipment. The second, preferred way to deal with reliability problems is to use reliability growth proactively, as is shown in Table 6.2. The new product development process should have testing with data gathering (FRACAS) that can help the designers find these failure mechanisms well before the design is finalized. FMEAs and Highly Accelerated Life Test (HALT) are two very useful tools for identifying potential reliability issues early in the design process. The reliability person should learn of and track every failure with the design person until the root cause is found and the design fix is implemented and validated.

Table 6.2 Advantages of proactive reliability growth.

Proactive	Reactive
Rapid problem discovery	Data more difficult to obtain
Failure data can be more accurate because the design team collects it	Data clouded by less-skilled service personnel
Equipment is easier to modify in-house	More costly to modify (recalls)
Can react to the first incidence	May need several occurrences
Manufacturer's reputation maintained	Manufacturer's reputation decreased

6.6 Reliability Demonstration Testing

RDT is often confused with reliability growth. In the last section, we showed how reliability growth is used to bring a design to maturity before the first customer shipment. The more mature a design is at product launch, the lower the warranty cost will be. This is achieved by investigating all failures early in the design and keeping track of the company's corrective action to closure.

RDT is a process that statistically ensures that the reliability goal set early in the program is met. There can be confusion regarding the order in which reliability growth and RDT takes place. Reliability growth always precedes RDT. We can start reliability growth measurements once the design is defined. FMEA and HALT identify problems; these surface and then get tracked using reliability growth techniques. RDT cannot start until the design is finalized and there is a final product to be tested. Put another way, reliability growth is a tool that improves the MTBF of the product and the fruits of that effort are verified through RDT.

Earlier in this chapter, we showed that even though a system has a specified MTBF, systems in the field will demonstrate a variety of different MTBFs. The MTBF we specify for the fleet is a measure of the "average MTBF" for all the systems in the fleet. Individual systems in the fleet will experience failures at different times, at different rates and have different MTBF experiences. The fleet MTBF is the accumulated runtime for the entire population in the field, divided by the sum of all the field failures for that same population. An individual single system MTBF varies because of the random nature of failure events. The random nature of these events can be mathematically modeled as a confidence interval. The confidence interval represent the range of expected values that are likely. As the confidence interval increases, i.e. greater confidence that the measured value will be between confidence interval, the range of expected values will also increase. The MTBF calculation assumed an exponential distribution and the confidence interval is based on a chi-squared distribution that is not symmetrical. The upper and lower limits will likely not be symmetrically distributed around the mean MTBF value.

RDT is the tracking of the accumulated product runtimes and the number of failures to verify that the product has achieved its MTBF goal. RDT is a way to show, through product testing, that the product indeed achieves the specified MTBF promised. RDT is a common method used to verify that the design meets a contractual reliability requirement. Reliability engineers are responsible for tracking the accumulated runtime hours

Table 6.3 RDT multiplier for failure-free runtime.

Confidence bounds (%)	Failure-free multiplier
95	3.0
90	2.3
85	1.9
80	1.6
75	1.4
70	1.2

against the number of accumulated failures. To verify that a system has achieved its stated MTBF, we will need to accumulate more product runtime without a failure. This total time is always more than the desired MTBF time.

There is a rule of thumb that can be applied to RDT. The rule is used to verify that a system has achieved the specified MTBF goal at 90% confidence, that a system must run failure-free for 2.3 times the desired MTBF goal.

For example, we can say a system has a demonstrated MTBF of 1000 hour at 90% if it has functioned properly for 2300 hour without a failure. Recall that, when we talk about MTBF, there is some uncertainty specified around it. This rule assumes that the design maturity goal has been met with a 90% confidence level. If we wanted a higher confidence level, then the runtime without a failure will have to increase. Likewise, if the design maturity goal can accept a lower confidence level, the failure free runtime will decrease. Refer to Table 6.3 to see the effect of the confidence interval on failure-free runtime. A 90% confidence interval, typically, is widely accepted.

There is always a confidence level associated with RDT. The confidence interval is the way to account for system variability and determine if it is acceptable. Let's imagine that a product has an MTBF of 8760 hour. With an exponential failure rate, there will be one failure every year. Of course, not every product will experience one failure every year; life is not that simple. There will be some users who will have no failures in the first-year time period and some who have one, two, or more failures in the same first year. The variability is defined through a confidence interval. The confidence interval takes into account that not all users will have the same experience in the product's reliability. The way we deal with this random nature is through the use of confidence intervals to account for the variability in the reliability experience for RDT.

When a system or a number of systems in the field have demonstrated that their accumulated runtimes have reached at least 2.3 times the stated MTBF of 1000 hours without a failure, we can state statistically that there is a high degree of confidence (90%) that the product has demonstrated a 1000-hour MTBF. By using more than one system in the RDT, the manufacturer can speed up the process. No manufacturer can state, however, precisely how long any individual unit will operate properly before failure.

What do you do when there is a failure during the design maturity testing to verify the product's MTBF goal? If this happens, does it mean that you cannot achieve your product's MTBF goals? No. There are still more statistics that can be helpful.

This point is illustrated in Figure 6.5 for demonstrating a 1000-hour product MTBF.

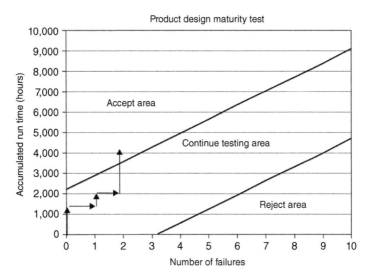

Figure 6.5 Design maturity testing – accept/reject criteria.

The *x*-axis is scaled for the number of failures. This test can be set up for one system or more. More is always better! The *y*-axis is in accumulated runtime hours. Again, this can be for one system or for more. When using more than one system to determine RDT, you must add the runtime hours and total the number of failures for all systems being used in the test. If you use two systems in the RDT, the accumulated runtime hours will collect at twice the rate as that of a single system. Using 5 or 10 systems will accelerate the RDT.

In Figure 6.5, small arrows depict what occurred when trying to demonstrate to perform a real RDT on a product that has a stated 1000-hour MTBF. The RDT begins at the origin, zero failures and zero accumulated runtime hours. For illustration, we show the first arrow as representing the test. This arrow lengthens and runs until there is a first failure at approximately 1500 hours. Then, the second arrow points to the first failure at the "number of failures" and extends until 1500 hours. The unit is repaired and the test continues, and the third arrow climbs toward the accept line to represent this condition. When the unit reaches approximately 2400 hours, there is a second failure. Again, the unit is repaired and the test is resumed. Then, there are no more failures and the unit passes through the "accept" or "pass" line at approximately 3600 hours. This is the statistically required point in accumulated runtime hours that a system must run with two failures to have "demonstrated" that it has an MTBF of 1000 hours with a 90% confidence interval. If the system had no failures, the accumulated runtime hours needed would only have to have reached 2300 hours. As you can see, there may be times when doing an RDT that you might experience failures before the arrow passes through the "accept" line. This is still statistically correct. If the failures collect too rapidly and the tracking line passes through the reject line, then stop the RDT because the test has statistically proven with a 90% confidence interval that the reliability of the product is not capable of operating at the specified MTBF.

If the arrows never punch through the accept line, then you have more work to do in reliability growth before you should again try to demonstrate the final system MTBF. If the arrows soon punch through the fail line, then you must have decided to do the RDT prematurely or you simply have greatly overstated the system MTBF.

Can you place five units in the RDT for a while and then remove a few units and still have a statistically correct result? Yes, as long as the accumulated runtime hours and the total number of failures are recorded. Five units can speed the process. Removing a few units from the RDT will slow it down.

Can you place five units in the RDT and halfway through remove some, install some new units and continue? Yes, statistically, but here it gets a little tricky. You must be certain that the infant mortalities have been removed from any of the new units. This is true for all the above examples. If the early failures are not removed, the RDT will, in all likelihood, fail. If you place units in and out too often, then the test may become invalid.

Can you perform a continuing RDT to ensure that the stated MTBF is remaining to specification? Yes, place a group of units in RDT, remove them after a period of time for shipment, place new units in the same RDT, remove them for shipment, and so on. The idea is not to consume too much life from the units while testing them. The RDT test can be very challenging when the MTBFs are very long. For an MTBF of a million hours or more, you would have to demonstrate 2 300 000 hours without a failure.

What is so magical about the number 2.3? It is simply the statistical correction number that acts as a multiplier for the 90% statistical confidence interval. If you wanted more confidence, the number would be somewhat higher, and vice versa.

6.7 Maintenance and Availability

Electromechanical systems that are repairable will require maintenance at some point in time to keep them operating. Maintainability is the probability that an item that has failed can be successfully restored to an operational state. The maintenance can be either a preventative or a reactive/corrective activity. A preventative maintenance activity is intended to either prevent or delay the electromechanical system from malfunctioning. Preventative maintenance can be done based on a schedule, an opportunity (i.e. when other work is being done due to an unscheduled downtime), a warning indicator or based on a sensor/feedback system designed to detect degradation and wear. The maintenance method chosen will have an impact on maintenance cost and failure probability. A corrective maintenance activity is performed when there is a failure that needs to be repaired. Corrective maintenance activities can require part replacement or an adjustment to return the electromechanical system to an operational state.

Maintainability refers to the ease with which repairs can be made to keep the system operational and available. The MTTR is the sum of the time required to fix all failures in the fleet divided by the total number of failures in the fleet. The MTTR is the average repair time per unit that fails. The time required to fix the failure typically includes diagnostic troubleshooting, fault isolation, waiting period if replacement parts are not available, repair, and any testing that is required to verify that the problem has been fixed. Simply stated, repair time is the duration during which the customer could not use the product. Maintenance refers to an operations related activity to keep the electromechanical system operational or to restore it to an operational condition after a failure.

Availability is the measure of the total time when a electromechanical system is up and operational.

6.7.1 Preventative Maintenance

Maintenance is the activity of restoring a product or system to its nominal or operational condition. The result of the maintenance activity can either restore the system to a like-new condition or a state not as good as new, but more reliable than before the maintenance was performed. *Restorability* describes the degree to which the system can be kept in a working state and the probability of a failure in the future. It should be noted, that preventative maintenance can result in the system being less reliable. An example of this would be if maintenance is done wrong or replaced with faulty or counterfeit parts.

When maintenance is part of a scheduled activity (i.e. a maintenance schedule), it is considered to be preventative maintenance. The maintenance schedule can be based on time or operating time, distance, meter reading, or an event. Automobiles follow a maintenance schedule provided by the manufacturing for when service needs to be done, parts replaced, or items inspected. The philosophy behind preventative maintenance is that it reduces the likelihood of a failure. Often, preventative maintenance schedules are based on failure data and history. Replacing the oil in an automobile is another example of preventative maintenance. Oil is not tested to determine if it is good or degraded. The preventive maintenance schedule or event is typically based on collected failure data from a similar part. It can be based on past experience and is intended to be conservative so that the maintenance event occurs well before the risk of a potential failure.

As an example, consider a fan with a rated useful life of 25 000 hours. The fans do not all fail at 25 000 hours, but that is the mean useful life for the fan. If you assume the failure probability follows a Gaussian distribution as a function of time, then the probability of failure as you approach wearout might look like Figure 6.6. *Note:* Many mechanical wearout failures follow a log-normal distribution because there is a long failure-free time before wearout-related failures start to occur.

Using a preventive maintenance strategy, you need to decide a replacement point for when the fan would be replaced. The determination for when the fan is replaced is

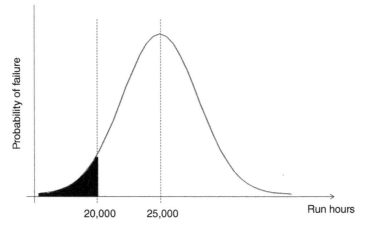

Figure 6.6 Number of fan failures vs. run time.

based on the risk tolerance you are willing to accept for allowing a fan fail before it was replaced. If failure could cause catastrophic collateral damage, be a safety issue or the unscheduled down time result in significant lost revenue, the risk tolerance will be low. The lower the risk tolerance for a failure, the sooner the schedule for fan replacement.

If the maintenance strategy is to replace the fan after 20 000 hours, the probability that a fan could fail before it is replaced is illustrated in the shaded-in black area. The fan may have another 5000–10 000 or more hours of useful life left in it, but preventative maintenance schedule is based on replacing the fan while it is still in a good state. It should be clear that this type of approach results has a higher maintenance cost. To lower the maintenance cost, a predictive or a prognostic and health management (PHM) approach is needed. If failure can't be tolerated at all, then a redundancy strategy is required.

6.7.2 Predictive Maintenance

Preventive maintenance can result in replacing components before they need to be replaced because the replacement decision is based on a schedule and not the condition of the component or system. For example the automobile manufacturers may recommend an oil change every 12 000 mi or 1 year, whichever comes first. The oil may still be good and not in need of replacement. To know if the oil is starting to degrade, a breakdown requires an analysis. If you are able to analyze the oil to determine if it has degraded, you can make a decision regarding when to change it that is more cost-effective. This is the theory behind predictive maintenance.

Predictive maintenance is a technique used to determine the condition of a component or system. If you can sense that something is starting to degrade you can make a better-informed decision regarding when it should be replaced or maintenance should be performed. This provides a much more cost-effective approach to maintaining a system and reducing unscheduled downtime.

Predictive maintenance uses techniques to determine wear, degradation, and faults like cracks and embrittlement. Predictive maintenance requires the user to make measurements to determine the health of the component. The measurements could be a thermal image to look at areas of hotspots, high friction wear, or signs of thermal runaway. It could be vibration analysis as a way to figure out if bearings or gears are wearing and might fail soon. But there are many other ways that can be simpler and less expensive. Inspecting your tires and using a gauge to measure tread wear is a predictive maintenance technique to determine when the tires should be replaced.

6.7.3 Prognostics and Health Management (PHM)

Predictive maintenance offers a significant cost advantage over preventive maintenance to minimize the cost of maintenance and unscheduled downtime, but there is a third approach to further minimize unscheduled down time and better manage maintenance costs. The method is called PHM and is based on the theory that most electromechanical systems do not go from a good state to a failed state instantaneously when operating under acceptable use conditions. Instead, they start to wear or degrade, and over time, the degradation becomes sufficient to cause a failure event. The PHM approach provides a methodology to detect degradation for a component, subsystem, or system. Once you have determined that it has degraded to an unacceptable level, PHM defines a process and methodology to determine when and how much degradation has occurred and estimate the time to failure or remaining useful life (RUL). To determine degradation and the time to failure requires an understanding of how it will fail, the failure mechanism, and a way to measure or sense degradation. The process is the same for electrical, mechanical, and electromechanical systems. On-board sensors can measure real-time operation and performance. PHM provides the following advantages over a preventative maintenance or predictive maintenance methodology:

- PHM provides an early indication when degradation is starting to occur.
- The number of unscheduled maintenance events is reduced and availability improves.
- The cost of ownership is reduced due to fewer maintenance cycles, better managed part inventory cost, and fewer or possibly no unnecessary inspections.

As an example, consider the fan described in Section 6.7.1. Using a preventative maintenance approach, the fan would be replaced after 20 000 hours. If the system was not designed to track actual fan run hours, an alternative method for estimating fan run hours would be made that is likely more conservative and results in the fan being replaced even earlier.

Using a predictive maintenance technique, you would test the fan under load to see how it is performing. If the fan performance has degraded below a threshold level (i.e. the fan output in cubic feet per minute [CFM] decreases), it would trigger a warning for a replacement event. This approach requires a maintenance schedule to monitor fan wear. The maintenance schedule interval to monitor fan wear needs to be set so the fan is unlikely to fail in between maintenance intervals. As the fan run hours increases, it may be necessary to reduce the maintenance intervals to prevent a fan failure. The PHM approach uses real-time sensor data and analysis to monitor and detect fan wear. There are many ways to detect fan wear resulting in degradation. If the fan wear is due to the bearings degrading, it will cause the fan to vibrate more. Fan vibration can be monitored using an accelerometer mounted in an area where it can detect bearing vibration. As the bearings wear, there will also be an increase in the amount and frequency spectrum of acoustic noise emitting from the fan. An acoustic sensor can be used to monitor the amount of noise and the frequency spectrum as a strategy to detect fan-bearing degradation. When the fan starts to wear, the speed of the fan slows down, so depending on the type of fan sensor it could be used to determine fan degradation.

PHM provides a process and methodology to compare real time data to an expected good set of data so a determination can be made regarding its health. The PHM process can indicate if there is a change in the operational health and provide a prognosis for future health. Developing a PHM process for an electromechanical system requires determining how it may fail or degrade over time. This can be accomplished through a failure modes, mechanism, and effects analysis (FMMEA). The FMMEA provides a systematic methodology to investigate how it will fail, what could cause it to fail, and the mechanism or mechanisms that can cause it to go from a good to a bad or degraded state. The failure mechanism is the process that physically causes it to change in health and can be due to an overstress or wearout. The overstress can be mechanical, thermal, or electrical radiation, or a chemical reaction (Figure 6.7). For example, an electrical overstress could cause dielectric breakdown in a device. Likewise, physical wearout can

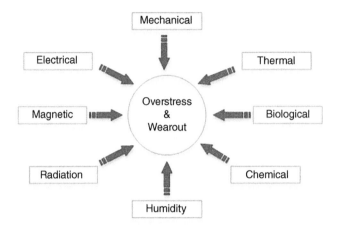

Figure 6.7 Mechanism that can cause degradation and failure.

Table 6.4 FMMEA for fan bearings (detection omitted).

	Failure	Failure mode	Failure cause	Failure mechanism	Occurrence	Severity	RPN
Fan	Ball bearing	Brinell mark	Mechanical overload	Material yields	3	3	9
		Furrows	Moisture	Corrosion	2	2	4
		Spalling	Cyclic overload	Fatigue	1	2	2
		Seizure	Thermal overload	Lubricant break-down/deterioration	5	4	20

also be due to the same mechanisms, namely: a mechanical, thermal, electrical, radiation, or a chemical reaction. The FMMEA helps determine which failure modes and mechanisms are good candidates for a PHM approach and provides insight on how to monitor its health and determine when a maintenance action is needed. For each item in the FMMEA that surfaced as a high-priority concern, a method for fault detection, diagnostics, and prognostics needs to be developed.

For the fan example earlier, the FMMEA would identify all the ways the fan can fail, what would cause it to fail, and the mechanisms at play for each failure cause. An example of this is shown for fan bearings in Table 6.4.

PHM requires a method to capture the health of the system in real time. Typically, sensors are used monitor the health of the system. There are many different types of sensors that can be used to monitor the health of a system (Table 6.5). The FMMEA can be used to determine the most likely components that will degrade over time, what will cause it to degrade and the mechanism causing degradation. After the degradation mechanisms are determined, an accelerated life test is developed to capture data for a good healthy state, transition to a degraded state and eventual failure.

Alternatively, sensors can be placed on an existing fleet to collect baseline and degraded failure data. The development of an accelerated life test plan would include

Table 6.5 Sensors to monitor for overstress in wearout degradation.

Overstress and wearout mechanism	Sensors
Biological	Electrochemical
	Magnetic
	Piezoelectric
	Thermometric
Chemical	Electrochemical
	Mass
	Optical
	Thermochemical
Electrical	Capacitive transducer
	Hall effect strain gauge
	Inductive transducer
	Thermal transducer
Humidity	Capacitive
	Resistive
	Thermal conductivity
Magnetic	Hall effect
	Magnetometers
	Magneto-optic
	Magneto resistive
Mechanical	Capacitive strain gauge
	Inductive impedance strain gauge
	Piezoelectric strain gauge
	Piezo resistive strain gauge
Optical	Fiber optic
	Photoconductor
	Photo emissive
	Photovoltaic
Thermal	IC sensor
	Resistive ther+A3:B9
	Thermocouple

the selection of a sensor or multiple sensors to monitor component health. Choosing different types of sensors to monitor health will allow determination for the best type of sensor and if multiple sensors are needed to monitor component wear and degradation. It may be necessary to use data from multiple sensors to get a better representation for how the component will degrade and eventually fail. An example of this is shown in Figure 6.8, where the circles represent the baseline data for a good system over time. The squares and triangles are real-time sensor data. PHM uses mathematical techniques and algorithms to process the data. The PHM methodology can involve a lot of data, and this data may need signal processing and conditioning before it can be compared to the baseline data. Typically, the first step is to clean and normalize the raw data into a more user-friendly data set. The square sensor data in Figure 6.8 requires

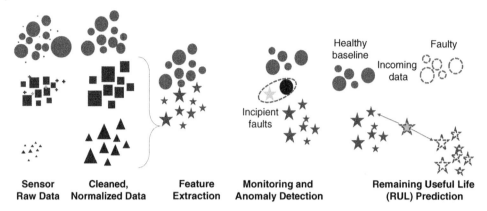

| Sensor Raw Data | Cleaned, Normalized Data | Feature Extraction | Monitoring and Anomaly Detection | Remaining Useful Life (RUL) Prediction |

Figure 6.8 PHM data collection and processing to detect degradation. Source: courtesy Anto Peter.

signal processing to eliminate the noise that is in the data. The multiple "+" signs represent the noise in the square sensor data. In this example, in addition to removing the noise in the square data, additional signal processing is needed to normalize the triangle sensor data. In this example, the square and triangle sensor data are used to create new set of data, illustrated as "stars," to compare against the good healthy state. The extracted data can be compared against the healthy baseline data looking for an anomaly. Once an anomaly is detected, a determination is made regarding the amount of degradation in the time remaining before failure. The time until failure is referred to as the "remaining useful life" (RUL). The time until failure is used to manage maintenance costs and prevent unscheduled downtime.

Returning to the fan example, the FMMEA identified bearing failure due to lubricant breakdown and deterioration as the highest-priority failure mechanism. Lubricant deterioration in the bearings can be physical (i.e. a loss of lubricant due to evaporation) or chemical deterioration (i.e. oxidation of lubricant, polymer separation breakdown consumption of antioxidant). This can occur due to mechanical loading, temperature, humidity, or electric fields. A model for ball-bearing degradation and failure due to lubricant breakdown and deterioration is then developed using mechanical loading and temperature in an accelerated stress life test. As the ball bearing wears out, there will be an increase in vibration, audible noise, heat generated, and power consumed. Sensors can be selected to monitor each of these characteristics (Table 6.6). The vibration and noise sensor data are evaluated for changes in power-level frequency spectrum. It may be possible to detect bearing degradation in the frequency domain as well as power level.

Table 6.6 Sensors to monitor bearing degradation.

Failure characteristic	Sensor to monitor health
Vibration	Accelerometer
Noise	Sound transducer
Temperature	Thermocouple
Power consumption	Power transducer

6.8   Component Derating

Earlier, we stated that the reliability of most components has improved 10–100 times in the last two decades. There is anecdotal evidence showing as much as a 10 000-fold improvement. Improvements of four orders of magnitude may be true in some cases, but using components properly and not overstressing them will pay large dividends in terms of system reliability. This will make components a very small part of the unreliability picture.

A design practice that selects parts to be stressed (in circuit) to a value well under their individual rated limits is called *derating*. An important specification for capacitors is breakdown voltage. It is well advised to select capacitors that have voltage and temperature ratings well above the specific needs of the circuit design. Diodes may have a peak reverse voltage rating. Selecting diodes that have a rating well above that of the circuit application will lower the stress on the diode, and it will perform longer in the application. There are so many different electronic components that they cannot be enumerated here. The three components noted here often have several parameters that should be derated, not just those mentioned. Examples of such derating systems include MIL-STD-975.

Derating can be seen everywhere and is similar to the factor-of-safety used in a mechanical design. Sometimes it is for public safety. The maximum weight requirements for an elevator is one form of derating.

In the simplest terms, it means that if a component, such as a resistor, can dissipate 1 W, use it in a circuit that never requires more than half of that, and it will withstand the stresses of the circuit longer. Many components are subject to accumulative fatigue due to applied stresses. Derating reduces the impact of these stresses and greatly extends the life of the component. When derating is also applied to the load, the performance may improve as well.

In all engineering specialties, there are specifications for parts of every sort. Valves have pressure limits, cables have load limits, and materials have temperature limits. Design the use of these and all components such that, by design, they are not the critical part of the reliability picture.

All electrical components should be derated to optimize the life of the component, reduce component failure rate, improve manufacturing yield, and reduce the potential for a safety hazard. Every electrical component should have a manufacturer's datasheet that identifies the component's electrical performance specifications when used within the operating limits specified in the datasheet. The component manufacturer should also list the device's absolute maximum limits. If the absolute maximum limits are not stated in the datasheet, the manufacturer should be contacted to find out that information.

In general, an electrical component should meet all the electrical specifications in the datasheet when it is used within its stated operational limits. When a component is used above the operational limits but below the absolute maximum limits, it may no longer meet all the electrical specifications called out in the datasheet. Using a component above the recommended operational limits but below the absolute maximum limits can result in an increase in the component failure rate and a reduction in its useful life. A component should not be used above its absolute maximum limits for any period of time. There is a process for qualifying an electrical component for use above the rated operational limits but below the absolute maximum limits called *uprating*, which is discussed in the next section.

Unfortunately, component manufacturers may use different terminology to describe the electrical performance and limits in their data sheet specifications which can lead to misinterpretation. As an example, consider the maximum temperature rating for an electrical component. If not specifically stated, the temperature rating can be ambient temperature, package temperature, or junction temperature. Even die junction temperature might mean either bulk die temperature or the peak junction temperature. The bulk die temperature can be 10–20 °C lower than the peak junction temperature.

If your organization does not have a set of component derating guidelines, there are industry standards that can be used. The Reliability Analysis Center (RAC) has a book called *Electronic Derating for Optimum Performance*. There are many other derating guidelines, each targeting different types of industries. Derating guidelines are a guideline for best practices and should not be used as a design rule. When a component violates the derating guidelines, it should trigger an investigation to determine the severity and risk from a system perspective. The first step is to contact the manufacturer to see if they can provide guidance about the impact to reliability and performance. It is possible that the manufacturer has done testing and is able to answer the question. It's also possible that the manufacturer derated their product specification so there may be some margin. The manufacturer may also have derating charts that will provide better guidance for component derating compared to the generic industry standard you are using for that component. If after you investigate you still cannot meet the derating requirements, you have two choices left. You can select a different component that meets the derating requirements or use engineering resources and money to qualify the use of the part outside the supplier-recommended operating limits (uprating).

6.9   Component Uprating

There is a process to evaluate and qualify a component for use above its operational limits when all alternative solutions have been exhausted and none were acceptable. The process is called uprating and requires testing and characterization with sufficient sample lots at accelerated stresses to determine acceptability. However, many manufacturers will not take responsibility or assume liability when a component is used outside its specified operational limits.

Electronic components are available in many different classifications. The majority of the electronic components used today can be classified as "commercial off the shelf" or COTS parts. Commercial grade parts are typically the lowest-cost version for that component. Other component grade classifications for electronic parts are industrial, military, and automotive (Table 6.7). It is not uncommon for an electronic component to be offered in several different component classifications, described by the component part number, as part of the available options. It is important to know what the differences are for the same component available in two different classifications. Sometimes the only difference is the temperature range for which the part can be used. It is possible for a semiconductor component that is available in a commercial and industrial classification to have slightly different parametric specifications. The parametric differences will be called out in the datasheet, often in a table so they can be compared. Sometimes the only

Table 6.7 Component grade temperature classifications.

Component classification	Temperature range (°C)
Commercial	0 to 70
Industrial	−20 to 85
Military	−55 to 125
Automotive	−40 to 105
	−40 to 125

difference between a commercial and industrial version part by the same manufacturer is that the industrial version is tested at a higher stress by the manufacturer.

Uprating should be considered only after all other alternatives have been exhausted to find an acceptable solution that complies with the manufacturer's recommended operating limits and the derating guidelines. Uprating can be expensive; it requires significant resources and time to qualify a component for use above its recommended operating conditions. It is recommended to discuss the requirements with the manufacturer, because there is no guarantee that the component can be qualified for use above its recommended operating conditions. Uprating requires determining which parameters do not conform to the design requirements. These parameters will need to be recharacterized when used above the recommended operating limits. The qualification process requires determining an appropriate sample size, the number of different production lots, the testing required, and the definition for success to achieve the desired confidence level. Before testing begins, determine which parameters are critical and which parameters can be relaxed. Then define the test margins required for each of the parameters to be tested. The qualification process needs to be completely documented, including all data sheets that were part of the evaluation, equipment used including model number, serial number and calibration, types of sensors used and where the placed, the test performed, including dates and test results.

If uprating is outside the company's core competency, the qualification can be accomplished using consultants and a test house with experience in qualifying components for use outside their recommended operating limits. The component manufacturer may also be willing to evaluate the component for use above its recommended operating conditions. This obviously would be desirable since they have the expertise, experience, knowledge base, test equipment, and test programs to perform functional and parametric evaluations beyond the recommended operating limits. One of the desired outputs from qualification is the component's failure rate and useful life when used outside the manufacturer's recommended operating limits.

Reference

1 Duane, J.T. (1964). Learning curve approach to reliability monitoring. *IEEE Transactions on Aerospace* 2: 2.

Further Reading

O'Connor, P.D. (2002). *Practical Reliability Engineering*, 360. Wiley.

Reliability Growth

H. Crow, P. H. Franklin, N. B. Robbins, *Principles of Successful Reliability Growth Applications*, 1994 Proceedings Annual Reliability and Maintainability Symposium, IEEE (1994).

J. Donovan, E. Murphy, *Improvements in Reliability-Growth Modeling*, 2001 Proceedings Annual Reliability and Maintainability Symposium, IEEE (2001).

L. Edward Demko, *On reliability Growth Testing*, 1995 Proceedings Annual Reliability and Maintainability Symposium, IEEE (1995).

G. J. Gibson, L. H. Crow, *Reliability Fix Effectiveness Factor Estimation*, 1989 Proceedings Annual Reliability and Maintainability Symposium, IEEE (1989).

D. K. Smith, *Planning Large Systems Reliability Growth Tests*, 1984 Proceedings Annual Reliability and Maintainability Symposium, IEEE (1984).

J. C. Wronka, *Tracking of Reliability Growth in Early Development*, 1988 Proceedings Annual Reliability and Maintainability Symposium, IEEE (1988).

Reliability Demonstration

P. I. Hsich, J. Ling, *A Framework of Integrated Reliability Demonstration in System Development*, 1999 Proceedings Annual Reliability and Maintainability Symposium, IEEE (1999).

Lu, M.-W. and Rudy, R.J. (2001). Laboratory reliability demonstration test considerations. *IEEE Transactions on Reliability* 50: 12–16.

K. L. Wong, *Demonstrating Reliability and Reliability Growth with Environmental Stress Screening Data*, 1990 Proceedings Annual Reliability and Maintainability Symposium, IEEE (1990).

Prognostics and Health Management

Pecht, M. (2008). *Prognostics and Health Management of Electronics*. Wiley.

7

FMEA

7.1 Benefits of FMEA

The tool that is second only to Highly Accelerated Life Test (HALT) and Highly Accelerated Stress Screening (HASS) is failure modes and effects analysis (FMEA). FMEA is an extremely powerful tool that can be applied without expensive equipment. In the late 1960s, the practice of using FMEA as a way to improve product design began to surface. It is a systemized series of activities intended to discover potential failures and recommend corrective actions for design improvements. These potential failures would otherwise not be discovered until the product was fully developed. The most important result of the process is that it will reveal a shortcoming before it is unintentionally designed into the product. In that respect, it is exactly like HALT and HASS in that it precipitates or identifies things that need changing in the design before the design is finalized. FMEA, like HALT and HASS, should be an integral part of the design process.

The FMEA process supports the design process by:

- Objectively evaluating the design through a knowledgeable team
- Improving the design before the first prototype is built
- Identifying specific failure modes and their causes
- Assigning risk-reducing actions that are tracked to closure

In addition, the output of the FMEA can provide inputs to other key tasks. These include:

- Test and troubleshooting documentation
- Service manuals
- Field replacement unit (FRU) identification

Successful implementation of FMEA will do the following:

- Improve the reliability and quality of products while identifying safety issues.
- Increase customer satisfaction.
- Reduce product development time.
- Track corrective action documentation.
- Improve product and company competitiveness.
- Improve product image.

FMEA utilizes a team generally composed of the following sections:

- Design Engineering (mechanical, electrical, software, thermal, etc.)

Improving Product Reliability and Software Quality: Strategies, Tools, Process and Implementation, Second Edition. Mark A. Levin, Ted T. Kalal and Jonathan Rodin.
© 2019 John Wiley & Sons Ltd. Published 2019 by John Wiley & Sons Ltd.

- Manufacturing Engineering
- Test Engineering
- Materials Purchasing
- Field Service
- Quality and Reliability Engineering

7.2 Components of FMEA

The FMEA is comprised of three parts: a functional block diagram (FBD), a fault tree analysis (FTA), and the FMEA spreadsheet.

7.2.1 The Functional Block Diagram (FBD)

The FBD is a step-by-step diagram that details the functionality of a development process. The process is broken down into three parts – input, process, and output (see Figure 7.1). The FBD is a high-level diagram detailing the high-level processes that take place for each input, process, and output. The steps identified under input, process, and output should not be highly detailed (see Figure 7.2). Each of the steps that are identified in the FBD becomes a process that is later evaluated using an FTA. For that reason, three to five steps are usually adequate to describe any input, process or output. Ten or more steps may be too detailed for the exercise and can bog down the subsequent FMEA.

The FBD details the outputs that are produced as a result of the processes that take place with the given inputs. The output is the result of transforming the inputs via the process. Therefore, in Step 1, the process is described as a sequence of events. For example:

1. For a radio receiver, you turn the dial to a radio frequency and through an electronic process you hear the sound. (The electronic components comprise a series of circuits that, one by one, convert the signal at the antenna to an audible sound from the speaker. This series of signal conversions is the process in the FBD.)
2. A DC power supply has an AC power input, and through an electronic process it has a DC output voltage, e.g. +12 V. (The AC from a wall outlet is converted into a varying DC; it is filtered, regulated, and sent out from the power supply as steady DC. This is the FBD of a simple power supply.)
3. An automobile transmission has a rotational force along an axis for the input, and through the drive shaft and mechanical differential it delivers this rotational force to the drive wheels. (When the engine rotates, the force is transferred through a shaft and is coupled to the transmission through a clutch. The gears of the transmission select the amount of power to deliver to the drive wheels. Then, from the transmission drive shaft the power is delivered to a differential that finally connects the force to the wheels. This is an FBD of a transmission.)

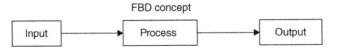

Figure 7.1 Functional block diagram.

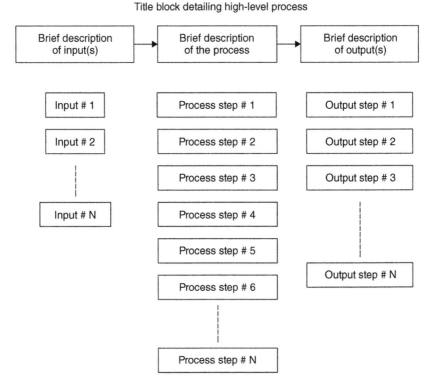

Title block detailing high-level process

Figure 7.2 Filled-out functional block diagram.

A very basic FBD would be to describe the process needed to fill a glass tumbler with tap water. On the other hand, an extremely complicated FBD would be to convert nuclear energy into electricity using steam turbines. Even though these two examples are very far apart in complexity, they can be defined through an input, process, and the output needed to get the results.

The FBD can be as simple or as complicated as needed. However, the FBD should include all significant processes that are involved. A word of caution – it is not always desirable to define processes down to the component level. Keep the processes at a fairly high level; often, the simpler it is, the better.

7.2.1.1 Generating the Functional Block Diagram

The FBD cannot begin until there is a technical understanding of the design (design FMEA) or process (process FMEA) by all the FMEA team members. Here is where the team leader can provide the necessary detailed information, i.e. schematics, mechanical drawings, theory of operation, bill of material, and so on. For our example, we will use a simple flashlight and its schematic (Figure 7.3).

The schematic illustrates the components that make up a simple flashlight. There are two batteries, slide switch, bulb housing, bulb or lamp, housing spring, reflector, conductor from the lamp housing to the positive terminal of a battery, the spring for the batteries, and the flashlight housing (not shown).

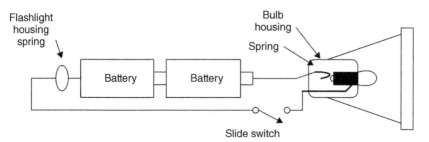

Figure 7.3 Schematic diagram of a flashlight.

The team must be sure that they agree that they understand the device described by the team leader and the documentation.

The FBD commences with writing the three high-level labels: Input, Process, and Output. The labels can be written on Post-its[1] and placed on the wall. The Post-its are part of a toolbox that helps facilitate brainstorming exercises. Some additional items that should be part of the toolbox are the following:

1. Several packages of 3″ × 5″ Post-its (they are useful in brainstorming)
2. A large blank wall or large paper flipchart on an easel
3. Several multicolored felt-tipped markers
4. A roll of masking tape

The FMEA team leader first instructs the group to identify every significant process around which they wish to do an FMEA. This is a team effort and is done as a brainstorming exercise. The schematic flow diagram should be used as an aid. From the schematic flow diagram, identify each of the significant processes involved. (If a schematic flow diagram is unavailable, then the team will need to identify the major processes for the design through a brainstorming exercise.) Using Post-its, create labels for all significant processes. Then, appropriately group the labels into common thoughts. Finally, review each of the grouped labels and agree that an FMEA should be done on that process.

7.2.1.2 Filling in the Functional Block Diagram
There are several possible ways to approach filling in the FBD; we will look at two. Both approaches start the same way. First, identify the high-level processes that take place in the design. A schematic flowchart of the design can aid in identifying these high-level processes. Next, detail the process steps first; then, identify the inputs and outputs required for the process to take place. The second approach works backward, starting with the outputs for each high-level process. We will briefly describe the essential details of the two different approaches. They both work well, and the appropriate choice is personal preference.

For the first approach, place each high-level process identified earlier as a title block for the functional block diagram. Then, write three FBD labels (Input, Process, and Output) on Post-its and place the labels on the wall beneath each FBD title. Next, describe

---

1 Post-its is a 3M registered trademark.

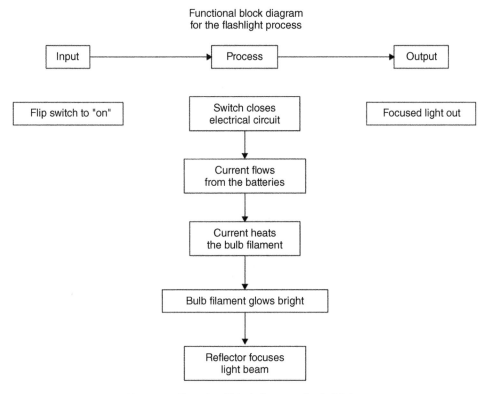

Figure 7.4 Functional block diagram of a flashlight.

the processes that take place under the high-level title. For each high-level process, identify the process steps or sequence of events involved. Once the process steps have been identified, identify the necessary inputs for the high-level process to take place. The inputs are the ones necessary to support the process. Align them under the "Input" label. Finally, write down the outputs that result from the process, placing them under the "Outputs" label. An example of an FBD for the simple flashlight is shown in Figure 7.4.

In the alternate approach, we identify what the desired outputs are for the high-level process, write the desired output statement on yellow Post-its and place it under the "Output" label. In the next step, we identify all necessary inputs required to achieve the desired output and align them just under the "Input" label. Finally, begin writing down the process steps needed to take the inputs and generate the desired outputs. Place these labels under the "Process" label.

The FBD is an interactive task that needs everyone's participation. Team members write labels on Post-its and place them under the appropriate FBD blocks (Input, Process, and Output) as shown in Figure 7.5. The labels can then be moved around and rearranged with ease. As the activity progresses, you'll find team members rearranging many Post-its until there is agreement on the FBD for each high-level process. After the FBD is completed, review each label with the team and make sure everyone understands the label and agrees with it. This step is often called *scrubbing the wall* and is intended to ensure that everyone understands and agrees with every item on the FBD. When

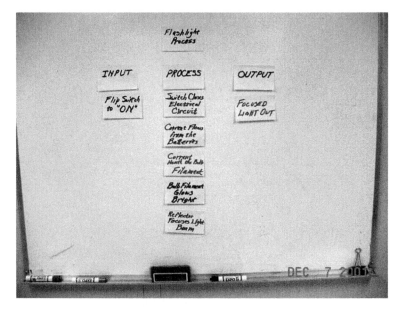

Figure 7.5 Functional block diagram of a flashlight using Post-its.

everyone is in agreement with the functional block diagram, it is time to begin the fault tree analysis.

The FMEA process can consume significant engineering resources. One way to reduce this time is to prepare the FBD in advance. The top designer or team leader can prepare the FBD prior to the first team meeting. This will save time and speed up the process. It is best to circulate the FBD a week or two before the team's first meeting for team members to review. If the FBD is developed in advance, it should be reviewed at the first meeting to gain team agreement around which aspects of the design the FMEA will be performed.

7.2.2 The Fault Tree Analysis

The FTA is a logical, graphical diagram that describes failure modes and causes. The FTA diagram graphically shows all failures for a system, subsystem, assembly, printed circuit board (PCB), or module. The FTA uses standard logic symbols (Figure 7.6), commonly found in flowcharting for process control, quality control, safety engineering, and so on, to tie together the sequence of events. The output from the FTA provides a better understanding of the causes that can lead to a failure mode. The results of the FTA can then be transferred to an FMEA spreadsheet. The FMEA spreadsheet uses the failure modes and their causes from the FTA and determines the effects of each failure cause on the design. The FMEA spreadsheet is also used to identify the most likely failure modes that will occur, as we will show later. The FMEA spreadsheet sets action plans in place to either reduce or eliminate the possibility of failure.

7.2.2.1 Building the Fault Tree

Begin the fault tree by stating the first failure mode from the functional block diagram. The failure mode is defined by taking the first item from the FBD (start with either inputs, processes, or outputs), and turning it into a failure statement. This is usually called the top event or high-level failure mode. If the FBD has as an output of "24 V

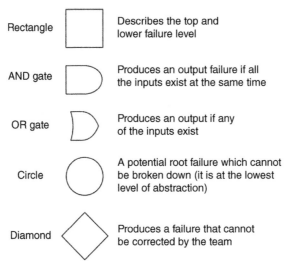

Rectangle		Describes the top and lower failure level
AND gate		Produces an output failure if all the inputs exist at the same time
OR gate		Produces an output if any of the inputs exist
Circle		A potential root failure which cannot be broken down (it is at the lowest level of abstraction)
Diamond		Produces a failure that cannot be corrected by the team

Figure 7.6 Fault tree logic symbols.

provided to the output," then the failure statement would be: 24 V is not present at the output. Place the label "24 V is not present at the output" at the top of your fault tree. Then begin the brainstorming process to create a set of inputs that would be contributors to a failure that could cause the 24 V to not be present at the output.

7.2.2.2 Brainstorming

Brainstorming is a process in which everyone can contribute equally. The basic concept is that everyone sits quietly and writes down ideas on yellow Post-its. Obviously, several members of the team will have similar ideas. Initially, this could be considered counterproductive or inefficient. However, the benefit that results from engaging everyone's participation is that more ideas will surface from the group. Set the ground rules for the brainstorming session as follows:

1. There are no bad ideas.
2. Everyone writes two to three labels and places them on the tree.

In brainstorming, there are no bad ideas. Everyone needs to feel comfortable about submitting their thoughts. Begin by having everyone write his or her two to three ideas on Post-its. When the team is satisfied that everyone has done this, it is time to start the FTA process.

Place the top-level (system-level) failure mode on top of the fault tree. Next, begin identifying failure causes associated with the above failure mode. You can go down several levels associating a second-, third-, and possibly fourth-level failure cause that is associated with the above failure mode. Place each subsequent failure cause beneath the previous failure cause using the Post-its statements. To get to the next lower level of failure cause, ask the question, what event would have to occur to cause the higher-level failure? Usually, two to three levels of abstraction in failure causes are adequate. The goal is not to drive to the root cause but to bring to the surface failure causes in the design cycle that cannot be tolerated. Leave enough room between the levels for interconnecting lines and logic gates. Continue to do the process until the desired level of abstraction has been reached. See Figure 7.7 for a sample of a fault tree for the flashlight example.

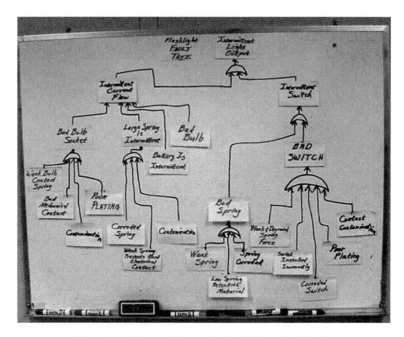

Figure 7.7 Fault tree diagram for flashlight using Post-its.

The level of abstraction will significantly influence the amount of time needed to develop the FTA. If the team believes it is necessary to drill down to the most fundamental aspects of the design, then this is probably what the team should do. However, the team leaders should use their expertise and knowledge of the design to prevent the group from going down an endless sequence of failure scenarios. All failure modes have causes. We circle the lowest-level failure cause on the FTA, and this will be the failure cause that is transferred to the FMEA spreadsheet (Figure 7.8).

It is possible, depending on the level of design complexity, for several failure modes to have the same cause. This is normal and is easily handled in the FTA exercise. Create several identical cause Post-its and place them appropriately as inputs in the several failure mode statements. As you continue to build the FTA, it is easy to see why we recommend using a large wall to paste up the many Post-its.

In building the FTA, you eventually reach a point where a decision needs to be made. The decision is that you have reached a sufficient level of failure cause description to evaluate its effect on the design. These failure causes are circled. However, you may reach a point where you cannot go further because the team lacks the expertise, knowledge, or understanding of the failure cause. These failure causes receive a diamond, because it will require outside expertise to resolve. All the lowest-level failure causes on the FTA should be either circles or diamonds. At this point, you are done with the FTA. Refer to Figure 7.9 for a simple failure mode and cause logic diagram.

7.2.3   Failure Modes and Effects Analysis Spreadsheet

The FMEA spreadsheet is a form that captures the process steps from the functional block diagram and investigates how each process step can fail (fault tree) including root cause.

Brief logic flow diagrams

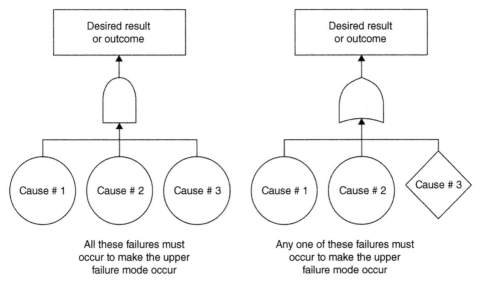

All these failures must
occur to make the upper
failure mode occur

Any one of these failures must
occur to make the upper
failure mode occur

Figure 7.8 Logic flow diagram.

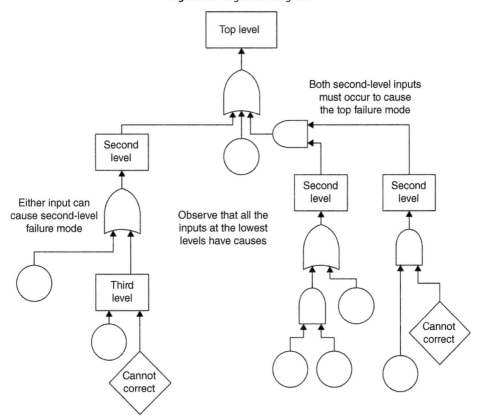

Figure 7.9 Fault tree logic diagram.

A sample FMEA spreadsheet is shown in Table 7.1. A team scribe is selected to enter the information into the spreadsheet form. It is important that the scribe be familiar with the product design, nomenclature, and three-letter acronyms (TLAs) so the information is clear and will not be misinterpreted.

The next stage of the FMEA process is to insert the failure modes and causes into the FMEA spreadsheet. This is actually the easy part of the FMEA process because the Post-its rectangles from the fault tree are the failure modes. You merely paste them into the failure mode column. Insert the highest-level failure mode from the FBD in row one. Then, enter the causes from the fault tree under the next column. There may be several causes, so be sure to include them all for the specific failure mode. Remember that the rectangles from the fault tree are the failure modes and the causes are the circles and diamonds that feed into the rectangles. In the next column, enter the effects that were caused by the failure mode. Here too, there may be more than one effect. Do not fill any more columns to the right; it is best to do that later. Take the next failure mode and enter it in the next line. Again, add causes and the effects. Continue until all the failure modes have been addressed. At this point, your rectangles, circles, and diamonds will have been completely consumed. See Figure 7.10 and Table 7.1.

Ask, "What effect does this failure have on the rest of the system or process?" In the flashlight example, an effect might be: no light output, or dim light, or the light gets weak very soon after turning the flashlight on, and so on. Complete the effects column fully. Then, move on to the fault detection column.

Now, beginning at the top, under the fault detection column, enter the mechanisms by which the failure modes could have been detected. An example might be: customer complains that the light gets dim too quickly, or life testing in a laboratory, validation testing or supplier qualification for a design FMEA, and so on. Complete this column to the very bottom of the form.

Move on to the severity column. If the scale does not suit your specific need, then change it accordingly (refer to the risk priority number (RPN) ranking in Table 7.2). If the severities are small, they are to be assigned small numbers. Severity rankings set impact. Severely impacted customer satisfaction gets larger numbers. Continue until the column is complete (refer to the RPN ranking in Table 7.2).

We will pause here to emphasize the importance of approaching the data entering one column at a time. As the FMEA team judges the levels for the various failure mode occurrences, if they stay in the occurrence frame of mind, their interpretation of what each number means will tend not to drift. If the team goes from occurrence, severity, and then to detectability, one parameter will tend to confuse the other. Because this is a very subjective measurement, it is best to avoid anything that may tend to impact the team's judgment.

Now in the Occurrence column, assign numbers between 1 and 10 that, in your judgment, describes as the frequency of this specific failure mode (refer to Table 7.2). If the scale does not suit your specific need, then change it accordingly. Remember, it is the scale you'll use uniformly throughout this FMEA. Stay with this column until all the occurrence rankings have been entered.

Move through the Detection ranking columns in a similar manner. Things that can be easily detected get smaller numbers. If there is a very low probability of detection, the numbers will be higher.

Table 7.1 The FMEA spreadsheet.

Failure Modes & Effects Analysis

FMEA#:   Company/Organization name:

Assembly:

Owner:

Date:

Team members:

#	Failure mode	Cause	Effects	Fault detection	S	O	D	RPN	H	FRU	Recommendations	Who?	When?	A
1	2	3	4	5	6	7	8	9	10	11	12	13	14	15

D, detection ranking; S, severity ranking; O, occurrence; RPN, risk priority number; H, safety Hazards; FRU, field replaceable unit; A, audit.

1. *Line or row number.* Provides a way to reference issues.
2. *Failure mode.* A brief description of the low-level failure mode.
3. *Cause.* What could cause failure to occur?
4. *Effects.* What effect does this failure have on the top-level design or process?
5. *Fault detection.* What could have been put in place to minimize or prevent the failure mode from occurring?
6. *Severity (S).* A metric in units from 1 to 10, with 1 as minor and 10 as major. Severity is thought of from the point of view of the customer or end user.
7. *Occurrence (O).* A metric in units from 1 to 10 with 10 the most frequent and 1 the least frequent. It is an estimate of the probable period before observing an occurrence; generally thought of as a field failure issue.
8. *Detection ranking (D).* A metric in units from 1 to 10, with 1 as a very high probability that the failure mode will be detected and 10 as a very high probability that it will not. (This can be confusing. The larger number represents a measure that is more difficult to detect.)
9. *Risk priority number (RPN).* A metric that is the product of occurrence, severity, and detection ranking (just multiply the three together to get the RPN), this number can range between 1 and 1000. The higher the RPN number, the higher the risk of the failure mode.
10. *Hazard or safety (H).* Does this failure mode create a hazard? Does this failure mode create a safety problem?
11. *Field Replaceable Unit (FRU).* Used to generate a recommendation for FRUs to your field service department.
12. *Recommended action (What).* A brief description of what the FMEA team recommends will have to be done to mitigate the failure mode.
13. *Who.* The person or persons assigned to the recommended action.
14. *When.* The date on which the recommended action is to be completed.
15. *Audit (A).* A check-off placeholder that indicates that the recommended action has been completed to the satisfaction of the FMEA team.

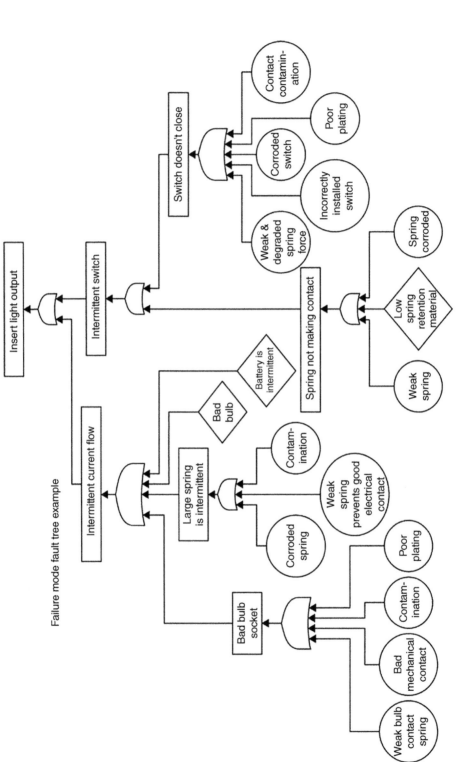

Figure 7.10 Flash light fault tree logic diagram.

Table 7.2 RPN ranking table.

Occurrence ranking	Severity ranking	Detectability ranking
1 Failure is unlikely or remote	1 Essentially no effect	1 Certain detection
2 Less than 1 per 100 000	2 Not noticeable by customer	2 Very probable detection
3 Less than 1 per 10 000	3 Noticed by discriminating customer	3 Probable detection
4 Less than 1 per 2000		4 Moderate detection probability
5 Less than 1 per 500	4 Noticed by typical customer	5 Likely detection
6 Less than 1 per 100	5 Slight customer satisfaction	6 Low detection probability
7 Less than 1 per 20		7 Very low detection likely
8 Less than 1 per 10	6 Some measurable deterioration	8 Remote detection likely
9 Less than 1 per 5	7 Degraded performance	9 Very remote detection
10 Less than 1 per 2	8 Loss of function	10 Uncertainty of detection
	9 Main function loss, customer dissatisfaction	
	10 Total system loss, customer very dissatisfied	

When the severity, occurrence, and the detectability columns have been completed, the next step is to calculate the RPN by multiplying the three metrics together. The RPN number can range between 1 and 1000. After you have completed entering all the RPN numbers, you will observe that the FMEA is beginning to take shape. Usually, there will be many numbers below a certain level or baseline, say 200. There will be a few numbers above that baseline as well. The magnitude of the RPN will highlight the top areas that need to be considered for improvement.

The next column, Hazard or Safety, is used to consider if the failure mode could harm or cause injury to personnel. If the FMEA team considers this failure mode to be a safety issue, place a "Y" for yes in this column. Continue down the column with a "Y" or an "N" for no until the column is complete. Do not assign numbers here. This is not a metric that is to be multiplied together as part of the RPN. This is either a safety issue or it is not, and it should be treated appropriately. Some recommended action to remove this as a safety issue is needed.

In the next column, FRU, you can help determine if this failure mode can be best handled by fixing the problem in the field. If so, place a checkmark here under the FRU column. The FMEA process is not intended to be a tool that generates a list of FRUs. Here it is merely a tool that can be used to generate a recommendation for FRUs to your field service department.

The next column, Recommended Action, is where the FMEA can get bogged down. It is where the team makes a recommendation for change that will mitigate the failure mode. The team is not to determine a design change then and there. Each recommendation is to be assigned to a person who is an expert and can most efficiently deal with the failure mode. Usually, that person is a member of the team. Three columns can be addressed at the same time for they are the next improvement actions.

The "Recommended action," "Who," and "When the task is to be completed" are to be discussed at the same time. Here is where the FMEA team "assignors" comes to agreement with the "Who," and sets a completion date agreed upon by all the FMEA team members. The FMEA team leader can manage the activities of each of those members who were assigned tasks in a normal fashion. The date is usually the date when the project needs completion of the recommended action.

The Audit column is for the reliability department so that they can track the status of the recommended actions from the FMEA. This ensures closure of all the recommended actions to the satisfaction of the FMEA team. Note in ISO 9001 companies and biomedical companies, it is important to show a closed-loop corrective action system.

A sample FMEA spreadsheet can be found in Figure 7.11 that matches these descriptions. The team leader should fill out the appropriate sections of the form. They are self-explanatory.

7.3 Preparing for the FMEA

To start an FMEA, the team needs a leader. This individual is most likely to be someone familiar with the FMEA process who can guide the team to success. It need not be a technical person. The first goal of the leader is to form a team that will collaborate to identify potential failure modes and their causes and correct the problems before the design is released. In preparation for the first meeting, the leader will assemble documentation that describes the design or process. The documentation is then distributed to each of the team members either before or at the first meeting. Often, it is desirable to submit this documentation a week or more in advance to give members a chance to familiarize themselves with the design. The following is a general list of the documentation needed in preparation for a design FMEA:

- Mechanical drawings
- Electrical/electronic schematics
- Design process algorithms/software
- Process documentation that identifies inputs and outputs
- Miscellaneous items that describe the product and its function(s)
- An operational or functional block diagram

The FMEA should be completed before the scheduled pre-production design release date. By the time the product is released for production, all the recommendations and actions assigned by the FMEA team should be completed, closed, and documented. Therefore, it is important to allow sufficient time in the design schedule for the FMEA process to be completed. Depending on the size of the assembly, the FMEA process typically takes (per assembly) 12–28 hours to complete. It will take much longer to conduct FMEAs for large and complex products such as airplanes. The FMEA team should consist of cross-functional members who have been trained in the FMEA process. It is inadvisable to include team members who have not had FMEA training. Team members who are unfamiliar with the FMEA process will typically cause delays, as they do not understand the process. If there is an occasion where the whole team needs FMEA training, then a skilled coach who fully understands the FMEA process can complete this training task in approximately 4 hours.

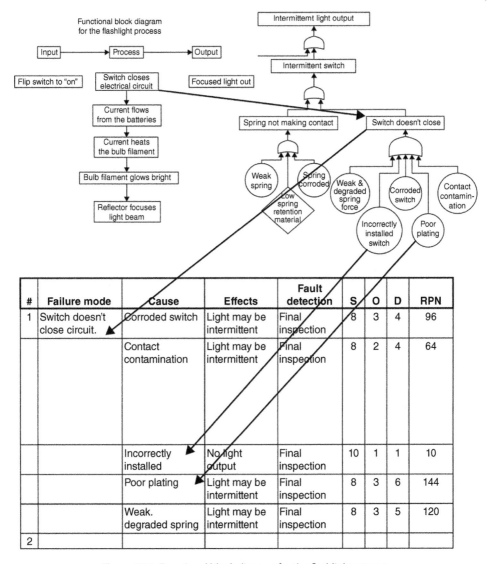

Figure 7.11 Functional block diagram for the flashlight process.

The first step of an FMEA is for the team leader to describe what the assembly is designed to accomplish. The simple way of doing this is to provide an FBD to the team. This diagram has basically three parts: inputs, outputs, and the functional process. The team is then tasked to review the FBD and to come to an agreement on how the assembly works.

The next step is for the team to identify failure modes and their causes. This is best achieved by using the brainstorming process. Any one of a number of process tools can be used to capture the team's findings:

- Fishbone or cause-and-effect chart
- Flowchart process
- Fault tree analysis

The FTA is the most common method used to identify failure modes. We, therefore, will describe the process for using a fault tree to identify failure modes and their causes. The design team may already be familiar with the FTA method as a way to determine possible root causes for a known failure.

The fault tree process begins with top-level failures, then second-level failures, third, and so on. The top-level failure represents the highest or most fundamental failure level. It can be as basic as turning on a light switch and the light does not go on. The second- and third-level failure represent events that take place as a result of attempting to turn on the light. Logic gates (i.e. AND, OR, NOR logic used in binary process flow) are then used to interconnect the lower-level failure modes to the highest failure level, linking the entire process. At each level, the failure modes may have one or more causes. These causes can be mitigated by design change, manufacturing process steps, improved material selection, and so on. In some cases, the causes cannot be addressed by the team and should be set aside for outside expertise. The identified failure modes and their causes have an effect on the product. The failure mode is always something that will dissatisfy the customer to various degrees. These causes, when identified and corrected, greatly improve customer satisfaction.

Upon completion of the fault tree, a list of the failure modes and their causes has been constructed. At this point, more than half the FMEA process is complete. The next step is to fill out the FMEA spreadsheet. An example of the spreadsheet is shown in Figure 7.11.

In Figures 7.4 and 7.10, we built an FBD and fault tree for the flashlight FMEA. Next, we will illustrate in Figure 7.11 how easily the failure modes and causes can be taken from the FTA and placed onto the FMEA spreadsheet.

The determination of the RPN number comes as a result of the team assigning appropriate weightings for severity, occurrence, and detectability. For brevity, the remainder of the FMEA spreadsheet is not illustrated.

When the team has completed the FMEA spreadsheet, the team leader either assigns actions to be addressed for each of the high RPN results and any hazard- or safety-related entries or determines that no action is required. The decision is based on the engineering resources available to fix problems and the severity of the problems found. We have found that an 80/20 rule is a good guide in deciding what gets addressed. The 80/20 rule assumes that the top 20% of the RPN issues identified represent 80% of the potential problems. Keep in mind that all potential safety issues need to be addressed. All recommended action items should be completed, closed, and documented before the scheduled production release.

At the beginning of the chapter, it was noted that the FMEA process is second only to HALT and HASS. It is important to note that FMEA has several advantages over HALT and HASS. The HALT process is expensive. FMEA can be accomplished with very little expenditure other than the time used by the FMEA team. All the documentation used during the design review process can be used again in the FMEA.

The cost to introduce and perform the FMEA process is the same, regardless of the size of the company. The best part of an FMEA is that the cost to implement it is small and independent of the size of the business. In addition, the resources expended on performing the FMEA will be recovered by reducing the total development time.

7.4   Barriers to the FMEA Process

Often, the greatest barrier to implementing the FMEA process is getting the design community to accept the concept of reviewing someone's design for reliability (DfR) in a systematic and detailed fashion. Most development engineers believe that they design highly reliable products. One reason for this is confidence in their skills. Another reason is that most product development engineers are rarely aware of the field failures that resulted from their previous designs. Typically, design engineers go from one design task to another. Other engineers, often referred to as *sustaining engineers*, are responsible for resolving production and field problems. The disconnect between sustaining engineering and product development engineering is why mistakes from past designs are repeated in new designs. If product development engineers are involved in resolving the original design problems, these design problems will not be repeated.

On one occasion, I pointed out to a design engineer that he was incorrectly mounting a component vertically (to save circuit board real estate space). This eventually leads to a failure caused either by shipping or by long-term vibration fatigue. The engineer replied that he had been designing that particular component the same way for the past 20 years without any problems. I asked the engineer if he had reviewed the failure reports that resulted from his past designs. He said he never had and that it wasn't his job. Then I asked him how often failures are reported. He said, "So far, never." This is a very large barrier called, *I've always done it this way*. This is a barrier that exists deep within most experienced design engineers. You'll have to overcome this barrier, one engineer at a time. The more experienced senior engineers will be the most difficult to change into rethinking the way we review designs for reliability. This does not imply that you should work on the less experienced engineers first. On the contrary, work on the senior engineers first. They will change and when they do, you will have allies that are already highly respected by the rest of the staff. It will make the changes easier to implement when you have their support.

Less experienced engineers and recent college graduates are much easier to persuade. They are not entrenched with their own particular set of tools and accept the FMEA process more readily. The experienced engineer tend to be more difficult, but their support will persuade other engineers to use the new process. There is one subtle advantage within the FMEA process that usually occurs around the first or second day of the FMEA process. Psychologists call it the "aha experience." Let me pause for a simple example.

Almost everyone has had to change a flat tire at some time in his/her life. When this happens, they will jack up the car and begin to loosen the lug nuts. When they do, the wheel will turn. Then, they have a fight on their hands. With one hand, they will hold the wheel to keep it from spinning in one direction, while they try to loosen each lug nut in the other direction. Eventually, they will get the lug nuts off, remove the flat tire, install the spare tire and refasten the lug nuts. The reason this usually happens is that we are rarely trained at replacing a flat tire. We are on our own the first time. Then one day, we have the "aha experience." When watching someone else loosening the lug nuts, we observe that this time the wheel is left low, touching the ground and then each lug nut is loosened just a little. This way they can use both hands on the lug wrench and let gravity hold the tire in place while it is still on the ground. Once all the lug nuts are loose, the car can be raised so that the tire no longer touches the ground. Now the lug nuts can be spun off with fingertip ease. "Aha." This will happen in the FMEA process, too.

Somewhere in the middle of the process, one engineer will observe a failure mode and its cause which he had never considered. It will come as a surprise. This is the anticipated "aha experience." It often comes from the most experienced engineer on the team. This golden moment is when that engineer is converted. This doesn't mean that he is completely on board, but his attitude has changed toward a more favorable direction in accepting the FMEA process.

Another barrier will be from management saying that the FMEA process will delay product introduction. This perception is not reality. If the FMEA is done early in the design cycle, it should not impact the design completion date. Some of the issues that will be uncovered in an FMEA would have surfaced during design verification anyway, and led to project delays. In the past, when the FMEA process was not used, these issues hopefully surfaced in design verification. Then the engineering change process kicks in and the product is delayed while the fix gets implemented. The FMEA process can also save time because design engineers can spend a greater portion of their time on product development and less time fixing previous design problems. Design problems eventually get fixed. They can be fixed before the first prototype is built when the cost is minimal or the company can wait until your customer drives the fix.

Another significant barrier is that, during the initial implementation phase, the FMEA process will take a long time to complete. This is normal. The process is complex and there can be a significant learning curve associated with implementation. At the end of the FMEA, have the team members note the strengths and weaknesses in the process so that improvements can be made. After you have implemented a couple FMEAs and implemented the process improvement suggestions, the process will be faster and proceed more smoothly.

Some feel that a good design review serves the purpose of finding all the design oversights and, as such, consider the FMEA to be redundant. Program managers are tasked with meeting delivery dates and will argue that the two processes are not necessary. Others consider the FMEA process as a replacement for the design review process. They argue that they should do one or the other, but not both. Design reviews and FMEAs are two completely different processes with different goals and objectives. Both are needed. A design review is intended to ensure that the design requirements are met and that the documentation is complete and correct. The FMEA process is designed to discover potential failure modes and safety issues that cannot be allowed and to implement design changes to ensure that they will not surface. To develop highly reliable products, both need to take place.

FMEAs must be and hardwired into the product development schedule so there is no confusion or ambiguity where in product development process they take place and when they need to be completed. A design FMEA can start as early as phase 2, when the product architecture is being developed. If the hardware or software platform and/or architecture is new then an architecture FMEA is scheduled in phase 2. The FMEA allows for early discovery of potential failure modes. Discovering potential failure modes earlier in the development cycle means they can be readily addressed at minimal cost. Waiting until the product is completely designed will result in compromises that would not have been made if addressed earlier in the development cycle. Discovering failure after the product has been designed, results in compromised solutions because development cost and time to market become significant factors in the decision. How many

FMEAs and where they are performed will become clear as the user develops experience in the process and measure the return from the effort.

7.5   FMEA Ground Rules

Keep the FMEA *team engaged.* Because the FMEA process usually takes several days or longer, many of the team members see a need to go back to their desk to tackle other tasks. There should be adequate time set aside for team members to get other work done. By keeping the FMEA meetings to two or three hours at a time, the team should be able to stay engaged and focused. If members leave from time to time, questions will arise during the FMEA process that cannot be answered because of their absence. Murphy's law dictates that they will be away when they are needed to explain something. This delays the process. In the long run, the process will go faster if everyone commits to just doing the FMEA without interruption, distractions, or multitasking. To achieve this, you may need to set ground rule to minimize distractions, like all cell phones off and no laptops.

Minimize interruptions. In some companies, where most of the engineers are very busy and are often interrupted, it is best to perform the FMEA off-site, thus ensuring a minimum of interruptions. Even breaks and lunch periods can be optimally managed. There should be scheduled breaks. Break times should be short enough so that the members don't wander. Lunches should be catered so that everyone can begin again without the delays caused by stragglers.

Use the data available to assist in determining the level of importance. But don't stop if data is not available. Use the experience of the FMEA team. When discussions come down to opinions, seek outside information or help. Don't get caught up in endless opinion-dependent discussions.

Remember to complete one column at a time until you reach Recommendations. This will help to avoid jumping to conclusions or making incorrect recommendations and it greatly speeds the process.

Maintain focus. It is easy to get distracted on a nonrelevant direction, get bogged down in an argument, or jump to solving a problem. None of this should be allowed in an FMEA. The purpose of the FMEA is to capture potential failure that the team can later investigate based on priority. It is also recommended that you select a site free from outside interruptions.

Use the 80/20 rule. An FMEA can identify hundreds of potential failures and then prioritizes them based on the results of the RPN number. Use the Pareto rule (top 20%) to determine the minimum number of issues that need to be investigated. The FMEA team may decide to go deeper down the list and investigate more of the issues. This should not be discouraged.

Create a "parking lot" for issues. When brainstorming failure modes, causes, and effects with a cross-functional team, issues and problems will surface that are really important but not part of the FMEA. It is very important to capture these issues but not have it derail the FMEA meeting progress. This can be satisfied by setting up a "parking lot" at the beginning of the FMEA (Table 7.3) for issues that surface that are important or need to be addressed but are not associated with the current FMEA. For example, when doing a design FMEA, an issue might arise regarding special custom tooling that

Table 7.3 FMEA parking lot for important issue that are not part of the FMEA.

Parking lot items:	Who	When
1		
2		
3		
4		
5		
6		
7		
8		
9		
10		

will be necessary for manufacturing. The idea should be captured as a parking lot item to be included in the process FMEA.

Limit meeting times to shorter time periods. Marathon FMEA efforts are unproductive. The FMEA process can be mentally demanding; after four to six hours, the team is likely to become mentally fatigued and unproductive. It has worked well to limit FMEA sessions to two- to four-hour time periods and then break. This allows the team to address daily business matters without having to distract from the FMEA process.

7.6 Using Macros to Improve FMEA Efficiency and Effectiveness

An FMEA can take a considerable amount of time and staffing resources to complete. For that reason it should be part of a continuous improvement program where the organization is continuously looking for ways to make the FMEA process more efficient. Look for ways to streamline the process without sacrificing the effectiveness of the exercise. When I first implemented design FMEAs, the process took a full week to complete. Over the years I found ways to streamline the process; now it typically takes 9–10 hours to complete an FMEA. I have found that Microsoft Excel is an effective tool to run FMEAs when you create macros to improve efficiency. Using the macro developmental tool, macro can be design and macro buttons created to automate redundant and repetitive processes, enable filtering, and improve the efficiency of documentation. Here are some of the efficiencies that we've added over the years to improve FMEA efficiencies:

- During the voting process for severity, detectability, and occurrence, it is recommended that you focus on the voting on one item at a time (i.e. severity, detectability, or occurrence). Create a macro button so that when you select one to vote on, i.e. severity, it hides detectability and occurrence. This will prevent the team from making a judgment based on the weighting that was placed on the other factor. The team should not see what the RPN results are until the voting has completed. This also minimizes the risk of judgment creep as you vote on each issue.

- Voting can be done by either typing in a number or using a pulldown menu that is created in each cell. In addition to the numbering scale, whether it's a 1–5 or 1–10 scale, there should be a method to streamline the voting process if the team has decided to fix the failure issue or decided that the failure issue cannot happen. We use the letters Y and X as a voting option to address this issue. It is not uncommon during the voting process that when voting on an item the team decides that an issue is so important that needs to be fixed no matter what the RPN score is. In these cases, we change the voting from a number scale to Y; this indicates that the team has decided to fix this issue. Once you select Y as the weighting factor the remaining voting scales (e.g. detectability and occurrence) are eliminated. This way the team does not spend any additional time ranking of this issue for the other RPN classifications. It is also possible when voting that the team learns the issue has been fixed, the design changed so it no longer applies, or it is decided that is not a real issue. In these cases, we select the letter X to ensure that the team doesn't spend any more time voting or discussing this issue.
- It may be possible, depending on the product and the type of FMEA, to automate the voting process for detection, severity, and probability. Let's start with detectability. A macro button would pull all the fault detection responses into a single column in a new worksheet. It then runs a delete duplicate routine so any duplicate fault detection responses are deleted. With this reduced list, the team then groups the fault detection responses into different buckets for the weighting factors. If the team decided that they would use a 1–10 voting scale, then there would be 10 buckets for the team to move each of the detection responses into. After the team does a first pass at placing the fault detection responses into the 10 buckets, the team then reviews the results to make sure everyone agrees with the weighting. Once this is done, a macro can then cross reference each of the fault detection responses and enter the appropriate detection scale weighting factor. It may also be possible depending on the product, to define all the possible fault detection options in advance and in a weighted bucket form. Then the fault detection cell for each failure issue can be a pulldown menu with the available detection options. When the detection information has been entered for each issue the voting process can automatically fill in the detection weighting value.
- The same process may be possible for severity voting as well. To determine the weighting factors for severity, we move the list of all the effects responses into a single column in a new worksheet with each response having its own cell. Like before, we delete all duplicates and then start the process of moving each unique severity response into one of the 10 buckets. After this is completed the team reviews the 10 buckets and makes any tweaks necessary. Once completed, the voting process is automated and the severity values are place into the cell for each issue.
- The same process may be repeated for occurrence voting. In this case, you're using the causes associated with each failure mode to sort them into the buckets of likely probabilities. This sometimes can be easier than using a predefined scale for probability of occurrence.
- When brainstorming "causes" for a single failure mode, it is not unusual for the team to identify more than one cause. Each failure cause needs to be its own line item with its own voting value for severity, detectability, and occurrence. When brainstorming failure causes, it is desirable to be able to capture all failure causes in a single cell and not be bogged down by inserting a new row for each new failure cause. The FMEA

creativity process should not be stifled due to the logistics of managing the FMEA spreadsheet. Once solution is to enter all the causes into a single cell with each cause separated by unique character like a pipe character "|." This allows rapid entry of all the brainstorm ideas without being bogged down with continually adding additional rows to the spreadsheet. Once all the ideas are captured an Excel macro can search for the "|" and inserted in the next row below until there are no additional failure causes.

7.7 Software FMEA

Design FMEAs have been applied to product development with great success for a very long time. The design FMEAs have primarily focused on the physical hardware, architecture, safety, and manufacturing process for the product development. However, few have applied the success of design FMEAs to the software development process. This is especially true in organizations where the product being designed has been predominantly electrical or electromechanical systems that required minimal software and/or firmware. The amount of software and firmware required for product development is rapidly changing. Organizations are finding that the software and firmware required to develop next generation products is rapidly increasing, but the software development resources have not increased at the same pace to meet this increasing demand. The change can be so significant that there no longer is the right mix of software and hardware design engineers for product development. As the level of software and firmware required for product development continues to increase, the number of software bugs also increases, but by a significantly greater proportion. The ratio of software to hardware bugs can easily be 50 or 100:1 or even higher. This is partly due to the fact that it is easier to apply DfR guidelines, lessons learned, and decades of well-documented circuit theory to make hardware designs reliable. It is also possible to write hardware requirements so the hardware can be fully functionally tested. The same cannot be said for software quality.

So why don't organizations apply the success achieved with hardware design and process FMEAs to software product development? One reason is that there's not a lot of training material on how to implement a software FMEA. This may seem strange, since software FMEAs have been around for quite a while, but they are primarily performed by companies whose main product is software.

A software failure mode is defined as any departure from the expected results that can be described at the subsystem, module, or component level. The failure mode describes how the subsystem, module, or component might not meet the design intent or produce the desired output. To identify potential failure modes, ask the question, "How could the subsystem, module, or component fail?" The failure mode of the component likely falls into one of four categories: not functioning, partially functioning, intermittently functioning, or resulting in an unintended function. A list of common failure modes is listed in Table 7.4.

The software can also fail if it is used in a way that was not intended, encounters an unexpected situation, unplanned environment, or an inexperienced user (pilot error). There can be many different causes for a software failure mode; a list of common software causes is captured in Table 7.5. A list showing how a failure mode can be broken down into failure causes is shown in Table 7.6.

Table 7.4 Common software failure modes.

Software failure modes	
Algorithm error	Memory bit error
Computation error	Memory error
Data handling error	Memory overflow
Decision Logic flaw	Order fault
Error detection fault	Recovery fault
Fail to alarm	Reset error
False alarm	Rounding error
Faulty data	Sequence fault
Functionality fault	Synchronization error
Invalid operation	Timing fault
Invalid state	Typing error
Label error	Logic wrong
Logic error	Logic extra
Logic missing	Linkage error
Loop logic error	

Table 7.5 Common causes for software failure.

Software failure causes	
Calculation error	Multitask deadlock
Counter rollover	Multithread deadlock
Divide by zero	Non-reentrant code executed reentrant
Finite precision error accumulation	Numerical overflow
Infinite loop	Output out of sequence
Initialization error	Output too early
Invalid code executed	Output too late
Invalid operation	Stack or heap size insufficient
Logarithm of zero	Unhandled exception
Logic defect	Unprotected critical sections where data in use may be interrupted
Memory leak	Unprotected critical sections where data in use may modified by other task
Memory not cleared	Wrong code executed
Missing data	Wrong formula
Multiprocess deadlock	Wrong units

Table 7.6 Failure modes and associated possible causes.

Failure mode	Failure cause
Data input error	Data reversed and misplaced
	Data redundant
	Deficient data
	Data precision error
	Data out of range
	Data error format
	Right inputs refused
	Right range but wrong value
Data output error	Numerical output inaccurate (too small, too large, noisy, or corrupt).
	Unexpected non-numerical value
	Pointer or index output incorrect (wrong address/index, null)
	No/missing output
Execution flow failure	Early return/completion
	Late return/completion
	Fail to return/completion (hang, halt, terminate)
	Diverted execution (e.g. incorrect branch, function call, raised exception or interrupt)
Time-sequence error	Data/signal too early
	Data/signal too late
	Data/signal too overtime
	Data/signal too frequency abnormality
	Out-of-sequence outputs

The software FMEA is a top-down approach where the software is decomposed from subsystems to modules and then components. These components are treated like *black boxes,* describing the components' functionality.

The software FMEA can start once the architecture is defined (for a software architecture FMEA) or when there is sufficient information available to construct a flowchart for the subsystem, module, and components being developed.

The software FMEA team must include the software developers and designers, plus some or all of the following functional groups:

- Application engineer
- Customer
- Design engineer
- Reliability engineer
- Software design engineer
- Software test engineer
- System engineer
- Hardware test engineer
- Others as required

Before starting a software FMEA, the following information is needed:

- A functional block diagram of the subsystem to be evaluated, broken down from the subsystem to the module and component level.
- Detailed flowchart or sequence diagrams of the software being developed.
- A description for each method or function in the FBD or class diagram including the required inputs and expected outputs at the component level.
- A description for how the software flowchart and FBD interface and interact.

The software design detail should be down to the lowest level of decomposition that the team will consider. This would include information regarding sequencing, data and control flow, state transition diagrams, messaging, logic description, and communication protocol. The software design detail should include information regarding the interfaces between the software items and the external systems. This would include the hardware-software interface.

7.8   Software Fault Tree Analysis (SFTA)

The software fault tree analysis (SFTA), like an FMEA, is a tool to anticipate failures and their underlying causes. The SFTA is an effective tool as part of a quality assurance program for defect prevention. The SFTA can be applied to requirements, design phase, architecture, safety, and to improve the effectiveness of the software verification/validation test plan. The SFTA is not typically applied to low-level code; it is applied at the component/module level and higher. It is an effective and efficient tool to identify weaknesses in the design and stated requirements, safety hazard, and potential failure. A SFTA can identify single points of failure, missing components, ambiguities, uncertainties, and issues with variable values or conditions. If probabilities are assigned to all the low-level causes, SFTA can be used to calculate the probability that a fault will occur. An example of this is shown in Figure 7.12 and based on Table 7.6 for an execution flow failure.

7.9   Process FMEAs

The first step in developing a process FMEA is to state desired outcome. What is expected to be achieved or produced from the process that is being evaluated? List the process steps in sequential order that take place in order to produce the desired output. For each process step, list the inputs and outputs that are required for the process. Each process step must have clearly defined requirements for success of the desired output. Assume that each input comes in a good state and should not be the cause for a failure mode. If an input could be the reason for a bad output, then that input should be included as one of the process steps so it can be evaluated for potential failure modes. It is often desired to keep the analysis of the process at a high level of abstraction.

The process FMEA should include all critical processes, including anything that is new or has not been done before. These critical processes can include customer requirements, industrial standards, product and operator safety, product and operator liability, and regulatory and compliance requirements.

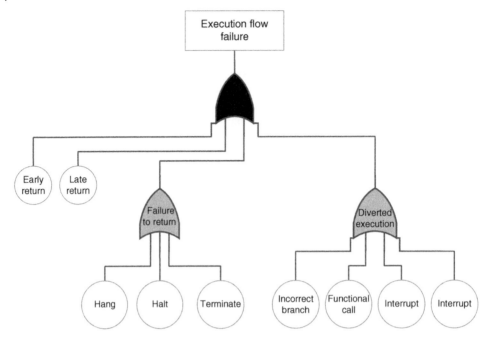

Figure 7.12 Example of a SFTA for an execution flow failure.

The process FMEA starts in phase 4 of product development and can begin prior to the first build of prototype assemblies. However, it is beneficial when doing a process FMEA to start the process FMEA after the first prototype has been built. It's important that all the team members that will participate in the process FMEA are participants in the first prototype build. One of the advantages to starting an FMEA after the first build is that manufacturing problems and lessons learned can be incorporated into the FMEA process. The objective of the process FMEA is to demonstrate the key attribute requirements, including quality, reliability, cost, productivity, maintainability, and serviceability that are necessary for success through the life of product manufacturing.

The process FMEA is very similar to the process followed for Ishikawa fishbone diagram. The process FMEA can include the same fishbone headings as part of the process evaluation. The process FMEA considers failure modes that are associated with materials, measurements, methods, environment, manpower, and machine. The process FMEA investigates how a process, function, tooling, or facility can fail. Often, it is the manufacturing engineer that leads the process FMEA.

The process FMEA team may include any or all of the following functional groups:

- Customer
- Design engineer
- Industrial engineering
- Line foreman or supervisor
- Manufacturing engineer
- Material engineer
- Operators
- Process engineer
- Reliability engineer

- Safety engineer
- Software engineer
- Tooling engineer
- Test engineer
- Others as required.

7.10 FMMEA

Failure modes, mechanisms, and effects analysis (FMMEA) is very similar to an FMEA with the added step of identifying the failure mechanism associated with each failure cause. The process is the same as an FMEA. Begin by identifying potential failure modes. The failure mode is the way the failure is physically observed by the user. The failure cause is the driving force that causes the failure mode to manifest itself. The failure cause is the stress, either internal or external, that is the driving force behind the failure mechanism. The failure mechanism is the process driving the failure cause. The failure mechanism can be chemical, electrical, mechanical, physical, thermal, or a radiation stress. It can be a single stress mechanism or multiple stress mechanisms working together to cause the failure mode. Finally, the failure effect is how the failure mode effects its use from a user's perspective. The FMMEA spreadsheet is that same as the FMEA spreadsheet with an added column for failure mechanism.

An FMMEA can be used as part of a prognostic and health management (PHM) process to develop a method to monitor for degradation and estimate how much time before failure occurs. PHM is covered in Section 6.7.3. The failure cause and mechanism is used to develop a prognostic (PHM) strategy, determine the type of sensor or sensors to monitor for wear and degradation, and create the test plan for an accelerated life test to develop a data set for how it transitions from a good state to a failed state. The accelerated life data set is then used to develop a methodology for detecting wear and degradation using real time sensors. It is important to determine what is causing the damage and what parameters should be monitored for prognostics. Knowing the potential failure mechanism or mechanisms causing deterioration might make it possible to mitigate or manage the mechanisms to prolong useful life and reduce downtime. The failure mechanisms can be used to improve the design and part of the design and validation process.

Several mechanism can cause overstress and wearout, including chemical, electrical, mechanical, radiation, or a thermal stress:

- Examples of chemical stress are corrosion, contamination, dendrite growth, deploymerization, and intermetallic growth.
- Examples of electrical stress are dielectric breakdown, conductive filament formation (CFF), electrical overstress (EOS), electrostatic discharge (ESD), electromigration, hot carrier injection (HCI), hot electron trapping, interdiffusion, secondary breakdown, radio frequency (RF) spikes, surface charge, time-dependent dielectric breakdown (TDDB), and trapped charge.
- Examples of mechanical stress are coefficient of thermal expansion (CTE), creep, fatigue, interfacial de-adhesion, shock, wear, and yield.
- Radiation stress includes alpha particle, gamma particle, trapped charge in oxide and ultraviolet.
- Thermal stress includes glass transition (T_g), phase transition, resistive heating, stress-driven diffusion, and thermal runaway.

8

The Reliability Toolbox

8.1 The HALT Process

Without a doubt, the most important tool available to the product development and manufacturing process is a Highly Accelerated Life Test (HALT). Many other methods have been applied throughout the years to improve product reliability, but the HALT process has become the most effective and fastest method to improve product reliability.

HALT is an accelerated test designed to identify field failures before the first product is shipped. It is a method of applying stresses to a product while still in the design phase, which will reveal imperfections, design errors, and design marginality. After these design issues are identified, they can then be corrected through redesign. The HALT process is then repeated to verify that the design changes worked and that no new design issues resulted from the design changes. The HALT process is very simple, yet few companies have fully implemented the process.

In fact, many companies do not have a reliability program in place. They consider their quality programs sufficient to achieve product reliability. These companies use the traditional approach of product development. That is, products are designed with "checks and balances" in place like design reviews. Design reviews check to verify the design is complete. The design review will verify, for example, that the parts list needed to build the design is complete. The design review may also verify that the material list is in a "standard format," usually defined by the manufacturing process. Design reviews typically verify design completeness through a concurrent activity involving all involved functional groups to review the documentation package and verify that the design is complete. Examples of some of the areas that are covered in a design review are as follows:

Engineering

- Schematics
- Block diagrams
- Theory of operation
- Outline drawings
- Input/output descriptions
- Thermal design
- Component derating
- Power descriptions

Improving Product Reliability and Software Quality: Strategies, Tools, Process and Implementation, Second Edition. Mark A. Levin, Ted T. Kalal and Jonathan Rodin.
© 2019 John Wiley & Sons Ltd. Published 2019 by John Wiley & Sons Ltd.

Manufacturing

- Design for manufacturing (DFM) guidelines
- Printed circuit board (PCB) guidelines
- Material list (bill of materials [BOMs])
- Assembly drawings
- Assembly instructions
- Manufacturing cost

Test

- Design for test (DFT) guidelines
- DFT cost
- Test software
- Fixtures

Supplier

- Approved suppliers
- Material costs
- Delivery lead times
- Alternate sourcing

Software

- Software debug and validation

The list of items covered in a design review can be extensive. Most companies also use some form of continuous improvement to improve and streamline the design review process. The one component typically left out of design reviews is reliability, especially field reliability information. Design reviews may cover some reliability issues, but the issues are generally based on lessons learned. Issues such as derating, DFM, and DFT improve product reliability and are sometimes covered in design reviews. However, reliability should be a bigger part of the design review process. The best way to reveal the reliability issues in a design is through HALT.

After the initial design is complete, a prototype is fabricated to test and verify that the design meets specification. Usually, not all the requirements are met and redesign is needed. Later, after the changes have been made, the redesigned prototype units are tested again to "prove out the design." The process of verifying that the design meets specification is referred to as a design verification test (DVT). At this point, the design is considered complete and ready for production. If you perform HALT *before* the DVT, there is strong likelihood that you will pass the DVT the first time. HALT is not intended to replace DVT. By performing HALT on the first engineering units that are functional, reliability issues are identified and fixed early in the development cycle. The end result is a faster time-to-market and passing DVT the first time.

In the traditional approach, products are manufactured and shipped to the customer. In the first year or two of production, there is an accumulation of field failures that consume warranty dollars and often create dissatisfied customers. Teams are then formed that are dedicated to investigating the field failures, determining their root cause, and developing corrective action. This is followed by an endless stream of corrective engineering change orders (ECOs), to eliminate the problem. The design problems can take years to resolve and delay the product from reaching design maturity. This long delay will have an adverse effect on profitability and customer satisfaction. The HALT process speeds up the product design cycle and significantly reduces the number of field failures typically experienced by early production.

The HALT process is an accelerated test, which will precipitate field failures in a relatively short time period, well before any product is in the field [1]. Once these failure modes are identified, they can be removed through redesign. Then, by applying HALT after redesign, the design fixes can be verified with the assurance that no new failure modes have been designed into the product. The end result is a final product that is free from defects while having a significant reduction in lost dollars to warranty claims. HALT is a stress process that accelerates failures so that they can be corrected before first shipment. HALT yields design maturity before the first unit is shipped. So HALT can be considered a "design maturity accelerator."

HALT requires that the device is powered up and operational while diagnostics monitor the device for normal operation. By monitoring the device under stress, failures can be detected along with the point-in-time and environment conditions when the device fails to meet specification. If a failure occurs, the device under test (DUT) is removed from the HALT chamber and the next device is tested. Similar stresses are then applied to the next device to learn if it fails in a similar manner. While HALT is performed on the next device, the previously failed device is evaluated to determine the root cause of the failure. That device is then fixed and returned to the cycle for the next step in the HALT process. After five or six devices have been HALT stressed, a Pareto chart can be created of the failures. The Pareto chart (Figure 8.1) graphs five (hypothetical) precipitated failures and how many of each was discovered in the five test units.

In Figure 8.1, the design team might consider not dealing with failure E because it only happened on one unit out of five. This is often referred to as a *single event or anomaly* and is of little importance. Not true. If 30 units had received HALT, there would probably be more failures in column A through E. There may well be even more failure modes. The point here is, that with more units, failure E would no longer appear as trivial; there could be many more assemblies with failure mode E. Because of the very small sample of five units in HALT, even the single failure mode E is significant. Failure analysis to

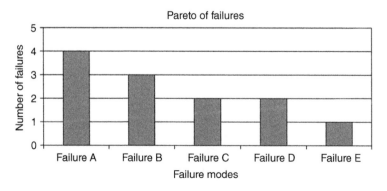

Figure 8.1 Pareto of failures.

the root cause is needed for every HALT failure because they are all likely field failure modes.

The full intent and purpose of the HALT process is to drive units to failure. Investigative techniques such as failure analysis, which drive down to the root cause of the failure, will reveal the true physics of failure. Once these failures are identified, they can be remedied by redesign.

Finding the root cause of the failure is critical. Just fixing the failure is of little value. A major automobile manufacturer discovered that some vehicles had completely dead batteries upon delivery to the dealer while others, on the same truck, were fine. Replacing the battery fixed the problem and allowed the dealer to sell the product. But what caused some of the batteries to fail? The solution turned out to be simple. Some cars, when loaded onto the delivery trailer, were at a significant angle. This caused the trunk lid sensor to activate the trunk light, but just for those cars at a steep incline at the rear of the delivery trailer. The trunk light went on because the sensor in the trunk lid turned the light on because the angle of sensor was correct for an open trunk lid. The solution was to disconnect the batteries before shipment to eliminate this problem. The failure mode was known and the root cause was discovered. The corrective action was acceptable for the short-term and a long-term solution was forthcoming. It was a lot easier to remove one battery cable and reattach it later than to replace batteries at the dealer.

Many years ago, radios and televisions had vacuum tubes. They were relatively unreliable in that their filaments would burn out, a major failure mode. The service person who replaced the tube fixed the problem but did not find out the root cause. Over the years, filament design extended the life of the tube technology by finding the root cause of filament failure and making changes that improved the product. Still, the tubes failed. Then, the transistor was invented and the failed filament root cause was solved. There are no filaments in transistors to fail.

8.1.1 Types of Stresses Applied in HALT

The HALT chamber is capable of applying two different types of stresses to the product, vibration and temperature. For HALT to be effective, these two stresses (at a minimum) are required. However, these are not the only stresses that can be applied to accelerate a product to failure. Examples of other stresses that can be applied in conjunction with temperature and vibration are as follows:

- Voltage margining
- Clock frequency
- AC supply margining (voltage and frequency)
- Power cycling
- Voltage sequencing

The stresses just described can be applied individually at first and then in combination with the other stresses. The decision of which stresses to apply is based on experience and what is feasible. At a minimum, vibration and temperature are required. This point is illustrated in Figure 8.2 [2].[1] The graph is from the work done at Qualmark Corporation, a HALT testing facility, and is a summary of its testing on 47 products

---

1 HALT stress test failure breakdown.

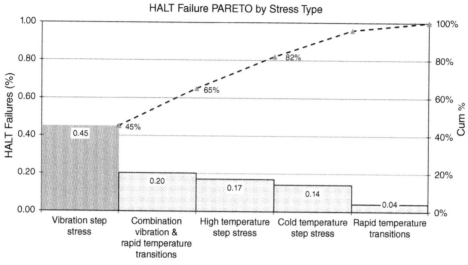

Figure 8.2 HALT failure percentage by stress type.

from 33 companies and 19 different industries. The testing started with cold step stress and proceeded around in a clockwise direction ending with the combination of vibration and rapid temperature transitions. If only temperature testing was performed, 35% of the design failures would be identified. Likewise, if only vibration testing was done, 45% of the design failures would be identified. The power of combination stresses to identify design failures is evident. Temperature or vibration alone identify less than half of the reliability design issues. That is why it is important to apply both temperature and vibration to achieve the goal of accelerating the greatest number of field failures in a relatively short time.

Using accelerated stresses, first singularly and then in combination, will reveal reliability design issues that can be eliminated through redesign. The redesign is performed early in the design cycle where it is the least expensive to implement. After the design is corrected, it is necessary to retest the product with HALT. This will ensure that the fixes worked and that no new failure modes were designed into the product as a result of the redesign.

8.1.2 The Theory behind the HALT Process

When a product is designed, it is tested to verify that it meets all of its design specifications. Design specification may include: output performance, temperature, vibration, shock, power supply levels, duty cycle, frequency, distortion, power source limitations, altitude, humidity, temperature, and many more. We will illustrate this point with a single specification (i.e. temperature) and refer to it as having an upper limit and a lower limit (refer to Figure 8.3). This is the design operating range the product must meet in order to function normally and meet design specifications.

The design operating range, described in Figure 8.3, can be applied to all the design specifications. It is in the operating range where the end product is designed to properly

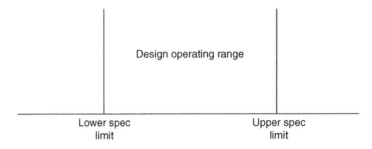

Figure 8.3 Product design specification limits.

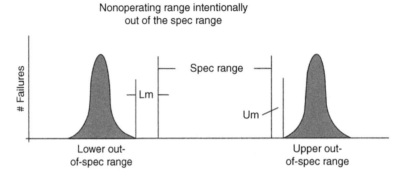

Figure 8.4 Design margin.

function. This is the range where the development team tests the product to ensure acceptance through DVT. Ideally, the product will function beyond the upper and lower specifications limits, commonly referred to as *design margin*. This design margin provides a safety range that allows for component, design, and process drift that would otherwise reduce the yield of the product in production. Over time, in the field, product performance begins to drift, leading to system failure. Design margin helps maintain the product operation over time. This point is illustrated in Figure 8.4. The shaded curves show the distribution of where a sample of products fails. The graph shows that when the sample is stressed to the limits of the specification, there are no failures. In fact, the first product failures begin to occur at a point beyond the upper and lower design margins.

You may be wondering why there is no shaded coned distribution beyond the upper and lower design limits. To explain this, consider that we are going to stress test 1000 production units to find the upper and lower limits where the product fails. What we will find is that they will not all fail at the very same point. There will be one point where most of the units fail. Then, as we go above and below that center point, we will find the number of units that fail decrease and eventually go to zero.

When testing the first prototypes, oftentimes not all the design specifications are met. Sometimes, there is little or no margin for safety. Figure 8.5 illustrates a specification where the upper and lower "out-of-spec ranges" have fallen inside the design spec range. In this situation, not all the products are able to meet the design specification. This problem will manifest itself in manufacturing as a low first-pass yield and early field failures.

HALT will improve product reliability, DVT acceptance, and first-pass production yields by increasing the design safety margins. To illustrate this point, refer to Figure 8.6.

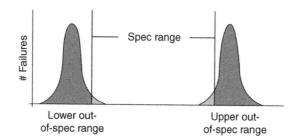

Figure 8.5 Some products fail product spec.

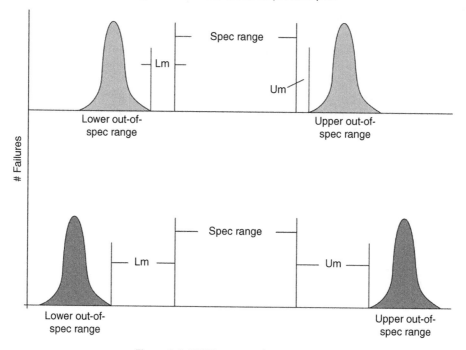

Figure 8.6 HALT increases design margin.

In HALT, we stress the product beyond its design specifications. At some point, the stress becomes so great that the product no longer operates. This is referred to as *a failure*. However, there are two types of failures that are possible, *soft* and *hard failures* as shown in Figure 8.7. They are referred to as *recoverable* and *nonrecoverable failures*. To determine which type of failure you have, reduce the stress level to its normal specification range. If the product returns to normal operation, then it is a soft failure. If the product still does not operate, it is a hard failure. The product may need to be reset before it can return to normal operation. If so, this is still considered a soft failure because no rework was required to fix the product. Hard failures require troubleshooting to determine what failed. All hard failures are later investigated to determine the root cause of the failure.

In HALT testing, the product is stressed to hard failure, the root cause of the failure is determined, and appropriate design changes are implemented. The soft failures are also designed out. To verify that the design change has been successful, we perform a second HALT. If the failure modes are removed and no new failure modes surface, then the design fix is considered good. So how does HALT improve product reliability, DVT acceptance, and first pass production yields?

The hard and soft failures that were precipitated in HALT are designed out of the product. In doing so, it increases the "design margin" where the stress causes product failure. Fixing the hard and soft failures causes the design margins to widen. By correcting the failures found in HALT, the gray shaded areas in Figure 8.6 are pushed out, leaving greater design margin between the design specifications and the stress points where failures occur. The end result is an improved product reliability and improved production yield due to improved sensitivity to process variation.

HALT intentionally stressed the product beyond the spec limits in order to cause failures. Then, the failures are corrected. The resultant product is more reliable, because the field failures surfaced in design and were fixed. Also, the product can now operate to specification limits beyond the design specifications. This point is further illustrated in Figures 8.7 and 8.8.

The product spec range lies between the upper and lower operating points where the product will remain in specification. Ideally, the product will have some margin where it will still function without failure. The measure of this is the upper and lower operating margin. When stresses are applied to the product beyond these margins, then soft

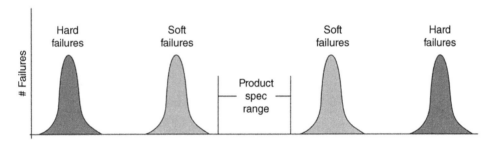

Figure 8.7 Soft and hard failures.

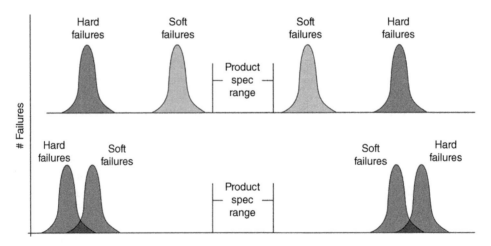

Figure 8.8 Impact of HALT on design margins.

failures will start to occur. Continuing to increase the stresses will drive the product to hard failures.

8.1.3 HALT Testing Liquid Cooled Products

A HALT chamber uses turbulent forced air flow to heat and cool the product under test. Heating is typically achieved with resistive coils. The HALT chamber uses fans to force air over the resistive coils to heat the air. Cooling is achieved by converting liquid nitrogen which is stored in a tank or Dewar to a vapor form. Using PID (proportional, integral, and derivative) control circuitry the chamber switches between heating and cooling forced air to regulate the chamber air over the product to get it to the desired temperature. This works quite well for products that can be regulated with forced air. This does not work when product to be tested uses liquid cooling to control the product. The liquid cooling could be through a cold plate that is in intimate contact with the circuitry through a thermal interface pad. The product could also be designed for immersion cooling where an inert liquid is circulated through the product to regulate the temperature. In these two situations using forced air is not an acceptable or feasible way to thermally stress the product.

One solution is to use the chamber air to regulate the liquid coolant that is thermally regulating the product under test. The liquid coolant runs through a heat exchanger, and the heat exchanger is placed in front of the chamber forced air output port (Figure 8.9). The chamber forced air changes the temperature of the circulating liquid temperature to the chamber set temperature by means of the heat exchanger. The liquid coolant is circulated through the heat exchanger using a pump that forces the liquid from a reservoir tank. The external pump must be rated for the temperature range of the chamber. A coolant distribution unit (CDU) can be converted for this purpose. It's important to

Figure 8.9 Two heat exchangers placed in front of chamber forced air.

make sure that the liquid used will not boil within the temperature range to which the product will be stressed. You can work with your coolant supplier to select a higher temperature rated liquid for this application. We use HFE 7500, which boils at 128 °C.

8.1.4 Planning for HALT Testing

Before the product can go through HALT, some planning is in order. There can be a significant amount of preparation work required before HALT testing. The reliability engineer and the lead engineer should work together to have everything in place for the day the HALT process is to begin. The following is a list of items that should be ready for the test:

1. The product, hopefully, three to five working units, and a spare. The spare is often called a *golden* unit because it is not intended for stress testing; it is used when there are subtle testing issues and it is difficult to tell if the DUT or the test instrumentation is at fault. Inserting the golden unit will verify if the problem exists with the DUT. This can speed the troubleshooting process greatly. Thus, the information learned by using this unit is "golden."
2. Test instrumentation. This is probably the most important item on the list after the product itself because the failures have to be discovered and corrected. Poor monitoring will miss some failures and render the HALT process less effective than it might have been.
3. The output specifications that will be monitored and the monitoring instrumentation.
4. Documentation, i.e. schematics, assembly drawings, flowcharts, and so on.
5. A mechanical fixture to affix the DUT to the HALT table.
6. Input and output cabling.
7. Special devices, i.e. liquid cooling apparatus, air ducts, power sources, other support devices, and so on.
8. Software, where required.
9. The stress levels intended to be applied to the DUT (established and agreed to by the HALT team). What are the stress limits that can be applied to the product? If the product has built-in protection circuitry for temperature, can it be overridden to allow for higher thermal stress? The HALT test plan should include verification that the temperature protection circuitry works before enabling a mitigation plan to test beyond those monitor limits. The agreed upon test limits should be documented in a table such as that shown in Table 8.1.
10. The time required for the testing.
11. The lead engineer needs to be available for the entire HALT process.
12. A test engineer assists in failure analysis.
13. The reliability engineer and a HALT chamber operator – the reliability engineer usually writes the final HALT test report.

The HALT team starts by placing the DUT in the HALT chamber and interconnecting it to the power sources, loads, and instrumentation. Then, the DUT is turned on and monitored to verify if it is operating properly. This step is needed to ensure the test setup is functioning properly. The runtime here is determined by how long it will take

Table 8.1 Agreed upon HALT limits.

Profile for standard HALT instrument		
Ambient temperature is considered to be 20 °C		
Enter diagnostics run time and stress limits	Temp. (°C)/Vib. G rms.	Run time (min)
Calibration		10
Checkers/diagnostics		5
Other diagnostics (describe)		0
Enter lower temperature limit (°C)	−40	
Enter upper temperature limit (°C)	120	
Enter upper vibration limit (Grms)	60	
Total diagnostic run time		**15**
Total mandatory test time/DUT (h)		**10.4**

Table 8.2 HALT profile for test setup checkout.

#	Process steps	Min/step	Sub totals (min)	Sub totals (h)
Mandatory **Test setup verification**				
1	Set chamber temp to 20 °C.	2	2	
2	Run diagnostic test software prior to HALT to verify the DUT and the test setup is functioning properly.	8	10	
3	Vibrate at 5 Grms while running diagnostics to verify adequate test connections.	15	25	
4	Stop vibration	1	26	
	Total Time		26	0.4

to verify that the unit is functioning properly. It is typically a function of the time it takes for the test software to run one or two complete diagnostic test cycles to completion.

Once the test setup is confirmed to be operational, the HALT chamber doors are closed and a low level of vibration is started with the chamber set for room temperature. The vibration level should be set at 5 Grms, random vibration over 6 degrees of freedom (Table 8.2 and Figure 8.10). The purpose of this test is to verify that the interconnections and test monitoring setup is hooked up properly and there are no loose connections. Again, one to two test cycles are run to verify that the hardware is ready for the HALT.

The first HALT test is typically temperature step stress and power cycling (Table 8.3 and Figure 8.11). It is recommended that you start this test going down in temperature until you reach the lower limit and then proceeding up to the upper temperature limit.

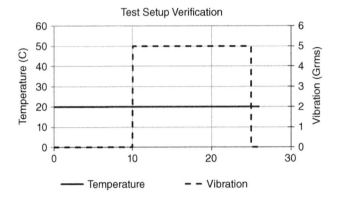

Figure 8.10 Test setup profile to checkout connections and functionality.

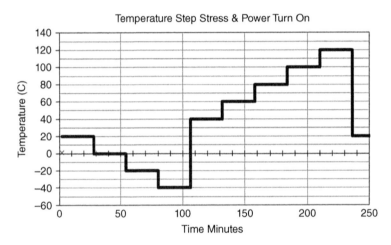

Figure 8.11 Temperature step stress with power cycle and end of each step.

If testing ends at a cold temperature (i.e. a hard failure at −20 °C), the DUT should be brought back to room temperature prior to opening the chamber doors. The chamber may get to room temperature quickly, but it is important to wait for the product to reach room temperature as well. This will prevent condensation from forming on the product if the chamber is open when the product is still at a cold temperature.

Temperature testing begins by stepping down to lower temperatures in 10–20 °C increments with dwell times of typically 10 minutes. The idea is to start with the weakest stress and move to stronger stresses as the stress testing continues. This way, subtle failures will not be lost with excessive stress testing. The time at dwell is driven by the time it takes the product to reach 90% of the target temperature and the test instrumentation diagnostics/checkers run time required to complete a full test.

Placing a thermal couple on the product under test in an area where there is high thermal mass will allow you to monitor the product temperature and ensure that the products reaches temperature before checkers/diagnostics start running. Most HALT chambers provide an option to control the chamber air based on the air temperature or the product temperature. Selecting product temperature is preferred when the product

Table 8.3 Temperature step stress with power cycle and end of each step.

Mandatory Combinational cold and hot power turn on (−40 to +80 °C) and temperature step stress (−40 to +120 °C)[a]	Time in minutes		
	Min/step	Sub totals (min)	Sub totals (h)
5	Set chamber temp to 20 °C.	2	2	
7				
8	Ramp the temperature 20 °C and soak for 10 min. During the temperature transition.	11	13	
9	Run diagnostics to verify proper operation. Toggle the power off for a minimum of 1 min and on.	15	28	
10	Repeat steps 8 and 9 until the lower temperature limit is reached.	78	106	
11	Ramp up in 40 °C and soak for 10 min.	2	108	
12	Perform power off/on test.	3	111	
13	Run diagnostics to verify proper operation.	15	126	
14	Increase temperature by 20 °C and soak for 10 min. During this time, toggle the power off and on.	11	137	
15	Run diagnostics to verify proper operation.	15	152	
16	Repeat steps 14 and 15 until the upper temperature limit is reached. Note: Do not exceed 120 °C.	130	236	
17	Return chamber to 20 and left the system stabilize for 10 min.	13.333333	249	
18	Run air purge before opening the chamber		249	4.2

a) *Profile process:* To reduce test time, it is recommended that you perform the cold and hot turn on during temperature step transition at the end of the temperature step stress. Once DUT is temperature stabilized (10 min typ), proceed with diagnostics to verify operational functionality. After functionality test perform power cycle and verify DUT turns onto a good state.

has high thermal mass. One reason for this is that the hot and cold thermal boost kick in when there is a significant difference between the temperature set point and the temperature of the product or air depending which one the chamber is set to control to. As the product or air temperature approaches the set point, the boost begins to decrease and eventually goes to zero when you reach the set point. If the chamber is set to control on the air temperature and the product has a high thermal mass (i.e. product temperature changes slowly), the chamber air reaches the set point quickly cutting off the boost even though the product still has significant time to reach the set point. That is why it is best to control the chamber based on how the product is responding to the environmental stress.

The product under test is power cycled either at the beginning or at the end of each temperature step stress. It is recommended that power cycling be performed at the end of each thermal step stress. Document the lower and upper limits where the DUT fails to power up. Investigating the root cause for the upper and lower temperature limits where the DUT does not power up can identify power supply, timing, and temperature sensitive issues. Continue the temperature step stress to the lower temperature limit (Table 8.1).

Upon completion of the cold temperature step stress, start the high-temperature step stress. Step the chamber to the first set point above room temperature. Begin to increase the temperature in a similar manner (10–20 °C increments) until you reach the high temperature limit. Record all soft and hard failures. If failures occur during temperature step stress, it is recommended that you stop and investigate the failures to determine root cause. If it is a soft failure, investigate ways to "work around" the failing element in order to continue temperature step stress. This usually involves bringing outside air into the chamber to regulate the temperature sensitive element so it will continue to operate. After completing high temperature step stress, set the chamber back to room temperature and allow the product temperature to stabilize. Before completing temperature step stress, run checkers/diagnostics to verify that the DUT is still working correctly. The first stress element for HALT testing, temperature step stress, is complete.

After temperature step stress testing has completed, start vibration step stress. Vibration step stress is run with the chamber set to room temperature, typically 20 °C. The stress test plan with upper limits (if needed) for vibration step stress should be defined before testing begins (Figure 8.12 and Table 8.4). Increase vibration in 5–10 Grms increments until you reach the limit of the chamber's capability or the defined limit called out in the test plan.

Note: The vibration test limit defined in the test plan typically is the chamber vibration limit. Some HALT practitioners will after each vibration step stress test has completed, lower the vibration to tickle vibration (5 Grms) for one test cycle. The reasoning is that vibration-caused failures may not reveal themselves to the test instrumentation at the higher vibration levels, but the failure may become detectable at low vibration levels.

Continue stepping the vibration level until the highest stress level that was established for the test has been reached. Record the failures and of course, troubleshoot to root cause since the design team is there support the HALT testing. *Side note:* Some components are known to be problematic under vibration and need to be planned for. Examples

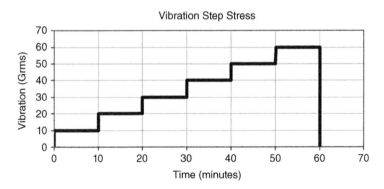

Figure 8.12 Vibration step stress.

Table 8.4 Vibration step stress.

Mandatory		Time in minutes	
Vibration step stress: Max limit 60 Grms/minute in 10 Grms steps.	Min/step	Sub totals (min)	Sub totals (h)
Set chamber temp to 20 °C.	5	5	
Set the vibration level to 10 Grms, run for 10 min with diagnostics running.	10	15	
Increase the vibration level in 10 Grms steps and diagnostics running. For 10 min. Continue increasing vibration level by 10 Grms until the upper vibration limit is reached. 10 Grms steps for 10 min per step. Loop on diagnostics during vibration.	50	65	
Stop the vibration. Verify proper DUT operation.	15	70	
Total time		70	1.17

of vibration sensitive components are: mechanical relays (they can chatter and resonate under vibration), piezo effect components (i.e. oscillators, filters, etc.).

Once temperature and vibration step stress tests have completed, the soft and hard failure limits for the product are known for each type of environmental stress. The limits determined from these two tests are used for the subsequent HALT tests that follow. The last of the required HALT test is the combinational temperature and vibration step stress (Table 8.5 and Figure 8.13). This stress test combines the temperature and vibration step stresses in combination and is the most stressful to the product under test. The upper and lower stress limits for this test are based on the upper and lower failure limits found during vibration and temperature step stress. The general rule of thumb is to set the upper and lower temperature limits at 10 °C above and below the hard failure limits (or soft failure limits) determined during temperature step stress.

The stress limits should always be below the hard failure limits. If the soft failure caused the product to not self-recover, fails to operate or shuts down, then use the soft failure limit to determine the maximum stress for combination step stress. If the soft failure was a parametric test and resulted in the product being out of spec, then that is OK. We do not expect DUT to meet all the parametric specifications when stress outside its operation specification limits, but we are expecting it to operate functionally. The same theory and process applies to setting the upper vibrational stress limit. The vibration step stress limits are set at 10 Grms below the hard failure limits or soft limits, as already described.

Many practitioners then move to a rapid thermal cycling stress test (Table 8.6 and Figure 8.14). This is where the chamber temperature is made to change as rapidly as possible. The temperature limits for this test should be 10 °C below the high and low temperature failure limits determined in temperature step stress (basically the same temperature limits used for the combinational temperature and vibration step stress).

Table 8.5 Temperature and vibration step stress.

	Mandatory Step temperature/step vibration stress: (−40 to +120 °C (20 °C steps), 20/40/60 Grms) (60 °C min⁻¹)[a]	**Time in minutes**		
		Min/step	Sub totals (min)	Sub totals (h)
32	Set chamber temp to low temp limit (i.e. −40 °C) and soak for 5 min.	5	5	
33	Perform vibration step stress in 20 Grms steps to the upper vibration limit. Dwell at each step for 10 min per step. Loop diagnostics during vibration.	30		
34	Increase temperature in 20 °C steps until the upper temperature limit 120 °C is reached. Repeat vibration step stress at each temperature step.	210	210	
39	Ramp to 20 °C and soak for 5 min. Verify proper DUT operation.	10	10	
	Total		225	3.8

a) *Note:* Set temperature boost to achieve rapid cycling (10 °C cold & 10 °C hot). Start at the low temperature limit (i.e. −40 °C). Set vibration at 20 Grms for each new step transition.

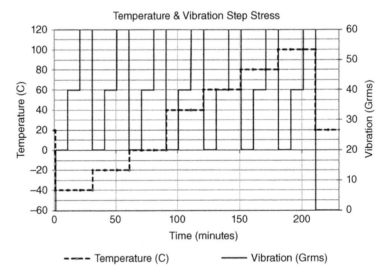

Figure 8.13 Temperature and vibration step stress.

There should be a minimum of three cycles, but five cycles is preferred. The goal is to rapidly transition from one temperature extreme to the other. It also is important to dwell, typically 10–15 minutes based on the products thermal mass, once the target temperature has been reached to ensure the product reaches the target temperature. This is especially important if the product being stressed has a large thermal mass. The chamber temperature will typically change much quicker than the product temperature.

Table 8.6 Rapid thermal cycling.

	Optional **Rapid temperature cycling: (60 °C min⁻¹)ᵃ⁾**	**Time in minutes**		
		Min/step	Sub totals (min)	Sub totals (h)
20	*Set temperature boost to achieve rapid cycling (10 °C cold, 10 °C hot).*			
21	Verify proper operation of the DUT.	15		
22	Close doors and set chamber to 20 °C.	5		
23	Purge chamber with nitrogen for 2 min (nitrogen purge).	2		
24	Ramp temperature cycle from upper limit −10 °C to lower limit −10 °C (limits identified in temperature stress test). Soak 10 min once product gets within 10 °C of set point (upper or lower limit). Loop on diagnostics during soak.	33		
25	*Enter the number of temperature cycles (repeat above step) ⇒*	5	165	
26	Ramp back to 20 °C and soak for 5 min.	7		
27	Verify proper operation of the DUT. (Run diagnostics.)	15		
28	Run air purge before opening the chamber.			
	Total		187	3.1

a) *Note:* When going cold, turn off any air being used to mitigate power boards. The airline can contain moisture that can lead to condensation.

So it is possible for the chamber to be thermal cycling at the temperature limits set, but the product does not change much because there was no dwell at temperature to allow the product to reach the temperature set point. There are several ways to avoid this error. You can set the chamber to be controlled by the product temperature instead of the chamber air. In this case, you would place the thermal couple for controlling the chamber at the place where there is high thermal mass. The other method is to run the test manually, place a thermal couple on the high mass section of the product and just wait until it reaches temperature before starting the next temperature ramp. The rapid thermal cycling test uncovers any design sensitivities that are thermal rate of change sensitive (Table 8.6 and Figure 8.14).

A slow temperature ramp is an effective way to discover instabilities that are temperature sensitive (Table 8.7 and Figure 8.15). The temperature limits for this test should be 10 °C above and below the low and high temperature failure limits determined in temperature step stress (basically the same temperature limits used for the combinational temperature and vibration step stress). The slow temperature ramp will uncover temperature-related instabilities that can be extremely difficult to reproduce if you don't know the temperature window they occur in. The instability may occur over a very narrow temperature range, so a slow temperature ramp may be the only way to observe the anomaly. In addition, failures reported by the end user that turn out to be due to temperature sensitivity will not contain the necessary information needed to reproduce

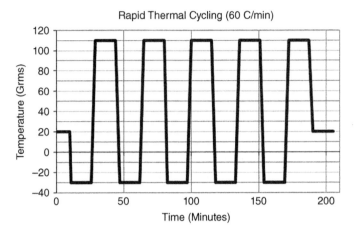

Figure 8.14 Rapid thermal cycling.

Table 8.7 Slow temperature ramp.

	Optional **Slow temperature ramp:**	**Time in minutes**		
			Sub totals (min)	**Sub totals (h)**
34	*Set temperature boost to achieve rapid cooling (10 °C cold)*	**Min/step**		
35	Set temp to 20 °C and soak for 5 min.	5		
36	Set vibration to 0 Grms. Loop on diagnostics during vibration.	0		
37	Ramp temp to lower temp limit (i.e. −40 °C) less 10 °C at 60 °C min^{-1}. Add 10 min for dewll at lower temperature (reach temperature stability).	11		
38	Soak at lower temp limit (i.e. −30 °C) less 10 °C for 5 min or till stable.	5		
39	*Set temperature boost back to 0 for hot and cold.*	0		
40	Ramp to upper temperature limit (i.e. 120 °C) less 10 °C at 1 °C min^{-1}.	142		
41	Loop on diagnostics during the entire ramp.	0		
42	Set chamber temperature to 20°C and soak for 5 minutes when there. (60°C min^{-1})	1	162	2.7
	TOTAL (mandatory and optional) test time =		162	2.7

the instability. The instability could be an oscillation, an increase in noise, a reduction in output gain, power supply instability, automatic gain control (AGC) instability, a loss of phase lock, or an increase in phase noise. These are just a few failure modes that can occur over a narrow temperature range. In addition, because the phenomena only occurs within a narrow temperature range, a step stress may miss the window where this undesired behavior occurs. The slow temperature ramp provides you an opportunity to observe the instability or other unacceptable behavior as you sweep through the temperature changes.

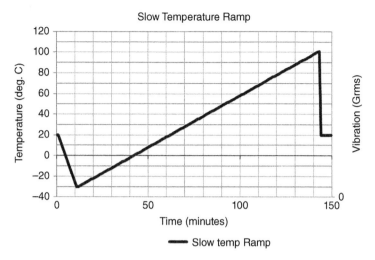

Figure 8.15 Slow temperature ramp.

Table 8.8 Slow temperature ramp with constantly varying vibration level.

	Optional Slow temperature ramp and ramping up/down vibration:	Time in minutes		
			Sub totals (min)	Sub totals (h)
34	*Set temperature boost to achieve rapid cooling (10 °C cold)*	**Min/step**		
35	Set temp to 20 °C and soak for 5 min.	5		
36	Set vibration to 10 Grms. Loop on diagnostics during vibration.	0		
37	Ramp temp to lower temp limit (i.e. −40 °C) less 10 °C at 60 °C min^{-1}. Add 10 min for dwell at lower temperature (reach temperature stability).	11		
38	Soak at lower temp limit (i.e. −30 °C) less 10 °C for 5 min or until stable.	5		
39	*Set temperature boost back to 0 for hot and cold.*	0		
40	Ramp to upper temperature limit (i.e. 120 °C) less 10 °C at 1 °C min^{-1}.	152		
41	Loop on diagnostics during the entire ramp.	0		
42	Turn off vibration and ramp to 20 °C. Soak for 5 min. (60 °C min^{-1})	1	174	2.9
	TOTAL (mandatory and optional) test time =		174	2.9

A variation of the slow temperature ramp includes constantly varying (up and down) vibration (Table 8.8 and Figure 8.16). The temperature limits for this test should be 10 °C below the high and low temperature failure limits determined in temperature step stress (basically the same temperature limits used for the combinational temperature and vibration step stress). The vibration step stress limits are set at 50% of the hard failure limits or soft limits as described above. The search pattern technique is valuable

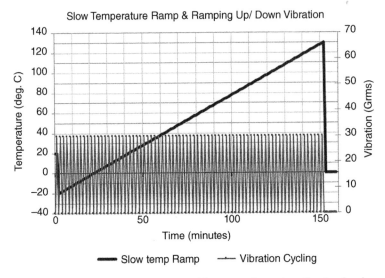

Figure 8.16 Slow temperature ramp with constantly varying vibration level.

where the soft failure is very close to the hard failure. The temperature increase slowly up at $1-2\,^{\circ}\mathrm{C}\,\mathrm{min}^{-1}$ with the vibration level is oscillating up and down all the while the product is being continuously monitored.

Repeat these stresses in the same sequence on the next DUT and carefully note where there are similarities and differences. Continue until all the DUTs have completed HALT and the data is recorded. It might be best, in some cases, to save some of the DUTs and stop testing early. This will allow you to work on the failures to their root causes, and to apply the fixes to new systems.

It is very important to record the stress levels where the soft and hard failures occurred. Later, when you have made design corrections, these stress levels should have increased, thus increasing your operating margins.

The stresses thus far to accelerate failure have been environmental in nature, namely temperature and vibration. The same methodology can be applied to electromechanical stresses to accelerate failure. For example, the power supply voltage can be margined up and down from its nominal value. This should be part of the hardware DV, but is usually done at room or nominal ambient temperature. During HALT, this is test is repeated combined with environmental stress, typically temperature. If the product design uses multiple power sources or voltage references, then HALT should include voltage margining and varying power supply turn-on time strategy. This test should include all allowable cases at the upper and lower temperature limits with guard band added. This includes varying supplies up and down alone and in combination to cover the range of allowable states. Create a test matrix for the combination of voltage supply upper and lower limits at the upper and lower temperature limits. Power supplies will not all turn on at the same time.

Some components require multiple voltages to operate. This can be especially true for application specific integrated circuits (ASICs). It is important to verify that the component will not power on in a bad state based on the order the power supplies turn on. Sometimes the manufacturer will state this in the datasheet but was overlooked by the design team. The component manufacturer may be unaware of a possible condition during turn-on where the component can turn on in a bad or locked state. If the device is an ASIC that uses multiple supply voltages, it likely will not be known if there any unallowed power on conditions. There are other electrical stresses that should be margined in combination and alone as well. These stresses are AC line input voltage and AC line frequency, clock frequency, and so on.

8.2   Highly Accelerated Stress Screening (HASS)

Once the reliability design issues identified from HALT have been designed out, the product is ready for manufacturing. However, a good design is only half the battle. Product reliability is achieved through both good design and manufacturing processes. Contrarily, design flaws and poor manufacturing processes result in field failures. The HALT process focused on the design issues that result in field reliability issues. There is also manufacturing process variation, which can produce product weaknesses that eventually lead to field failures. Can the HALT process be applied to manufacturing to prevent products from shipping which have an unacceptable process variation?

The HALT process can be applied indirectly to the manufacturing process. The process that is used in manufacturing is called *HASS*. HASS can prevent marginal and defective units from being sold. The purpose of HASS is to identify products that have process-related defects and manufacturing weaknesses before shipment. HASS is also helpful in identifying when suppliers provide potentially defective parts.

HALT and HASS use similar types of accelerated stresses to identify failures. Both processes also stress the product when it is powered up and operational. But the similarities stop there. HASS stresses the product in a similar way to HALT, but at reduced stress levels. HASS is a gentler form of HALT. HALT is a means to stress test the design and we emphasize the "T" for test. HALT is intended to reveal design-related failures. It is a proactive tool to improve product design and is performed by design engineering. HALT reveals failures that customers will experience in the field if the design is not fixed.

HASS is a stress screen with the last "S" used to emphasize screen. After products are manufactured, they pass through HASS to verify that the manufacturing process is in control. HASS is performed in manufacturing as a means of product acceptance. Process variations are flagged by HASS and can immediately be corrected to prevent an unsatisfactory product from being shipped. HASS is a reactive tool to assure that the manufacturing process stays in control.

The more complex the manufacturing process, the greater the opportunity for manufacturing defects to enter the product, rendering it less reliable. Control of the manufacturing process is critical. Even with the best manufacturing practices in place, that is, statistical process control (SPC), continuous improvement, electrostatic discharge (ESD) protection and training, manufacturing defects due to process drift and supplier issues can surface. HASS is intended to either pass or "screen out" nonconforming product. A HALT chamber, added to the end of the production line, which applies reduced stresses to the product can perform the vital HASS step. The HALT chamber and the HASS chamber can be the same. This is an option for small companies who outsource manufacturing or produce products in low volumes. Many companies prefer to avoid

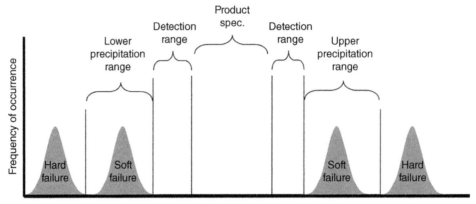

Figure 8.17 HASS stress levels.

the scheduling conflicts between design and manufacturing by having separate chambers for HALT and HASS. HASS chambers also tend to be bigger for batch processing.

From the HALT results, the product operating and destruct limits are learned. In HASS, the product is stressed at levels beyond its operating limits but below the destruct limits. The HASS profile consists of two parts, the precipitation screen and the detection screen. The test begins with the precipitation screen. The precipitation screen is a stress level that is below the destruct limit and above the operating limit. Refer to Figure 8.17. The HASS screening level applied to the product needs to be determined. A good stress level for temperature is between 80% and 50% of the destruct limits. The initial vibration stress level is set at 50% of destruct limits. It is important to stay below the destruct limits; otherwise damage to good product is likely. The purpose of the precipitation screen is to sufficiently damage anything that is a latent defect (meaning it is bad but not yet detectable as being bad) so it can be detected later in a detection test. However, the stress must not damage or severely degrade good product. Generally, if the right stress levels are applied, the defective assemblies will degrade at a significantly greater rate than good product. A proof of screen (POS) (discussed in Section 7.3.1), will identify if the precipitation stress is too severe or ineffective.

The precipitation stress was designed to sufficiently damage defective products, so it can be differentiated from good product. The way we identify bad products is through a detection screen. The detection screen applies temperature stress at levels that are between the soft failure limit and the product spec limit. Set the temperature stress midway between the spec limit and the soft failure limit. The vibration level is set between 3 and 5 Grms (often referred to as *a tickle vibration*). The HASS profile is usually short, typically three to five cycles of precipitation and detection is adequate.

8.2.1 Proof of Screen (POS)

The environmental stresses induced on the product by HASS will lower the life expectancy of the product. This is unavoidable. The goal of HASS is to provide a stress level high enough to precipitate identification of manufacturing defects without removing an excessive amount of product life. How much product life is removed in HASS can be estimated through a process called POS.

The POS process is simple; just repeat the HASS screen until the product fails. Applying the HASS stress repeatedly causes the product to degrade at an accelerated rate. Eventually, the product will fail because of the accumulated effect of the stress. If it takes 20 times to render the product nonoperational, then it is reasonable to estimate that 5% off the product life is removed with each HASS screen. If the DUT failed after only four HASS screens, it can be assumed that 25% of the life of the product was removed each time. There is no minimum number of stress cycles desired before a product fails. Some companies want at least 20 cycles without a failure. The test should be run on a large enough sample to assure that normal manufacturing process variation is accounted for.

If, on the other hand, you run the HASS test for 100 cycles without a failure, the HASS stress levels may be set too low. Some practitioners recommend seeding product to determine if the HASS screen is effective at detecting defective product. Seeding a board requires intentionally inserted manufacturing defects into the product. The product is then tested to determine if the defects are found during HASS screen. The problem with seeded defects is that it is difficult to insert seeded defects that are real representations of product defects (i.e. manufacturing process drift or supplier changes).

Are there alternatives to HASS? Is HASS the only way to screen manufactured products for defects? No, there are other techniques but HASS is the most effective. The alternatives are burn-in, environmental stress screening (ESS), and of course electrical test with no environmental stresses at all.

8.2.2 Burn-In

The burn-in process can be applied to components and final products as a final acceptance test. Recall the bathtub curve from Chapter 6; Figure 8.18. The early failure rate of a product is often higher than the failure rate during its useful life. It is the reason some people prefer to buy a year-old car because "the bugs have been worked out." Burn-in is designed to accelerate infant mortality so failures occur before the product is sold. Typically, infant mortality failures occur in the first year of product use. Product failures

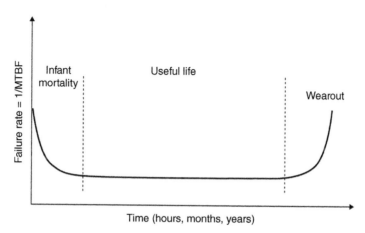

Figure 8.18 The bathtub curve.

are typically higher in the first year, when reliability and quality problems often surface. Theoretically, you could avoid the high infant mortality failure rates by operating the product in-house for a year before it is sold. This obviously is impractical for many reasons. However, if you can accelerate the products first-year use, that is, "burn-in the product," then the infant mortality failure rate will occur during the burn-in test. The burn-in must be long enough to remove most early life failures.

The burn-in process is performed on 100% of production and usually consists of powering on and off the product while running a test diagnostics. It is common for a burn-in test to run 24–48 hour. The product is also kept at an elevated temperature typically at the upper limit of its product specification range, through a temperature chamber.

The burn-in process typically takes, one- to two-days test time, but it can be longer. To get around this long test time, burn-in test chambers tend to be large so that many production units can be tested concurrently. This lowers test costs and increases throughput. Large burn-in chambers can cost several hundred thousand dollars.

The burn-in test, sometimes referred to as a *biased bake test*, is designed to accelerate the aging process. In the 1970s and 1980s, it was not uncommon for products to have a high infant mortality failure rate because the components they used had considerable variation in them. The advent of concurrent engineering, design guidelines, and quality programs such as total quality management (TQM), continuous improvement, and SPC component quality has changed all that. Components today have increased in reliability by orders of magnitude. Today, almost all component manufacturers deliver reliable parts. The reliability of the product is no longer driven by the quality of the components, but by the quality of the design and manufacturing process. This does not mean that part selection is no longer an issue. If you select the wrong part for the job, then expect to have a reliability problem. But the problem is no longer "bad parts."

In addition, studies have shown that burn-in is not an effective technique to remove early failures. Research indicated that burn-in at the component level tended to damage more good parts (via ESD, handling, and electrical overstress) than identify bad parts.

8.2.3   Environmental Stress Screening (ESS)

Another common form of a "burn-in" test is called *ESS*. An ESS test is an environmental stress test designed to accelerate the failure of faulty product. The test is performed while the product is operational and being monitored. The main difference between ESS and conventional burn-in is that ESS induces multiple stresses on the product. These stresses might include

- Temperature cycling
- Temperature soak
- Vibration

However, unlike HASS, the stresses applied to the product are generally below the product spec limits.

Conventional ESS testing can be as short as 2–4 hours or as long as 24 hours depending on the test. Completed products are placed into an ESS chamber (temperature/vibration chamber) and the power is turned on. A typical ESS starts with a temperature cycling profile where the temperature is increased to just below the design specification of the product. When the upper temperature is reached, the product is temperature soaked from 1–2 hours. Monitoring equipment will detect if the product goes out of specification. After a high temperature soak, the product is transitioned to low temperature while still biased and operating. When the lower operating temperature is reached, the product is again cold soaked 1–2 hours. After the temperature soak portion of ESS, a rapid temperature cycling test is performed. The temperature is raised and lowered in repetitive cycles while monitoring continues.

If through this sequential process the ESS total time is reduced to one day, then manufacturing process-related failures occurring beyond one day would probably not be detected if the process goes out of control again. Another shortcoming of the ESS burn-in process is that it may be a week before a manufacturing process error is discovered. This means that the manufacturer could produce many products that are not conforming and still need to be corrected.

This is a sort of a Catch-22. Short ESS cycles are desired to maintain low work in process costs and to reduce the size and cost of the ESS resources needed. Long ESS cycles are desired to catch manufacturing process defects that are undetectable with short ESS cycles. Reducing the ESS window is a large risk. It is easy to see that the ESS concept can provide a poor compromise.

Solder failures and connecting lead technology are high on the list of failures discovered in the field. The majority of these failures that can be detected by temperature cycling cannot be detected with these few cycles, even in a five-day ESS. This is why the customer discovers these failures after several months or years in the field. Often, several thousand cycles are required to precipitate this type of failure. You cannot afford to use the ESS process to discover these failures.

8.2.4 Economic Impact of HASS

HASS accommodates few units, simply because the chambers tend to be relatively small. (Some HASS chamber manufacturers will provide customized chambers for the specific needs of the manufacturer.) If the HASS process uses rapid temperature cycling, then the number of units in the chamber will necessarily have to be limited so that the internal temperatures of all the DUTs can be achieved rapidly. Because HASS can often be accomplished in a short time, such as 30–60 minutes, a high number of units can pass through the HASS process daily. Considering that the HASS process can also look at a smaller sample, it can be much closer in real time to when a defect might have been inserted into the process. This translates into a quicker recovery process when using HASS. Because most defects found in manufacturing are process-related defects and not design-related ones, process delay due to detection and improvement is considerably shorter. Once the HASS process has the manufacturing process in control, the cost of ensuring top-quality is significantly reduced.

The manufacturing ramp means that few units are produced at first, more are produced later and then full production volumes are reached. HASS needs to be applied to every unit until it is clear that there are no new process-related failures. It may be unnecessary to continue HASS because the process is in control, largely due to HASS discoveries. But it would be unwise to completely curtail HASS because process drift and change creep into the system with negative reliability impact. An audit process should be incorporated.

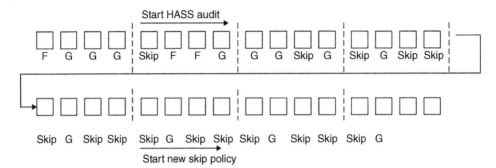

Figure 8.19 HASA plan. Source: Courtesy of James McLinn.

8.2.5 The HASA Process

Highly Accelerated Stress Audit™ (HASA) is a HASS audit process. It periodically examines the process and adds no significant cost when it is not auditing.

It is reasonable to assume that all the manufacturing processes will not stay in control forever, but only for a while. Once HASS has verified the manufacturing process is under process control, the screening can be moved to a skip-lot audit process. The number of units that can be manufactured without HASS has to be determined by the nature of the manufacturing process. If 100 units can be produced every day and placed in shipping hold, the HASS process can be an audit. Divide the 100 units into four groups of 25 and sample the groups with the following results shown in Figure 8.19.

The first lot has a failure and the next three lots are all good. Skip the next lot and sample the one after. This has a failure, go back and check the skipped lot. Next, sample the next two. The next lot has a failure and the one after is good. Sample the next two and both are good. Skip the next lot and sample the one after. This is good, so skip the next lot and sample the one after. It is also found to be good. So skip the next three lots. Sample the one after; it is good, so continue to skip the next three lots. Continue in this fashion until there is a failure and go back to every other lot until you find three good samples in a row. This is a standard skip-lot process.

As you can see, the number of units to be produced between HASS audits (HASA) is a function of many variables. Ideally, the number of units produced between HASA screens should be as high as possible and yet small enough to accommodate cost-effective correction action.

Summary of HALT, HASS, HASA, and POS Benefits

1. Electronic design/margin improvement
2. Packaging design improvement
3. Parts selection improvement
4. Production process improvement
5. Software implementation improvement
6. Rapid design and process maturation
7. Reduced total engineering time and cost
8. Lowered warranty costs
9. Higher mean time between failures
10. Rapid process corrective action

8.3   HALT and HASS Test Chambers

A brief description of the HALT chamber is in order. First, let's describe what it is not with a description of the typical burn-in type process performed by many manufacturers. Typical environmental chambers have the ability to raise and lower the temperature of the chamber interior. The chamber is heated by applying current through resistance wires. Cooling is usually accomplished using a form of air conditioner. (Some of these temperature chambers use liquid nitrogen.) The temperature of the oven can be increased rapidly using resistance wires, but it cannot be cooled rapidly. The air conditioner cooler system does not have the thermal capability of lowering the temperature in the chamber rapidly. This shortcoming of standard temperature control for environmental chambers leads to long test cycle times. A single temperature cycle from ambient to 140 °C and back down to −40 °C with dwells of one hour at the high and low temperatures may take five or six hours to complete owing to the slowness of the air conditioner and the thermal mass of the devices in the chamber. This inability of standard burn-in ovens makes temperature cycling of production units undesirable to the manufacturer.

HALT chambers are environmental chambers designed to quickly provide two environmental stresses. Typical HALT chambers can control temperatures from −100 °C to +200 °C. By using huge resistor banks and a specially tuned liquid nitrogen cryogenic system, these chambers can produce temperature rates of change of 60 °C min^{-1}. Some chambers with advanced cryogenic management can achieve rates of change in the 80 °C min^{-1} range. Through a special design, liquid nitrogen can be aspirated into the chamber very rapidly and the device in the chamber will respond much faster than it can in the burn-in type chamber. In the HALT chamber, a multitude of vents and hoses direct high-flow cooling or heating air toward the DUT to achieve very rapid temperature rates of change of the DUT. Thermocouples located inside the DUTs will ensure that the internal temperatures are achieved quickly to minimize overall cycle time.

The vibration table in the HALT chamber is also unique. It has the ability to move in three directions linearly and rotationally; hence the term *six degrees of freedom* (refer to Figure 8.20).

There are compressed air-driven piston actuators mounted under the table in many angles and directions. These actuators operate in a random sequence. They impart their energy to the table; thus, the table moves in harmony with the actuators. Control electronics randomly selects the actuators and controls when they operate and to what

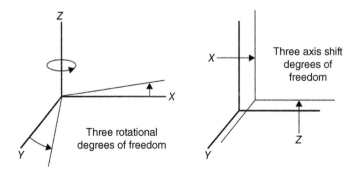

Figure 8.20 A HALT chamber has six simultaneous degrees of freedom (movement).

magnitude. It is easy to imagine that the underside of the table has 8–16 miniature air hammers mounted in all directions that are operated by miniature operators running randomly all at the same time. The table is mounted on a cushion of springs so that it can move in six degrees of freedom. Test devices mounted to the table move similarly. The frequency response of the actuators imparts a broad frequency range to the DUT. Typical table vibration frequencies range from 2 to 10 000 Hz. Tables typically can produce vibration levels to upwards of 60 Grms to devices under test that reach 1000 pounds or more.

8.4 Accelerated Reliability Growth (ARG)

The reliability activities through phase four are primarily focused on improving product design reliability. At the end of phase four, the product has gone through design verification and design validation. Product alpha and beta testing is under way and the focus is shifting toward manufacturing and product ramp. However, in all likelihood there will be hardware reliability escapes and software quality issues after product validation completes that will surface after the product is transferred to manufacturing. When hardware reliability and software quality issues are reported by your customers, engineering support teams (composing of both hardware and software engineers) quickly work to identify root cause and implement corrective action. This activity is reactive and based on the end user notifying you of the problem. The challenge thus becomes, what can be done in phase 5 to bring the customer experience in house providing the opportunity to identify the reliability and quality escapes before the end user experiences them? From my experience, I have found that many hardware reliability and software quality issues can be discovered in Phase 5 when the operations group is validating production ramp readiness. There are several tools that can help you achieve this goal. Two of the more effective tools to help achieve this goal are called ARG (accelerated reliability growth) and ELT (accelerated early life test).

In the early day's electronics, component quality was very poor due to a lack of quality control. Part of the problem was that manufacturers of electrical components were unaware of what were the critical processes, materials, and dimensions that controlled component quality. So it was not uncommon during this period for manufacturers to include component burn to precipitate and detect latent defects from the manufacturing and test process. However, component burn in is not a perfect process and defective components still found their way to the consumer. When consumers started demanding higher-quality components, it forced manufacturers to improve their quality control. Thus began the long period of quality improvement programs like: TQM, Six Sigma, zero defects (ZDs), quality circles, ISO 9000, and other industry standards, as well as quality function deployment (QFD), continuous improvement, mistake proofing a.k.a. poka-yoke, Kaizen, and plan-do-check-act (PDCA), to name a few of the more popular programs.

Component manufacturers over time figured out how to produce very-high-quality electrical components where the defect level was so low that component burn in precipitated more defects than it caught. That's when strategies like "get out of burn-in" (GOBI) became popular and manufacturers switched to a sampling plan like "acceptable quality level" (AQL). AQL programs are necessary to ensure that processes don't go out of control bands and protect counterfeit material and parts from working the way into the production stream.

Toward the end of Phase 4 product development, the focus shifts to product manufacturing and ramp readiness. There is the expectation that at the completion of Phase 5 the product is ready for high-volume manufacturing and design maturity has reached an acceptable level.

Reliability growth is the process of measuring product reliability improvements as failures are discovered and permanently removed through failure reporting, analysis, and corrective action systems (FRACASs). As mentioned earlier, reliability growth programs start when production material begins shipping to the end user. After that you have to wait for the end user to find and report reliability and quality problems. This triggers an investigation that can result in a significant retrofit cost, customer dissatisfaction, and product brand degradation. Reliability growth programs that start when production ships to the end user is a costly and inefficient reactive process. Identifying root cause for field failures generally takes longer because the design team has moved on to a new projects and typically does not get pulled back to support field failure issues. This can also result in the design team not learning from past mistakes.

Reliability growth does not need to be a reactive process and should start in Phase 4 with the first production units. The goal of an ARG program is to create the customer experience in house in a way that allows for the discovery of the design and quality escapes before high-volume production ramp. This can be achieved through an accelerated stress with sufficient run time and sampling from production material to identify the reliability and quality escapes. Reliability growth testing is performed on production material that has gone through the manufacturing and test process and is ready for shipment to the end user (Figure 8.21). Any quality or reliability issues found in this extended testing is treated like a field failure and is logged into FRACAS.

A burn-in is most effective when it is targeted to a known failure mode and mechanisms. The stresses can then be tailored to accelerate the failure mechanism (precipitation) and detection of the failure. If the failure mechanism is known, the acceleration rate

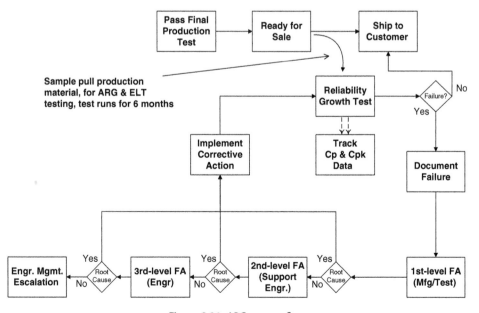

Figure 8.21 ARG process flow.

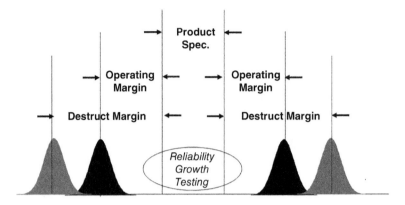

Figure 8.22 Accelerated reliability growth.

can also be calculated. The ARG protocol is an extended burn-in where the failure mode and mechanism is unknown and intended to be discovered. During HALT, we learned the soft and hard failure limits of the product. The ARG stresses are set to be outside the product spec but within the operating margin for the product (Figure 8.22). For derivative product designs, review of past design and quality escapes will provide insight into an appropriate acceleration protocol. An example of an ARG and accelerated ELT program is shown in Figure 8.23. The requirements for an ARG program should be different for high-volume versus low-volume products. The risk also needs to be part of the decision process for how many run hours and the number of production lots that need to be cycled through the ARG program.

The environmental acceleration stress is temperature, so the Arrhenius Equation (8.1) can provide insight to acceleration factor. The acceleration factor is a constant multiplier

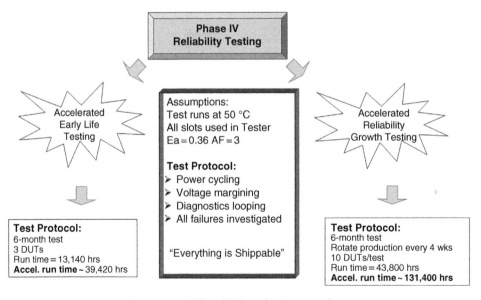

Figure 8.23 ARG and ELT acceleration test plans.

that describes the difference between the two stresses. In this case, the two stresses are temperature. The accelerated stress temperature is 50 °C and normal temperature the product operates at is 25 °C. When there are many possible failure modes, each will have its own failure mechanism and acceleration factor that may or may not be temperature dependent. So assigning a single acceleration factor to system that is thermally stressed needs to be conservative and only provides an estimate for the accelerated run time. In this example, I've chosen a conservative activation energy, E_a, equal to 0.36.

The Arrhenius Acceleration Model

$$AF = \frac{t_1}{t_2} = exp\left[\left(\frac{E_a}{k}\right)\left(\frac{1}{T_1} - \frac{1}{T_2}\right)\right] \tag{8.1}$$

where

A	=	Constant
exp	=	Natural logarithm, base 2.7182818
E_a	=	Activation energy for the reaction
k	=	Boltzman's constant = 8.625×10^{-5} eV/K
T	=	Absolute temperature (K)
t	=	Time

8.5 Accelerated Early Life Test (ELT)

An accelerated ELT is intended to identify unintended early life wear out mechanisms. These failures are systemic in nature so can potentially affect the entire field population. The unintended early life wearout mechanism is an escape in the reliability process. It can also be due to a component running hotter than anticipated, causing it to have a significantly shorter useful life. It can be a significant coefficient of thermal expansion (CTE) mismatch between a large component and the circuit board it is mounted to. The CTE mismatch can create significant strain energy as the product heats up causing it to develop solder cracks early in the product life. Once the solder crack is created, additional thermal cycles cause the stress intensity at the crack to be greater due to a stress concentration created at the solder fracture. The result is that the solder crack continues to propagate until there is a full separation of the solder connection. A component that is unintentionally used beyond its absolute maximum recommended limits is likely to suffer premature wear out. This may be due to a misunderstanding between the differences between absolute maximum limits and recommended operating limits. If you're lucky, these types of design errors get detected during HALT when you step stress the part. However, if it doesn't get detected in HALT, it is very likely to end up as an early life wearout failure.

The accelerated ELT can be the same test use for ARG testing (ARG) except that it runs for a significantly longer time. The sample size does not have to be large because the failure mechanism we are hoping to detect is due to wear out. Typically, ELT runs for a period of six months using the same test protocol as ARG.

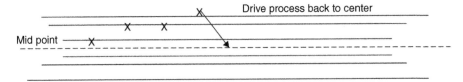

Figure 8.24 Selective process control. Source: Courtesy of James McLinn.

8.6 SPC Tool

Reliability is "quality over time." Early in this book, we discussed the difference between reliability and quality; here, we will point out how to use a well-understood quality tool to improve reliability. SPC is a tool used in manufacturing to minimize control process drift. The tool is used is to periodically monitor a given process to ensure that a product parameter has not drifted out of specification. The SPC process establishes high and low levels that are not to be exceeded. When the manufacturing process drifts near these limits, the operator is instructed to make an adjustment to return the manufacturing process to the center of these limits or to an ideal setting so that quality and consistency can be maintained as shown in Figure 8.24.

Consider that the entire manufacturing processes may be under SPC and at any given time many of the monitored processes have drifted too near the high level. If the finished product is to be assembled using all the processes at this time, this particular unit may still be within the quality standard, still conforming to all specifications, but is on the verge of falling out of tolerance in all the areas being monitored by the SPC process. Compare this unit to one that was manufactured when all the SPC controls were near their midpoints. This latter unit will obviously perform within the design specifications for a long time before any of these parameters go out of tolerance from normal process drift or accumulated wear. The former unit is on the verge of being out of quality specification initially, and very little stress and use may drive it outside its design specifications. This unit has a low reliability and may be caught by the HASA test. Wider design margins translate into improved reliability in the field.

Consider two units that meet the initial quality specification, but only one unit has high reliability because it was manufactured close to nominal tolerances and the other was not. The unit farther from nominal will cost more warranty dollars. This is the basis of the Taguchi loss model. The challenge is to identify the important few limits that can economically be controlled in the manufacturing process so that units stay well within acceptable quality and reliability standards. The SPC tool has been used to produce products to specification, and by tightening on the upper and lower tolerances for the critical few processes, you can produce products that also have greater reliability. When used as described above, SPC can be a reliability tool as well as a quality tool.

8.7 FIFO Tool

Rotating inventory so that units on the shelf the longest get used first is a well-established business tool. The accounting term, FIFO, first in, first out, can apply to material handling to improve reliability as well. Electrical components have leads that will eventually be connected to make a larger assembly. Usually, this means they will be attached to circuit boards, either by surface mount, plated-through-hole technology, or by the surface-to-surface (connectors) mating. The longer these components remain in inventory, on the shelf, the likelier it is that the electrical leads of these components will begin to oxidize and corrode. The solderability of these leads is greatly reduced by this oxidization process. The typical electronic assembly has hundreds and often thousands of electrical connections. If a small number of these connections are made unreliable by using components with questionable leads, the overall reliability of the assembly may be reduced.

One way to reduce lead oxidation is by placing all the components in bags or containers that are filled with nitrogen gas. Component lead degradation will be greatly reduced by reducing the oxygen environment around components. As part of the purchasing process, materials that are susceptible to oxidization can be purchased in nitrogen-filled containers. This is more expensive than using FIFO to control inventory.

By not using FIFO, quantities of older components and material collect in the back of stockroom shelves. Recently purchased material is placed on the front of the shelf and is often the first material used in manufacturing. There will be times when, to fill orders, all the material on the stock shelf will be required. In this case, there is a high likelihood that there will be solder reliability problems in manufacturing and later in the field. Solder and connection reliability can be maintained at high levels by controlling order quantities and using the oldest material in the stockroom. It is also important to note that you don't know how long those components may have sat on your distributor's shelves before they were shipped to your company.

Integrated circuits are also a critical component. They have leads that are susceptible to oxidation like most other components, but they have another problem. Plastic encapsulated integrated circuits absorb water from the atmosphere while they are waiting to be soldered into a final assembly. This can be a real problem in manufacturing. When a component goes through the soldering process, it is often heated to temperatures well above the boiling point of water. The moisture in the component changes to steam. The pressure inside the component from the steam is sometimes great enough to cause microcracks and even cause delamination. Many times, these potential failures will not be discovered during test. They will often manifest themselves as early field failures and high warranty costs. This failure mechanism is called *the popcorn effect* because the failure is caused just like popcorn.

It is recommended that integrated circuits and other plastic devices that are susceptible to the popcorn effect be purchased and stored in nitrogen containers. Additionally, these components should be preconditioned in a warm baking oven for 24–48 hours prior to passing through the soldering process. This preconditioning causes the moisture that has collected inside the integrated circuit to slowly migrate out. Thus, the popcorn effect is eliminated by process changes.

Sometimes, some or all the material selected for manufacturing may not be needed and may be stored to be used another day. It should be returned to nitrogen containers or placed in process so it can be preconditioned again before being placed through the solder process. Unused material often gets set outside the standard manufacturing loop, so attention to this detail is important. The amount of time between completion of preconditioning and placement in the solder process varies, depending on the components and the local environmental conditions. The desired temperature for the preconditioning oven varies but it is obtainable from the component manufacturer.

References

1 Silverman, M. (1998). *Summary of HALT and HASS Results at an Accelerated Reliability Test Center*. Santa Clara, CA: Qualmark Corporation.

2 Mike Silverman "HALT and HASS Results at an Accelerated Reliability Test Center," IEEE Proceedings Annual Reliability and Maintainability Symposium, (© 1998 IEEE)

Further Reading

R. J. Geckle, R. S. Mroczkowski, *Corrosion of Precious Metal Plated Copper Alloys Due to Mixed Flowing Gas Exposure*, Proc. ICEC and 36th IEEE Holm Conference on Electrical Contacts, Quebec, Canada, 1990, and IEEE Trans. CHMT (1992).

Gore, R., Witska, R., Ray Kirby, J., and Chao, J. *Corrosive Gas Environmental Testing for Electrical Contacts*. Research Triangle Park, NC: IBM Corporation.

FMEA

S. Bednarz, D. Marriot, *Efficient Analysis for FMEA*, 1998 Proceedings Annual Reliability and Maintainability Symposium (1998).

M. Kennedy, *Failure Modes and Effects Analysis (FMEA) of Flip-Chip Devices Attached to Printed Wiring Boards (PWB)*, IEEE/CPMT International Manufacturing Technology Symposium, IEEE (1998).

M. Krasich, *Use of Fault Tree Analysis for Evaluation of System Reliability Improvements in Design Phase*, 2000 Proceedings Annual Reliability and Maintainability Symposium (2000).

K. Onodera, *Effective Techniques of FMEA at Each Life-Cycle Stage*, 1997 Proceedings Annual Reliability and Maintainability Symposium, IEEE (2000).

Prasad, S. (1991). Improving manufacturing reliability in IC package assembly using the FMEA technique. *IEEE Transactions of Components, Hybrids and Manufacturing Technology* 14 (3): 452–456.

D. J. Russomanno, R. D. Bonnell, J. B. Bowles, *Functional Reasoning in a Failure Modes and Effects Analysis (FMEA) Expert System*, 1993 Proceedings Annual Reliability and Maintainability Symposium, IEEE (1993).

SAE International (2012). *Recommended Failure Modes and Affects Analysis (FMEA) Practices for Non-Automobile Applications*. SAE. https://www.sae.org/standards/content/arp5580/.

R. Whitcomb, M. Riox, *Failure Modes and Effects Analysis (FMEA) System Development in a Semiconductor Manufacturing Environment*, IEEE/SEMI Advanced Semiconductor Manufacturing Conference, IEEE (1994).

HALT

J. A Anderson, M. N. Polkinghome, Application of HALT and HASS Techniques in an Advanced Factory Environment, 5th International Conference on Factory 2000, April, 1997.

C. Ascarrunz, *HALT: Bridging the Gap Between Theory and Practice*, International test Conference 1994, IEEE (1994).

R. Confer, J. Canner, T. Trostle, S. Kurz, *Use of Highly Accelerated Life Test Halt to Determine Reliability of Multilayer Ceramic Capacitors*, IEEE (1991).

N. Doertenbach, *High Accelerated Life Testing – Testing With a Different Purpose*, IEST, 2000 proceedings (February, 2000).

General Motors Worldwide Engineering Standards (2002). *Highly Accelerated Life Testing*. GM.

R. H. Gusciaoa, *The Use of Halt to Improve Computer Reliability for Point of Sale Equipment*, 1998 Proceedings Annual Reliability and Maintainability Symposium, IEEE (1998).

Hnatek, E.R. (1999). *Let HALT Improve Your Product*. Nelson Publishing.

Hobbs, G.K. (1977). *What HALT and HASS Can Do for Your Products*. Nelson Publishing Inc.

Hobbs, G.K. (1997). What HALT and HASS can do for your products. In: *Hobbs Engineering, Evaluation Engineering*, 138. Qualmark Corporation.

Hobbs, G.K. (2000). *Accelerated Reliability Engineering*. Wiley.

Joseph Capitano, P.E. (1998). Explaining accelerated aging,. *Evaluation Engineering* 46.

McLean, H. (1991). *Highly Accelerated Stressing of Products with Very Low Failure Rates*. Hewlett Packard Co.

McLean, H.W. (2000). *HALT, HASS & HASA Explained: Accelerated Reliability Techniques*. American Society for Quality.

Minor, E.O. *Quality Maturity Earlier for the Boeing 777 Avionics*. The Boeing Company.

M. L. Morelli, *Effectiveness of HALT and HASS*, Hobbs Engineering Symposium, Otis Elevator Company (1996).

D. Rahe, *The HASS Development Process*, ITC International Test Conference, IEEE (1999).

D. Rahe, *The HASS Development Process*, 2000 Proceedings Annual Reliability and Maintainability Symposium, IEEE (2000).

M. Silverman, *Summary of HALT and HASS Results at an Accelerated Reliability Test Center*, Qualmark Corporation, Santa Clara, CA, 1998 Proceedings Annual Reliability and Maintainability Symposium, IEEE (1998).

M. Silverman, HASS Development Method: Screen Development, Change Schedule, and Re-Prove Schedule, 1998 Proceedings Annual Reliability and Maintainability Symposium, IEEE (1998).

Silverman, M.A. *HALT and HASS on the Voicememo II*™. Qualmark Corporation.

Silverman, M. (n.d.). *Summary of HALT and HASS Results at an Accelerated Reliability Test Center*. Santa Clara, CA: Qualmark Corporation.

Silverman, M. *Why HALT Cannot Produce a Meaningful MTBF Number and Why this Should Not Be a Concern*. Santa Clara, CA: Qualmark Corporation, ARTC Division.

J. Strock, Product Testing in the Fast Lane, *Evaluation Engineering* (March, 2000).

Tustin, W. and Gray, K. (2000). Don't let the cost of HALT stop you. *Evaluation Engineering* 36–44. www.evaluationengineering.com/dont-let-the-cost-of-halt-stop-you.

HASS

T. Lecklider, How to Avoid Stress Screening, *Evaluation Engineering*, pp. 36–44 (2001).

Rahe, D. (1998). HASS from Concept to Completion. *Quality Reliability Engineering International* 14: 403–407. https://vdocuments.site/hass-from-concept-to-completion.html.

D. Rahe, *The HASS Development Process*, 2000 Proceedings Annual Reliability and Maintainability Symposium, IEEE (2000).

M. Silverman, *HASS Development Method: Screen Development, Change Schedule, and Re-Prove Schedule*, 2000 Proceedings Annual Reliability and Maintainability Symposium, IEEE (2000).

Quality

Gupta, P. (1992). *Process Quality Improvement – A Systematic Approach*. Surface Mount Technology.

Carolyn Johnson, Before You Apply SPC, Identify Your Problems, *Contract Manufacturing* (May, 1997).

G. Kelly, SPC: Another View, *Surface Mount Technology* (October 1992).

Lee, S.-B., Katz, A., and Hillman, C. (1998). Getting the quality and reliability terminology straight. *IEEE Transactions on Components, Packaging, and Manufacturing* 21 (3): 521–523.

C.-H. Mangin, *The DPMO: Measuring Process Performance for World-Class Quality*, SMT (February, 1996).

Minor, E.O. (n.d.). *Quality Maturity Earlier for the Boeing 777 Avionics*. The Boeing Company.

S. M. Nassar, R. Barnett, *IBM Personal Systems Group Applications and Results of Reliability and Quality Programs*, 2000 Proceedings Annual Reliability and Maintainability Symposium (2000).

Oh, H.L. (1995). A Changing Paradigm in Quality. *IEEE Transactions on Reliability* 44 (2): 265–270.

T. A. Pearson, P. G. Stein, On-Line SPC for Assembly, *Circuits Assembly*, (October, 1992).

D. K. Ward, A Formula for Quality: DFM + PQM = Single Digit PPM, *Advanced Packaging* (June/July, 1999).

Burn-in

T. Bardsley, J. Lisowski, S. Wislon, S. VanAernam, *MCM Burn-In Experience*, MCM '94 Proceedings (1994).

D. R. Conti, J. Van Horn, *Wafer Level Burn-In*, Electronic Components and Technology Conference, IEEE (2000).

J. Forster, *Single Chip Test and Burn-In*, Electronic Components and Technology Conference, IEEE (2000).

T. Furuyama, N. Kushiyama, H. Noji, et al. *Wafer Burn-In (WBI) Technology for RAM's*, IEDM 93–639, IEEE (1993).

R. Garcia, *IC Burn-In & Defect detection Study*, (September 19, 1997).

C. F. Hawkins, J. Segura, J. Soden, et al. Test and Reliability: Partners in IC Manufacturing, Part 2, *IEEE Design & Test of Computers*, IEEE (October–December, 1999).

T. R. Henry, T. Soo, *Burn-In Elimination of a High Volume Microprocessor Using I_{DDQ}*, International Test Conference, IEEE (1996).

Jordan, J., Pecht, M., and Fink, J. (1997). How burn-in can reduce quality and reliability. *The International Journal of Microcircuits and Electronic Packaging* 20 (1): 36–40.

Kuo, W. and Kim, T. (1999). An overview of manufacturing yield and reliability modeling for semiconductor products. *Proceedings of the IEEE* 87 (8): 1329–1344.

W. Needham, C. Prunty, E. H. Yeoh, *High Volume Microprocessor Test Escapes an Analysis of Defects our Tests are Missing*, International Test Conference, pp. 25–34 (1992).

Pecht, M. and Lall, P. (1992). *A physics-of-failure approach to IC burn-in, Advances in Electronic Packaging*, 917–923. ASME.

A. W. Righter, C. F. Hawkins, J. M. Soden, et al. *CMOS IC Reliability Indicators and Burn-In Economics*, International Test Conference, IEEE (1988).

T. Sdudo, An Overview of MCM/KGD Development Activities in Japan, Electronic Components and Technology Conference, IEEE (2000).

Thompson, P. and Vanoverloop, D.R. (1995). Mechanical and electrical evaluation of a bumped-substrate die-level burn-in carrier. *Transactions On Components, Packaging and Manufacturing Technology, Part B* 18 (2): 264–168, IEEE.

ESS

H. Caruso, *An Overview of Environmental Reliability Testing*, 1996 Proceedings Annual Reliability and Maintainability Symposium, IEEE (1996).

Caruso, H. and Dasgupta, A. (1998). A fundamental overview of accelerated-testing analytic models. In: *1998 Proceedings Annual Reliability and Maintainability Symposium*, 389–393. IEEE.

M. R. Cooper, *Statistical Methods for Stress Screen Development*, 1996 Electronic Components and Technology Conference, IEEE (1996).

Epstein, G.A. (1998). *Tailoring ESS strategies for effectiveness and efficiency*. In: *1998 Proceedings Annual Reliability and Maintainability Symposium, IEEE*, 37–42.

S. M. Nassar, R. Barnett, Applications and Results of Reliability and Quality Programs, 2000 Proceedings Annual Reliability and Maintainability Symposium, IEEE (2000).

Up Rating

Das, D., Pecht, M., and Pendse, N. (2005). *Rating and Uprating of Electronic Parts*. CALCE EPSC Press.

9

Software Quality Goals and Metrics

9.1   Setting Software Quality Goals

It is not possible to remove every defect from a software release. Even if it were, it is not economically feasible to do so. Software quality goals should be set by examining the software quality needs based on a return on investment (ROI) analysis for the investment in quality.

Software quality needs are not the same for all products. It may not even be the same for all parts of a single product. So, how does an organization determine how much to invest in software quality and even which parts of the product to focus those efforts on to ensure the most bang for the buck?

When determining quality goals, one needs to take into account the impact of the defect and the cost to fix the defect. Acceptable quality targets might be lower for products (or product components) where the defects have minimal impact to the customer or are not costly to fix. On the other hand, quality targets should be high for products (or components) where the defects have a significant impact or are very costly to fix.

At one extreme, a typo on a web page might be an example of a low-impact, easy-to-fix defect. The developer can make an easy change and the defect can be remedied in a very short period of time. In many cases, this typo would have no material or financial impact whatsoever.

At the other end of the spectrum, a defect in a medical device might result in the death of a user. This defect escape is expensive, potentially resulting in lawsuits, loss of business, and can negatively impact brand equity. In the most extreme case, it might even lead to business failure. A different example might be a serious design defect found late in the development process. A defect of this nature might cause, in the extreme case, all of the product development work to date to be scrapped with the development organization losing its investment up to that point. Similarly, a defect in a mission critical component or function might cause a technology or product to fail irreparably. A defect in a Mars Rover, for instance, might result in the entire mission to fail.

There is a cost associated with improving software quality, but it should be viewed as an investment. The return on that investment is the positive impact from future revenue resulting from delivering high quality software. The cost to increase the software quality should be subjected to a cost benefit analysis, at least informally.

It can be costly to avoid or remove defects from a product. It is impossible to release a software product that is guaranteed to have no defects. The cost to find and remove software defects increases as the remaining defect density of the product is lowered.

Improving Product Reliability and Software Quality: Strategies, Tools, Process and Implementation, Second Edition. Mark A. Levin, Ted T. Kalal and Jonathan Rodin.

That is, it costs less per defect to remove the first defects than is does to remove the final defects. The costs become asymptotic over time.

Conversely, as the product defect density decreases, the financial impact of the remaining defects also decreases. A very buggy product might be a complete failure, a total loss of investment. For the most part, the financial impact of a single defect will be finite.

Therefore, over the course of the project, the financial impact of the remaining defects decreases while the cost to remove the remaining defects increases. Not all products are created the same and not all defects are equal. A single defect can have disastrous financial impact. Other defects may have no financial impact at all.

Organizations developing products and technologies typically have finite budgets. The cost of improving the software quality increases over the duration of the project while the financial impact of the potential escaped defects decreases as the expected number of escapes decreases. The financial impact of an escape is determined by assessing the following:

1. The cost to fix the estimated escaped defects. Remember that the cost to repair a defect increases as the project progresses and is highest for an escape found at the end of product development.
2. The potential lost revenue caused by the escaped defects. A very buggy product won't sell, customers might demand their money back, and future sales of other company products can be negatively affected.
3. Liability issues, lawsuits, etc. arising from malfunctioning software.
4. Possibility of mission failure

Hypothetically, a curve can be plotted that shows the potential financial impact of the remaining defects if they were allowed to escape. Another curve could be plotted that shows the cost to remove the remaining defects. Where those two curves cross is the sweet spot for a quality target for a given technology, product, or component. This is not the same for each product. In a product where the financial impact of a defect is very high (i.e. life threatening defect in a medical device), more investment in quality is warranted. In a product where the financial impact of the remaining defects is low, the value of investing in more quality is also low. The following two graphs show some hypothetical analyses of defect removal costs vs. cost impact of escapes. Figure 9.1 shows a situation where the financial impact of escapes is low (perhaps, the previously referenced web site example). Figure 9.2 shows a case where the financial impact of potential escapes is very high (for instance, the previously cited medical instrument software example). In these two graphs, the highlighted area around where the curves cross is the probable target for the software quality goals. In the first case, it is clear that less investment can be tolerated and the product can be released with a higher defect density. In the second case, a much greater investment is warranted for improving software quality to a much higher level.

9.2 Software Metrics

Peter Drucker is often quoted as saying, "If you can't measure it, you can't improve it." There is no way that software quality can be improved if the health and state of the

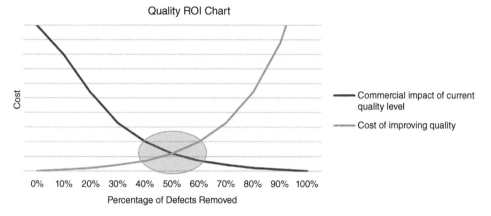

Figure 9.1 Quality ROI chart (financial impact of escapes is low).

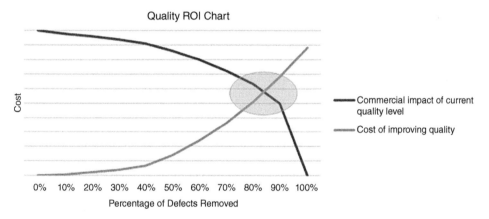

Figure 9.2 Quality ROI chart (financial impact of escapes is high).

software's quality is unknown. The same is true for software development processes. It is necessary to collect appropriate metrics over time in order to determine if quality goals are being met and to know if quality processes are effective. There are several important and useful metrics for managing software quality.

Metrics alone are not a panacea. Collecting and tracking software quality measurements is the first step in improvement process. The metrics are used to manage software releases and guide software process improvement efforts.

Some software developers believe that metrics are a distraction and misleading, and therefore should not be used. Others fear that metrics will be used to measure individual developer performance and thus are a demoralizing management practice. As with any idea, the naysayers will find flaws and this is true for the use of metrics as well. Most software development metrics are estimates. There is a danger when expectations are not set correctly and management expects actual project performance to adhere exactly to the metrics. Management time will be wasted trying to determine or explain why a project deviates in small ways from the metrics. A problem can arise when the wrong behavior is encouraged because the metrics are inappropriately used to measure individual developer productivity. This is such a common problem that it was immortalized in a famous Dilbert cartoon in which the boss offers a bonus for each bug found and fixed, which, in turn, encourages the developers to create more bugs [1].

Despite the undesired consequences that can come from using metrics inappropriately, one should not throw out the baby with the bathwater. Collecting data and tracking metrics are useful ways to determine the status of a project, indicate when and where a project goes astray, and evaluate process improvements. The following metrics are powerful tools when used appropriately.

9.3 Lines of Code (LOC)

Lines of code (LOC) is the number of lines of source code in a piece of software. LOC is a base metric used in generating other metrics such as defect density and is typically used as a factor in sizing software effort. LOC is often aggregated into units of 1000, typically referred to as KLOC (kilo lines of code).

LOC may seem like a simple number, but it can be counted several different ways. LOC can refer to:

- Total number of text lines in a source code (essentially a count of new lines).
- All lines of source code including comments and blank lines.
- Logical LOC. A logical line of code is a single software instruction statement irrespective of how many physical lines the statement uses in a file. The following pseudocode statement is a single logical line of code:

```
If (myVariable == TRUE) then
setSomeState = xyz;
```

- Physical LOC. A physical line of code is a line that contains an instruction up to the new line or carriage return. The pseudocode listed above would be considered two physical LOC.

It is important to choose one definition and count that consistently. For the purpose of calculating LOC, it is recommended that only noncomment, nonblank physical LOC be used.

There are many tools available to count LOC, and many IDEs also provide LOC counts. Most of the tools can be run over a folder or a set of folders to give a LOC count for an entire project. Most of these tools are capable of counting source code written in many different languages.

There are many other line counting tools available as well. The typical output of a line counting tools is show in Figure 9.3. In this example, the "code" column sum would be the nonblank, noncomment, physical LOC.

9.4 Defect Density

Defect density is the number of defects per measure unit of software, where generally units of software are measured in 1000 lines of code (KLOC). Predicting expected defect density for a software project is important for two primary reasons:

```
C:\tmp>cloc .
        7 text files.
        7 unique files.
github.com/AlDanial/cloc v 1.76  T=0.50 s (14.0 files/s, 3416.0 lines/s)
-------------------------------------------------------------------------------
Language                      files          blank        comment          code
-------------------------------------------------------------------------------
C                                 3             55            174           880
C++                               4              8             62           529
-------------------------------------------------------------------------------
SUM:                              7             63            236          1409
-------------------------------------------------------------------------------
```

Figure 9.3 Sample line counts.

1. Estimating the expected number of defects for a project provides input for a defect model, which can be used to determine how close the project is to achieving its quality goals.
2. The estimate for the number of defects expected in the software project is used to plan the effort required to produce the software.

Generally, the expected defect density is based on historical data. Software organizations should be tracking the number of defects for each project that are found prior to release and the number of software defects reported by users after product release (escaped defects). The sum of the discovered defects and escapes divided by the size of the release provides the defect density for that release. The defect density is generally stated as the number of defects/KLOC.

Defect counts and LOC should be tracked for each project or product release. This will provide the data to accurately estimate defect density. Tracking defects and code size over time provides a means to measure improvement in software quality over time.

Defect density could also be tracked per feature, product component, or technology. This more specific data can be helpful for both improving effort estimation as well as providing insight to areas of the product or process to improve. It is very informative, when creating effort estimates, to know that certain parts of the product may require more effort due to higher than average defect rates. Similarly, differentiated defect rate data can be very good input to consider when determining which areas of the product require improvement.

Defect density is not uniform for all projects. Three key factors can impact defect density:

1. *The language used to code the software.* Typically higher level languages, such as Java and C#, produce a higher defect density than a low-level language such as C or assembler.
2. *The code complexity.* More complex code typically has a higher defect density. For example, multitasking or real-time software tends to have higher defect densities. Older legacy software often falls in this category as well.
3. *The type of software application software being developed.* Higher-level software (e.g. database or web applications) tends to have lower overall defect densities.

Lower-level software (e.g. dealing with hardware or device drivers) typically has a higher defect density.

In most cases, a software group can collect historic defect density metrics for overall software releases and not worry about differentiating between different defect density rates for a given project. However, if different projects delivered by the same software organization have very different characteristics, it may be worthwhile using different defect density estimates for each type of project. Also, if forced to use industry data to create initial defect density estimates, it is useful to match that up with the type of project under development.

Historic defect density data may not be available when first starting to use defect models. If this is the case, it is possible to use generic industry defect density data as a starting point. The estimated defect density metric can be refined over time as actual data is collected. If historical data from your organization is not available, then industry data can be used in its place. According to research by Capers Jones [2], the average US software industry defect removal rate is 85%. That is, the average US software company removes about 85% of the software defects before release. As noted in Section 9.4, the best-of-breed software companies have an escape rate of 1.3–2.6 defects per KLOC, depending on language. Therefore, it is reasonable to assume that the average injection rate for a very good software development group would be between approximately 9 defects per KLOC for a low-level language and 18 defects per KLOC for a high-level language. One can assume that a best-of-breed software group would also employ good defect prevention techniques. So, these referenced industry numbers are too low and should be rounded up when using them to estimate defect rates for an organization that is just starting to deploy software quality control methods. If no historical data is available, I recommend using 10 defects per KLOC for low-level languages and 20 defects per KLOC for high-level languages.

9.5 Defect Models

Defect models estimate the expected number of defects that will be injected and/or found during a project. Some defect models can be very sophisticated and predict how many defects will be injected and discovered during each phase in the project. A simple defect model might just estimate the total number of defects expected to be created for a specific component or project.

Defect models are best if constructed based on historical data. If historical data is not available, initial defect models can be constructed using industry data.

A caveat about managing with defect models, in the words of George Box: "All models are wrong, but some models are useful."[3] The defect model should not be treated as gospel. It is an estimate and thus inherently inaccurate. A defect model cannot predict the exact number of defects that are in the code at any given time, just give a rough estimate. It should be used as a general guideline to approximate the quality of the project at any point in time. Defect models are a very useful tool to manage the project, so long as management does not expect any given project metrics to match the model exactly.

Defect models allow a development organization to more effectively plan the work required for a project. Using an estimate from a defect model enables managers to plan the work so that enough time and resources are available to find and remediate enough defects to achieve the desired quality goals. Managers can schedule the work so that the defects do not pile up. The most productive approach is to keep the number of open defects low and under control as the project progresses. This approach is recommended because if the number of open defects is allowed to get too high at any point in time, it becomes more difficult to test the software and harder to isolate individual issues. When individual defects take longer to determine the root cause and repair, overall productivity slips and the project will take longer and cost more. On top of this, if the defect count gets too high, then new development is occurring on a faulty foundation, which will lead to more issues as development progresses.

9.6 Defect Run Chart

A defect run chart shows the state of the software release quality over time during the development and testing of a software product and at product release. It allows managers to determine if the software release has achieved its target quality goals.

A defect run chart is composed of several data series. The first is the total number of defects expected to be found and fixed during the software release. This data is determined by the defect model. The second data series is the total number of defects discovered over time during the software development and test of a release. The third data series in the chart is the number of defects open at any given time during the release. A defect run chart is shown in Figure 9.4.

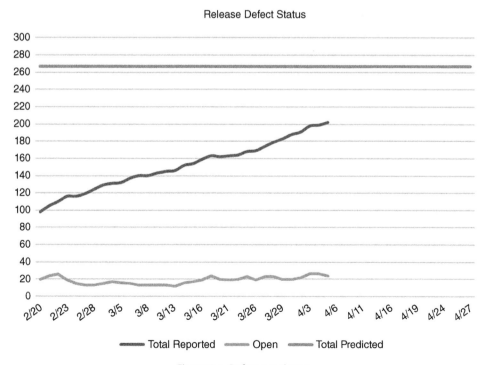

Figure 9.4 Defect run chart 1.

Assuming that new code development is complete for the project, the software is assumed to have achieved its quality goals when the total reported number of defects reaches the number of predicted number of defects and the number of open defects reaches zero. There are several things to keep in mind when interpreting the defect run chart. The current number of open defects is not sufficient by itself to show release readiness. Nor is it sufficient if the total reported count is equal to (or greater) than the predicted number of defects.

The total number of predicted defects is an estimate and inherently inaccurate. The predicted defects number should be used as a rough yardstick, not a definitive criterion. Also, the number of open defects can be zero well before the software is ready to ship if the development team is efficient at closing the discovered defects as they are reported.

When assessing software quality readiness using the defect run chart, in addition to the raw numbers, managers must take into account the status of the development and testing efforts. If development or testing is not complete, then the software is not ready even if the numbers look good. To assist in showing the status of development and testing, those milestones can be added to the defect run chart, as shown in Figure 9.5.

The defect run chart also provides a mechanism for assessing how well the development and testing teams are doing at meeting their schedules. If the metrics are not converging toward the expected targets, then more corrective action might need to be applied. For example, if during the system test phase (after coding is complete) the rate of discovered defects does not appear to be on a slope that will meet the total expected defects at the desired date, then mitigation plans should be applied. Similarly, mitigation

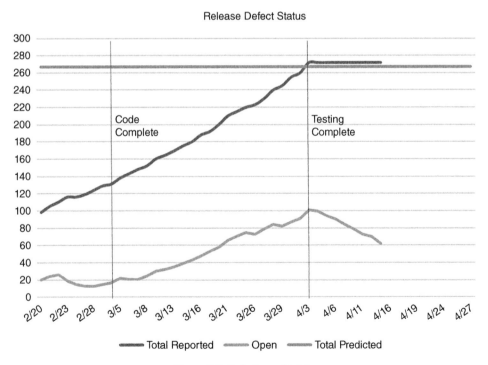

Figure 9.5 Defect run chart 2.

should be put in place if the defect closure rate does not appear to be on a slope to drive to zero open defects at the desired release date.

9.7 Escaped Defect Rate

Escaped defects are bugs that are discovered by end users after the product has shipped. While it is impossible to remove all defects from a software package, the goal of the software quality efforts is to minimize the number of escaped defects.

The escaped defect rate is the number of unique escaped defects per 1000 lines of code (KLOC). This should be tracked for each release. This means that the number of lines of new and modified code is recorded for each software release. The number of end user reported defects must also be tracked on a per release basis. Duplicate reported defects should not be incrementally counted.

The escaped defect rate is the primary metric recommended to determine if software quality is improving over time. Goals should be set to target an escape rate. On a release-by-release basis, the escaped defect rate should be improving if the software quality processes are improving.

Calculating that target escape rate is not trivial. One of the primary factors is the language with which the software is developed. High-level languages such as Java and C# pack a lot more functionality into each line of code than low-level languages such as C or assembly language. Thus, the expected escape rate for a high-level language would be higher than for a low-level language. To mitigate the problems with sizing per LOC, some researcher use function points (FP) instead of LOC. Function points are language neutral metric for stating the amount of functionality in a piece of software.

Research by Jones[2] shows that best-in-breed industry escaped defect rates are about 0.13 per function point. According to research performed by Quantitative Software Management (QSM) [4] there are on average 97 LOC per FP for C, 50 LOC per FP for C++, 54 LOC per FP for C# and 53 LOC per FP for Java. In general, the QSM data show that LOC per FP are roughly bimodal; around 50 LOC/FP for high level languages and around 100 LOC/FP for low level languages, give or take 10%–15%. Using the QSM and the Capers Jones data, the best software development teams in the industry deliver code that has defect escape rates of:

C	~ 1.3 defects per KLOC
C++, C#, Java	~ 2.6 defects per KLOC

Note that these defect escape rates are not to be confused with expected defect density. These metrics measure different defect rates.

Tracking the actual escape rate for each release needs to be thought through. There are several factors that will affect the reported escape rate. Those include the number of customers that adopt the new release and the length of time that the release has been in use. Normally, more usage results in more escaped defects being reported sooner. Thus, as more customers use the product, the rate of reported defects will increase. As time passes, the total number of escaped defects will increase. This will occur up to the point where a subsequent software release is available and adopted by customers, at which point the escaped defect reported rate will drop.

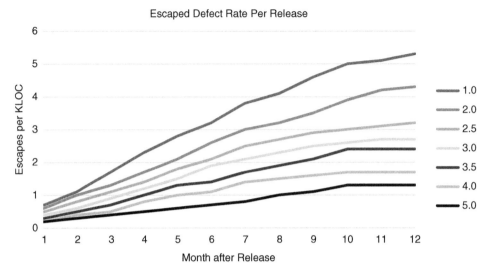

Figure 9.6 Comparative escape rates.

In order to compare improvements from release to release the rates must be viewed as a snapshot in time boxed period of time. The time box can vary, depending on release cadence and customer adoption rates; however, it is generally best to limit it to one year from release for accurate comparison. The number of reported escapes can be plotted monthly per release for a year after each release. An escaped defect rate comparison chart could look like Figure 9.6.

A software development group can measure its progress toward improving software quality by tracking the release to release escape rate. In this particular case, it is apparent from this chart that the escaped defect rate is improving from release to release.

9.8 Code Coverage

Unit tests should strive to cover as much of the product code as possible. In a perfect world, unit test code would test 100% of the product code. However, not all code is testable by unit tests. Most of the unit test frameworks can generate a report on the LOC covered by the unit tests when they are executed in the framework. The coverage metric should be used as a yardstick to determine how adequate the unit tests are. Setting a coverage target is a bit controversial. Achieving 100% coverage is not feasible. As a rule of thumb, getting 90% or higher code coverage from unit tests is excellent. Less than 80% code coverage is probably inadequate. There are diminishing returns derived from trying to achieve 100% code coverage.

One thing to consider is that unit tests alone are not sufficient to guarantee that the product code is free of defects. Since software is stateful, simply executing a line of code once is not sufficient to test that code in all states. So, even if it were possible to get 100% code coverage with unit tests, the code under test is probably not completely free of defects. However, measuring code coverage is strongly recommended because code coverage metrics are one of the best ways to determine how comprehensively the code is tested. Automated builds and test coverage can generate continuous reports on test code coverage without additional effort from individuals in the software development organization.

Code coverage is generally measured by tools that plug into the unit test framework while running unit tests. For Visual Studio, the code coverage analysis is built into the unit test functionality and does not require additional tools. There are many commercial and open source tools available that can be used with common unit test frameworks (including for Visual Studio). A few of the free tools are Jcov for Java, gcov (GNU code coverage tool) and Google Test (gtest), which can be used for C/C++. A quick search of the web will present many other tools. Choose the one best for your environment.

The code coverage tools will report which LOC are not exercised in addition to what percentage of code is covered. This allows the tester to add more unit test coverage until as much code as possible is tested. If the code coverage tool integrates into an IDE, such as the Visual Studio code analysis, it may visually display the specific LOC covered and not covered in the IDE. Other tools generate a report that shows code coverage for each analyzed module by line number.

It is unlikely, and probably not necessary, that the unit tests will exercise every line of code. Some code is very hard to test. The hardest code to test is usually error-handling code, where it might be very difficult to simulate or create all of the various error conditions. Risk must be considered when deciding which code could safely be omitted from the unit tests. Four general rules should govern the decision to skip unit testing of some code:

1. All code for nonerror use cases must be covered by unit tests.
2. All code for common error conditions must be tested.
3. Code for error conditions that require significant recovery processing should be tested.
4. Tests could be omitted for code to handle noncritical and hard-to-duplicate errors.

In addition to risk, the effort to add unit tests for hard-to-execute LOC should be evaluated in terms of ROI. It may not always be appropriate to add unit test coverage for this class of code if the cost of doing so is too high. However, this does not mean that code that cannot be covered by unit tests should go unverified. Other verification methods should be employed for all code that is not covered by unit tests. This code could be covered by system tests or could require extra code review attention.

References

1 Dilbert © 2018, Andrews McMeel Syndication, November 13, 1995, https://dilbert .com/strip/1995-11-13.

2 Jones, C. and Bonsignour, O. (2012). *The Economics of Software Quality*. Pearson Education Inc.

3 Box, G.E.P. (1979). Robustness in the strategy of scientific model building. In: *Robustness in Statistics* (ed. R.L. Launer and G.N. Wilkinson), 201–236. Academic Press.

4 Quantitative Software Management, Function point languages table, version 5.0 http://www.qsm.com/resources/function-point-languages-table.

Further Reading

Ann Marie Neufelder, Current Defect Density Statistics, SoftRel, LLC 2007 http://www
.softrel.com/Current%20defect%20density%20statistics.pdf.

10

Software Quality Analysis Techniques

10.1 Root Cause Analysis

Root cause analysis (RCA) is a procedure by which weaknesses in the overall development process can be identified with the intent to improve them over time. During the development process, defects are fixed when they are discovered. During this process, not much time is spent trying to understand how the defect was injected in the first place. Analysis should be performed outside of the process for a specific project development to identify any organizational, process, and technical weaknesses that exist. RCA is one of the most important analysis procedures to drive development process improvement efforts.

Many of the important RCA techniques originated from quality improvement efforts in manufacturing in the 1960s. They were subsequently adopted by other engineering disciplines and have been used successfully for decades. There are many tools and techniques that can be used for RCA. This chapter covers three powerful and more commonly used methods.

Adequate staff participation is important with all the root-cause techniques. Participants should include members of the team who worked on the project, additional subject matter experts, and process experts. Any RCA effort will likely be deficient if the team is too small or does not include representation from groups with the necessary experience and expertise.

10.2 The 5 Whys

The 5 Whys is an iterative technique for identifying the underlying root cause for any given defect. Unlike Pareto charts, which analyze a population of defects, the 5 Whys technique is used on a single defect at a time.

Just like it sounds from its name, in the 5 Whys method, the question "Why did this happen?" is asked five times iteratively while analyzing the cause of a defect. Each iteration of the why question digs a little bit deeper into the underlying root cause. An example of applying the 5 Whys to root cause a defect might be like this:

Defect: The robot picker stopped.

1. Why? The robot's vision system could not detect a part to pick up.
2. Why? The vision algorithm could not detect the edge of the part.

Improving Product Reliability and Software Quality: Strategies, Tools, Process and Implementation, Second Edition. Mark A. Levin, Ted T. Kalal and Jonathan Rodin.
© 2019 John Wiley & Sons Ltd. Published 2019 by John Wiley & Sons Ltd.

3. Why? The light was at an angle that caused a shadow on the part.
4. Why? The vision system was not tested under different lighting conditions.
5. Why? The vision system test plan was not reviewed.

Obviously, the why questions can continue ad infinitum; however, it has been found that generally five iterations are sufficient to get at the ultimate root cause.

The goal of the 5 Whys technique (and generally other RCA methods) is to ultimately identify process weaknesses that can be fixed in order to avoid future recurrences of similar problems. Notice in this example that the ultimate root cause was an inadequacy in a process, in this case the test development process. A process improvement that corrects this deficiency would improve all of the software quality, not simply prevent a single type of escaped defect.

10.3   Cause and Effect Diagrams

Cause and effect diagrams are effective for discovering the root cause of a defect. These diagrams show the relationship between an effect (the defect) and its possible causes. Cause and effect diagrams organize inputs by categories and allow users to explore possible root causes. Cause and effect diagrams are also known as fishbone diagrams (because they resemble a fish skeleton) or Ishikawa charts after their inventor Dr. Kaoru Ishikawa.

In a cause and effect diagram, the effect to be analyzed is placed on the right side of the diagram (as with the 5 Whys, the effect is the defect to be root caused). A horizontal line is drawn to the left of the effect. Categories of causes are drawn as slanted lines above and below the horizontal line and connecting to it. Detailed root causes are added to each category line. This is illustrated in Figure 10.1.

The cause and effect analysis process begins with a brainstorming session with the RCA team, where the root causes are identified. The team then sorts those root causes

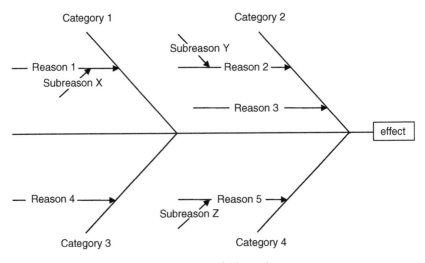

Figure 10.1 Generic fishbone diagram.

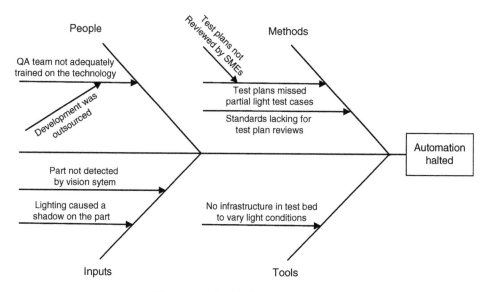

Figure 10.2 Sample fishbone diagram.

into a set of categories. Categories could include tools, infrastructure, processes, methods, and people, for example. A traditional method to do this is for each member of the team to write their ideas on sticky notes. The sticky notes are then stuck on the wall and organized by the team into categories by arranging similar items together. Once the team is in agreement as to the causes and the categories, they are documented into a cause and effect diagram. A completed cause and effect diagram example is show in Figure 10.2.

10.4 Pareto Charts

A key aspect of any task is knowing where to start. The biggest problems need to be tackled first to get the most bang for the buck. Pareto charts are typically used to identify either the most commonly occurring categories of defects (especially escaped defects) or the most common root causes of defects.

A Pareto chart is a method of organizing data in order to prioritize the improvement efforts. In a Pareto chart, the number of items of each category are sorted and put into a bar chart with categories where the most occurrences are closest to the y-axis and smaller categories are further to the right on the x-axis in diminishing order. Generally, a large population of defects is required to make a Pareto chart analysis meaningful. The percentage of each category is graphed, so that the sum of all categories adds up to 100%.

The categories in a Pareto chart can be organized by area of technology, component, or even by engineer. Multiple Pareto charts can be created with different categorizations to explore different areas of weakness. Also, for the largest categories of defects, derivative Pareto charts can be created to further understand the areas of weakness. An example of a Pareto chart is show in Figure 10.3.

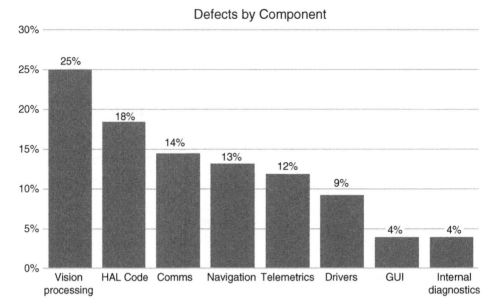

Figure 10.3 Sample Pareto chart.

The Pareto chart does not show root causes. It is used to determine where to drill down. After the defect data are analyzed with a Pareto chart, the largest categories should be root caused. This process should be repeated periodically. Often, several months after each software release, the escaped defect data for the previous release can be put into a Pareto chart to see where to focus next. If adequate process improvement activities are occurring, the category percentages will change over time.

After the root causes of the defects are analyzed (using techniques like the 5 Whys or cause and effect diagrams), those can be put into a Pareto chart too in order to determine the most significant process improvements to make. For instance, a Pareto chart showing the root causes of code review failures might look like Figure 10.4.

The lesson to be learned from this Pareto analysis example is that at least two changes need to be made to improve code review effectiveness: subject matter experts need to participate in code reviews and more time must be allocated in the schedule for code reviews. Perhaps, also, the developers should attend some training on holding effective code reviews.

10.5 Defect Prevention, Defect Detection, and Defensive Programming

Simply put, defect prevention consists of techniques to avoid introducing defects in the first place. Defect detection is a technique used to discover defects that exist in the software. From the perspective of productivity, cost, and quality, it is obviously far more efficient to not create bugs at all than it is to find and remediate them. Most of the techniques used for defect prevention have already been discussed. Those include failure modes and effects analysis (FMEAs), RCA, and reviews. The primary defect detection

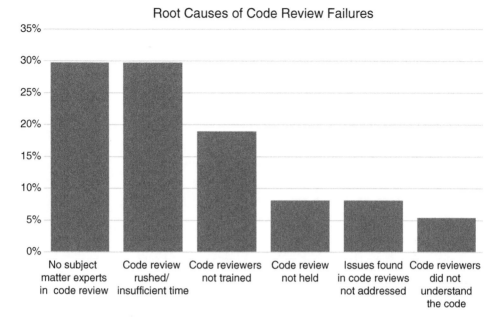

Figure 10.4 Code review root cause Pareto.

techniques consist of the various testing methodologies also previously discussed. An additional defect prevention technique is known as *defensive programming*.

Defensive programming is the art of anticipating what can go wrong during the software execution and adding in code to handle those errors and exceptions. A significant percentage of any well-written software package consists of code to handle errors and unexpected behaviors. There are no readily available definitive studies on the subject, but there are estimates that as much as 80% of all code written is to handle errors and exceptions. Errors or unexpected behaviors could include incorrect user input, hardware errors, bad data, buggy code, bit errors, etc. Another class of exceptions comes from failed allocation of system resources (e.g. if the software attempts to allocate a block of memory, but no memory is available).

Potential errors should be anticipated during code design and implementation and appropriate error handling must be put in place. The potential errors or exceptions can be identified using techniques such as FMEAs, fault tree analysis (FTA), and RCA.

Each time some code interacts with a user, with external data, with another piece of code or with hardware, there is an opportunity for that external entity to behave incorrectly or unexpectedly. To the extent possible, the code should be written to be able to handle an unexpected or incorrect response. The behavior of the error-handling code can vary, depending on the circumstances. The error-handling code might be as simple as to display an error message and wait for user instruction. At the other extreme, the code might be designed to attempt to recover from the error and continue operating as intended.

Some programming languages have specific constructs that can be used for error and exception handling. In other languages, developers have to implement their own

```
// the demoTryCatchMethod() calls another demoThrowMethod() which can
throw an exception

#include <iostream>
#define FrontSensorPin 10;

void demoTryCatchMethod()
{
    float pinValue = 0;

    try {
        demoThrowMethod( FrontSensorPin, pinValue );
    }
    catch( const char* msg ) {
        // code to handle the exception
        cerr << msg << endl;
    }
}

// demoThrowMethod() throws an exception if it detects an error

#define MaxPin 100;
#define MinPin 0;

void demoThrowMethod(int pin, float value)
{
    // check to see if pin has a legal value
    if (pin >= MinPin && pin <= MaxPin)
    {
        value = readPin( pin );
    }
    else // throw and error message
    {
        throw "Invalid pin number";
    }
}
```

Figure 10.5 Try-catch code example.

structures for detecting errors and handling them. In all cases, the developer must provide the code to actually do something when an error is detected.

Most modern object-oriented languages support "throw" and "try-catch" structures to make it easier for the developer to structure the error handling code. A method can "throw" an error if it detects an anomaly. The calling code can "catch" the error and handle it. This is called a *try-block*. The pseudocode in Figure 10.5 shows an example of the use of a try-block and throw.

Code that is intended to always recover from errors and keep functioning is known as *fault tolerant*. Most software does not have to be fault tolerant. Even software that must be fault tolerant does not necessarily have to provide fault tolerance in all aspects of its operation. Fault tolerance techniques include backups, redundant hardware and software, and code to reset hardware (and software) back to a known working state. Fault tolerance is not the subject of this book, so this subject will not be expounded on further.

In addition to including appropriate error-handling code, one less-obvious defensive programming technique is adequate code documentation. Most code is not worked on only once in its lifespan. Most software is used and modified for years, sometimes even decades. The code in an application may be extended or repaired by many engineers over a long period of time. Code that is well documented is easier to understand and thus is less likely to be broken when it is revised at some later date. Competent code documentation is a critical defensive programming technique to avoid future defect injection. This is important even if the original author of the code revises it at some later date. When the original engineer returns to some code that they wrote, their understanding of the code will have diminished. So, it is critical for the code flow and any method or function signatures to be adequately documented when initially written. Code reviewers should include code comment assessment as a key criteria during the reviews.

10.6   Effort Estimation

One might ask why a section on effort estimation is included in a book on product reliability and software quality. The answer is that if the project is not sized correctly, particularly if there is not enough time or staff allocated to deliver a product when it is needed, then quality will be sacrificed in order to make deadlines. Worse, the project might get so far behind or be so trouble ridden that the endeavor is canceled altogether. According to studies by the Standish Group [1], 31% of all software projects will be abandoned before completion and another 52% of software projects will cost nearly double their initial estimates. There are multiple reasons why the software project failure rate is so high. One of them is that the project was not sized correctly in the first place.

There are various ways to estimate the effort required to develop a software project, but they all use the same rule of thumb: count something. That is, the project estimator enumerates some criteria in the project and multiplies that count by a time factor. Some common project facets to count include estimated lines of code (LOC), number of requirements, and number of function points (FPs). These counts are then multiplied by the amount of effort historically required to implement and test previous projects of similar size. For example, if the new project is estimated to be five KLOC (kilo lines of code) of new code and historical data shows that it requires 40 person days of effort to implement and fully test 1 KLOC, then the new project would require 200 person days of effort.

On a side note, at first glance it might seem that 40 person days of effort seems like an awful lot of time to write 1000 LOC. After all, when developers are writing code, they often write hundreds of LOC per day. However, when all the work is considered that is required to produce a high-quality software product release, it is clear that actually writing the code is not the majority of time spent by a developer on the software. The other tasks, including writing requirements, designing the component, reviewing, testing, debugging, and fixing the software, all combine to take more time than just writing the code. All of these tasks need to be taken into account when determining the amount of effort required to develop a software package.

If historical effort data is not already being tracked, it is not too difficult to approximate a starting estimate. Assuming that the current software project is not the first software project implemented by an organization, then some historical data will be available.

Existing LOC can be determined by running a LOC counting tool on the current code base. The number of person days that went into developing that code can be estimated. This will allow the organization to derive a factor for person days per KLOC.

To improve future estimations, metrics for total labor and LOC should be tracked on a project-by-project basis. Over time, the accuracy of the effort estimation will improve as the input data becomes more accurate.

Another method for estimating effort for a project is the "like" method. With the like method, an engineer chooses a historical completed project that is like the new one and then uses the effort from the past project to estimate the effort for the new project. The estimate could be modified up or down, often combined with something counted. For example, an engineer might determine that new project X is like historic project Y, except that project X has 10 more inputs and outputs. Therefore, the effort for project X would be the same as the actual effort it took to do project Y plus the additional estimated effort to implement and test the additional 10 inputs and outputs.

As with all estimation, there is inherent inaccuracy in the prediction. Organizations that are determined to estimate more accurately will use multiple methods of estimation and will compare the different estimates to integrate into a combined effort estimate. Creating multiple effort estimates using different methods avoids having a single estimate that may be wildly inaccurate. As an example, a project might be sized by (i) estimating LOC and effort per LOC, (ii) estimating function points and effort per FP, and (iii) estimating by like projects. These three estimates could be averaged together to derive the effort estimate to be used for planning the project. As a general rule of thumb, unless a very high accuracy in effort estimate is required, it is probably not worth the work required to triangulate on effort estimate by using multiple different estimation techniques.

Reference

1 The Standish Group, CHAOS Report, 2014. This report can be purchased from the Standish Group from here: https://www.standishgroup.com/store/services/chaos-report-decision-latency-theory-10-package.html.

Further Reading

Cohn, M. (2006). *Agile Estimating and Planning*, 1e. Pearson Education.
Florac, W.A. and Carleton, A.D. (1999). *Measuring the Software Process: Statistical Process Control for Software Process Improvement*. Pearson Education.
Galorath, D.D. and Evans, M.W. (2006). *Software Sizing, Estimation, and Risk Management: When Performance is Measured Performance Improves*, 1e. Auerbach Publications.
Hill, P. (2010). *Practical Software Project Estimation*, 3e. McGraw-Hill.
Pfleeger, S.L., Wu, F., and Lewis, R. (2005). *Software Cost Estimation and Sizing Methods, Issues, and Guidelines*. RAND Corporation.
Stutzke, R. (2005). *Estimating Software-Intensive Systems*. Addison-Wesley.

11

Software Life Cycles

Software development is inherently iterative. Unlike hardware, software is never complete. As soon as a version of software is released, the developers start working on the new features and bug fixes for the next version of that software. All of the software development life cycles recognize this fact. The various life cycles primarily only differ in the duration of each iteration. Some life cycles favor short and frequent iterations. Other life cycles favor longer duration iterations. Within those iterations, most of the life cycles basically perform the same kinds of development and verification activities.

The following sections cover the two most common software development life cycles: waterfall and Agile. There is also a section describing Capability Maturity Model Integration (CMMI), which is a framework for determining the effectiveness of a set of processes that make up a software life cycle.

11.1   Waterfall

The waterfall model, sometimes referred to as the classic model for software development, was introduced in the 1970s. A waterfall starts at a higher level and flows downstream. The waterfall model for software development emulates a waterfall; it is a cascade of linear software processes. It starts with system-level requirements that flow down to software requirements, followed by software design, software implementation, software integration, acceptance testing, software release, and software maintenance. A waterfall only flows down, it cannot flow back upstream (Figure 11.1). The original waterfall method did not include a process or method to provide feedback to a previous stage for changes.

The waterfall method is widely used because it is easy to plan and implement since the software requirements and deliverables are defined in the beginning. However, it also requires that the software requirements are fully captured before proceeding to software implementation. Since there is no feedback, every stage of the process is supposed to be complete without issues, bugs, or changes before it can proceed to the next stage. In practice, stage transition is often allowed with provisos that open action items are tracked to completion. It tends to work best on projects where the requirements are well understood and unlikely to change or where there is a strict delivery time frame for a well-defined set of functionality.

Improving Product Reliability and Software Quality: Strategies, Tools, Process and Implementation, Second Edition. Mark A. Levin, Ted T. Kalal and Jonathan Rodin.
© 2019 John Wiley & Sons Ltd. Published 2019 by John Wiley & Sons Ltd.

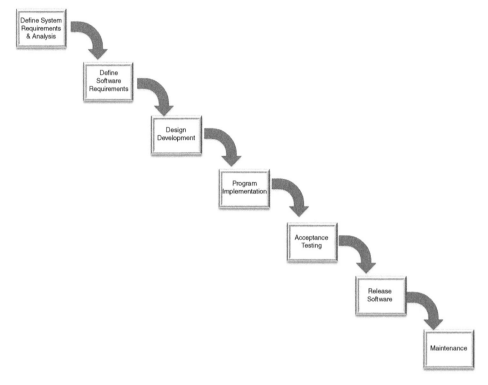

Figure 11.1 Waterfall life cycle.

Because requirements are not immutable, waterfall life cycles typically include formal change management procedures. The change management procedures allow for course correction during a waterfall project.

The waterfall process is easy to manage; the software development follows a logical sequential progression of events. Development progression is easy to measure and report its status to participants. The waterfall method follows a similar progression to hardware product development so it is easy to map their interdependencies.

However, it is almost impossible to guarantee the correctness of any of the stages before proceeding to the next. The requirements may not be fully understood or ambiguous and new customer requirements or market opportunities can lead to requirement changes. Testing happens late in the development cycle, when bugs can have more serious consequences and greater impact on the development schedule. You can measure the progress made from process stage to process stage, but you cannot measure the progress within a development stage. These inefficiencies lead to longer and less predictable development cycles.

As stated by Barry Boehm [1], the underlying assumptions of the waterfall model are:

- The requirements are knowable in advance.
- The requirements have no unresolved, high-risk issues.
- The requirements are unlikely to change much during the project.
- The right architecture for implementing the requirements is well understood.
- There is enough calendar time to proceed sequentially.

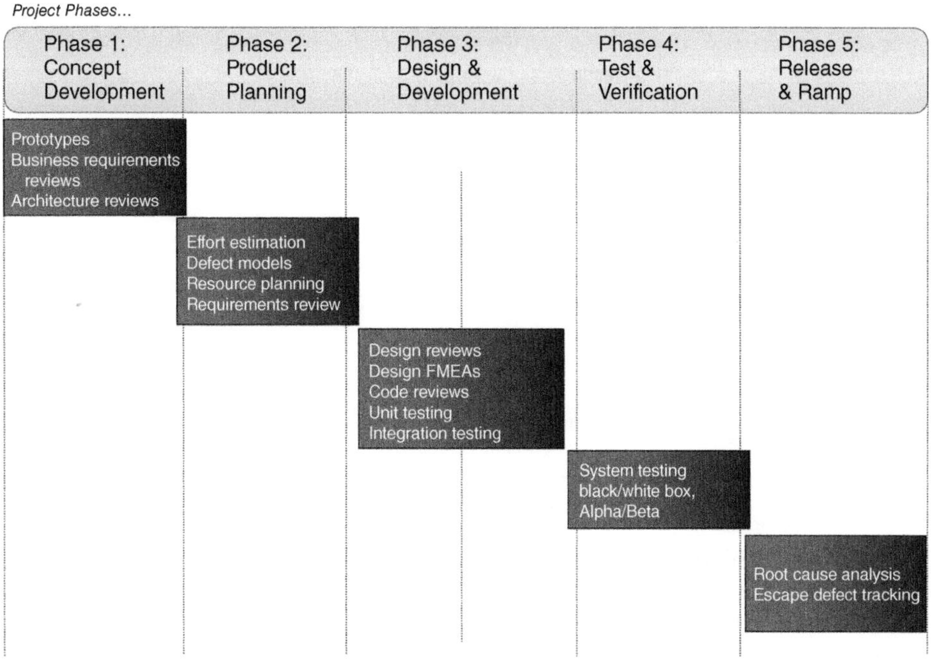

Project Phases...

Figure 11.2 Quality processes in a waterfall life cycle.

Figure 11.2 shows when the various software development quality activities occur during a waterfall life cycle.

11.2   Agile

Agile software development life cycles emphasize a series of iterative short development cycles called sprints that each deliver a fully functional set of software features ready to be used by an end user. Each sprint incrementally adds functionality to software delivered in previous sprints. Sprints are generally two to six weeks long, though there is no specific required sprint duration. Prior to the onset of each sprint, there is a planning period to prepare for the sprint. After each sprint is completed, there is a retrospective analysis session where the sprint is reviewed with the intent of determining how to improve subsequent sprints.

During each sprint, each team holds a daily brief meeting to focus on coordinating the day's work. This meeting is generally referred to as the daily stand-up meeting, and it should be kept to no more than 15 minutes. The purpose of this meeting is to make sure each team member knows the plan for intended work for that day and to identify any issues that might be impeding that work. In order to keep the meeting focused on coordination, it is critical to avoid diving into trying to solve any issues during the meeting. Other working sessions can be convened to deal with identified issues.

One of the general philosophies behind Agile processes is that formal documents and metrics are kept as lightweight as possible to enable quick development cycles. However, each sprint does include a full execution of requirements, design, code, and test

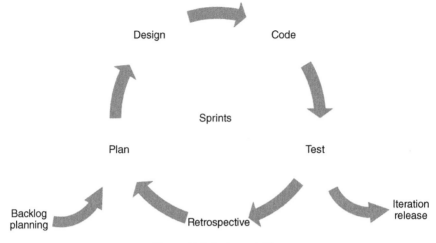

Figure 11.3 Sprint activities.

activities, as found in other life cycles such as waterfall. Figure 11.3 shows the activities in a typical sprint.

Several software development life cycles fall into the category of Agile processes. These include, e.g. Agile, Scrum, Kanban and Extreme Programming (XP). Each of these life cycles have slightly different practices, but they all share the general Agile principles of relatively short repeated sprints with fully functional deliverables at the end of each sprint.

Figure 11.4 shows when the various software development quality activities occur during an Agile life cycle.

11.3 CMMI

CMMI is not actually a software life cycle or set of processes. CMMI is a model or framework that describes what software processes are necessary in a mature software life cycle and specifies a set of attributes for each process type. CMMI can be applied to Agile or waterfall software life cycles. CMMI was originally developed by the Software Engineering Institute at Carnegie Mellon University and is now administered by the CMMI Institute. Initially defined for software development, there are now also CMMI models for service, delivery, and acquisitions.

The CMMI model for development describes 22 process areas and 5 levels of process maturity. Not all process areas are required for all maturity levels. Lower maturity levels require fewer process areas to be covered. The process areas are described in Table 11.1.

The lowest maturity level is an ad hoc approach to software development without much in the way of repeatable processes and procedures. The highest maturity level is applied to organizations that have defined and repeatable processes for all aspects of software development and employs sophisticated statistical controls to measure and manage all processes and process evolution. Each maturity level requires a subset of the process areas to be covered. That subset is incremental, with increasing levels of maturity. The five maturity levels are listed in Table 11.2.

- Product releases could occur after each sprint or after an epic (some number of sprints)
- Defect models and run chart can be applied to sprints or epics
- Backlog management is an ongoing activity outside of the sprints
- User story development occurs as part of backlog management
- Root cause analysis can be performed on an ongoing basis outside of sprints as part of backlog and epic management

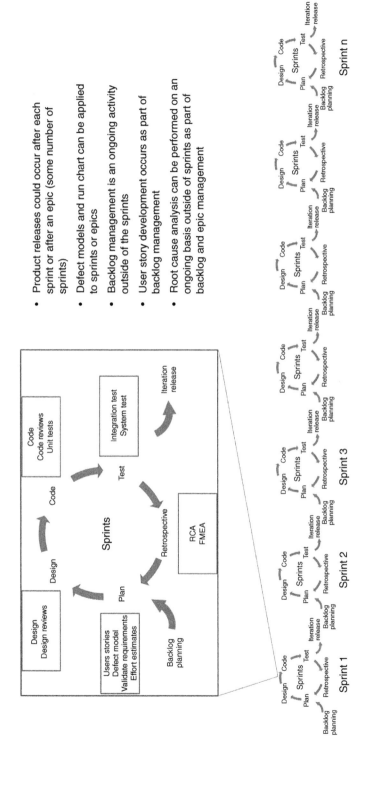

Figure 11.4 Sprint activities in an epic.

Table 11.1 CMMI process areas.

Abbreviation	Name	Description
CAR	Causal analysis and resolution	Processes for root cause analysis of outcomes with process improvements actions.
CM	Configuration management	Control and audit changes to work products such as requirement documents, source code, etc.
DAR	Decision analysis and resolution	Formal analytical decision-making processes.
IPM	Integrated project management	Manage the project with involvement of all the stakeholders.
MA	Measurement and analysis	Defining and tracking metrics to be used to measure project performance.
OPD	Organizational process definition	Processes for choosing life cycles and processes to be used for projects.
OPF	Organizational process focus	Processes for process improvement.
OPM	Organizational performance management	Processes to align and manage organization's performance with overall business objectives.
OPP	Organizational process performance	Analytical measurements and metrics of process performance.
OT	Organizational training	Processes to align organization and project skill sets with organizational and project objectives.
PI	Product integration	Processes for integrating product components.
PMC	Project monitoring and control	Processes for monitoring and managing project progress and status.
PP	Project planning	Processes for creating project plans, including creating effort estimates, schedules, etc.
PPQA	Process and product quality assurance	Processes to monitor compliance with process execution.
QPM	Quantitative project management	Quantitative project management processes.
RD	Requirements development	Processes to create requirements.
REQM	Requirements management	Processes to manage requirements including ensuring alignment between requirements and work products.
RSKM	Risk management	Processes to identify and manage project risks, create mitigation, and execute plans.
SAM	Supplier agreement management	Processes to manage suppliers, including outsourcers and third-party components.
TS	Technical solution	Processes to determine appropriate technical solutions (design, implementation, etc.) to meet project objectives.
VAL	Validation	Demonstrate that the defined product is suitable to purpose, i.e. it is the right product.
VER	Verification	Demonstrate that the product as built meets the defined requirements.

Table 11.2 CMMI maturity levels.

Level	Name	Description	Process areas
Level 1	Initial	Ad hoc approach to development in a generally very reactive organization.	None
Level 2	Managed	Basic project-oriented processes defined. Development organization still fairly reactive.	CM, MA, PMC, PP, PPQA, REQM, SAM
Level 3	Defined	Core development processes defined and managed. Development organization is proactive and can tailor processes to adapt to needs.	DAR, IPM, OPD, OPF, OT, PI, RD, RSKM, TS, VAL, VER
Level 4	Quantitatively managed	Development processes are measured and controlled.	OPP, QPM
Level 5	Optimizing	Strong focus on continuous optimization of development processes.	CAR, OPM

Development organizations are formally appraised by an appraiser certified by the CMMI Institute. An appraisal takes weeks of prework followed by a two-week on-site review. Results are posted on the CMMI Institute registry.

It can take several years for a software organization to evolve from Levels 1 and 2 to Level 3. A development organization is doing pretty well if it is operating at Level 3. There may or may not be a defendable return on investment (ROI) to improve to Levels 4 and 5. The right maturity level for any development organization is entirely dependent on the organizational goals. Familiarity with the CMMI model requirements is a useful means of determining improvement opportunities, even if a software development group chooses not to pursue an appraisal.

11.4 How to Choose a Software Life Cycle

Software life cycles are tools. As engineers, we know that no one tool is suitable for all projects. Likewise, no one software life cycle is suitable for all projects. Each life cycle has strengths and weaknesses. Note that there are many variants of each of the life cycles discussed here. Advanced practitioners of each life cycle have extended each in ways to make them more broadly suitable for different project types. Generally, the project criteria that can be used to determine which life cycle is most suitable are: (i) the size of the project, (ii) the size of the development team, (iii) the geographic location of the development team, (iv) how well understood are the requirements, technology and architecture, (v) how likely are the requirements, technology, or architecture to change, and (vi) how costly is rework.

Table 11.3 lists the strengths and weaknesses of the basic software life cycles discussed in the section. To summarize:

- Waterfall is best for larger projects where the requirements and technologies are well understood a priori and/or development teams may be large or dispersed.
- Agile life cycles are best for smaller projects where change is expected and teams are co-located.

Table 11.3 Life cycle comparison.

Life cycle	Strengths	Weaknesses
Waterfall	1. Can deliver projects with very predictable schedules and costs when the requirements are well understood or discoverable up front. 2. Emphasis on formal documentation makes it easier to work with large and/or geographically dispersed development teams. 3. Easier to manage for very large projects. 4. Better suited if the cost of rework is high.	1. Can devolve into churn if the requirements change frequently or are not particularly knowable a priori. 2. Can require too much overhead for small projects.
Agile	1. Can deliver functional software more quickly when the user can accept incremental releases. 2. Very suitable for developing software when the requirements are subject to change or not well understood up front. 3. Less risky life cycle if the ultimate software architecture is not known up front. 4. Can deliver software at lower cost for projects where the development team is co-located.	1. Requires ongoing communication with customer or surrogate. 2. May provide no real advantage if customer is unwilling to take frequent releases. 3. Can become very cumbersome and error prone if development teams are large or geographically dispersed. 4. Can be expensive if the cost of rework is high.

Reference

1 Boehm, B. (2000). *Spiral Development: Experience, Principles, and Refinements.* University of Southern California Center for Software Engineering. http://csse.usc.edu/csse/event/2000/ARR/spiral%20development.pdf.

Further Reading

Beedle, M. and Schwaber, K. (2002). *Agile Software Development with Scrum.* Pearson.

Boehm, B. (1988). A spiral model of software development and enhancement. *IEEE Computer* 21 (5): 61–72.

Chrissis, M.B., Konrad, M., and Shrum, S. (2011). *CMMI for Development: Guidelines for Process Integration and Product Improvement*, 3e. Addison-Wesley.

Kniberg, H. and Skarin, M. (2010). *Kanban and Scrum - Making the Most of Both.* C4Media Inc.

Martin, R.C. (2003). *Agile Software Development, Principles, Patterns, and Practices.* Pearson.

12

Software Procedures and Techniques

12.1 Gathering Requirements

In his book *Alice's Adventures in Wonderland*, Lewis Carroll writes of an exchange between Alice and the Cheshire Cat:

> "Would you tell me, please, which way I ought to go from here?"
> "That depends a good deal on where you want to get to," said the Cat.
> "I don't much care where — " said Alice.
> "Then it doesn't matter which way you go," said the Cat.
> "– so long as I get SOMEWHERE," Alice added as an explanation.
> "Oh, you're sure to do that," said the Cat, "if you only walk long enough." [1]

Carroll's quote has often been summarized in the following saying: "If you don't know where you are going, any road will get you there." In the case of developing software, documenting, and validating requirements is the only way to know where you are going. Requirements specify what the software must do. Getting the requirements right is the single most important task for delivering a successful product. If the requirements are wrong, then it does not matter how well implemented and tested the product is; the software will be a failure inasmuch as the software will not meet its functional purpose. Therefore, it is clear that quality software depends on having accurate and well-documented requirements.

There are different methods to document requirements, and different software development life cycles prefer or proscribe specific approaches to requirements documentation and review. Traditional waterfall life cycles tend to develop a requirements document at the start of the process that describes all use cases and requirements in detail. Agile processes create user stories with acceptance criteria. The user stories are created and managed on an iterative basis. In waterfall life cycles, the requirements document is intended to be definitive and is expected to be validated prior to proceeding with any software design or implementation. In Agile life cycles, an initial backlog of preliminary user stories is created. At sprint planning time, a set of user stories are selected for that sprint, and those are subsequently elaborated during the sprint. In Agile life cycles, new requirements are typically added to the backlog over time.

Irrespective of the life cycle, and however they are documented, the requirements should be reviewed with the stakeholders prior to implementation. Collecting and validating requirements is not an easy task. Often, users and customers are not sure what

Improving Product Reliability and Software Quality: Strategies, Tools, Process and Implementation, Second Edition. Mark A. Levin, Ted T. Kalal and Jonathan Rodin.
© 2019 John Wiley & Sons Ltd. Published 2019 by John Wiley & Sons Ltd.

they want. Even when they do know what they want, users tend to struggle to articulate their requirements. There are many techniques available to elicit requirements. These include surveys, interviews, observation of end users, brainstorming with subject matter experts, and prototyping. To get a complete set of requirements, it is recommended that several of these elicitation techniques be used and the results combined and cross referenced.

Internal requirements need to be identified, in addition to customer requirements. Internal requirements can consist of constraints (such as cost to implement or need to fit into an existing architecture) or nonfunctional requirements (such as buy-versus-build directives).

It is more difficult to choose appropriate requirements when there are multiple or many target customers involved than if the software is intended for a specific end user. Different users may have very different and even contradictory needs. A triage process must be deployed when deciding on requirements for a product with multiple intended customers. Typically, this would involve a requirements prioritization process such as weighting each requirement by assigning a return on investment (ROI) or a market share to each.

Requirements need to cover many aspects of the product. There are requirements that are external for end user needs and there are requirements for meeting internal needs. The following areas must be considered when determining the requirements for any given product, though not all may apply to every product:

- *Functionality.* The features and functionality delivered to the user is the most obvious and necessary set of requirements to get right. If the product does not do what the customer needs it to do, the product will not be successful. If the feature set is not correct, then the rest of the requirements simply will not matter.
- *Security.* The level of security required by the product must be determined. Some end uses clearly dictate high level of security, for example, a banking application. Other applications might not be so obvious. If the product is a medical device or a motor controller that can be connected to a network or remotely accessed, it might be critical that the connection to the device be secure to avoid malicious misuse leading to harmful outcomes. Even if the product is a connected device, such as an IoT (Internet of Things) appliance that cannot be harmful in and of itself, if it is not secure then it could be hacked used as a platform to attack other systems.
- *Performance.* Some products require that their functions are performed fast. In some cases, performance might be a competitive feature. For instance, a product that manufactures parts might not be competitive if it manufactures slower than other similar products. The time units of performance may vary greatly. For example, a network switch might need to transfer packets in microseconds, whereas a user interface may only need to be responsive in one second. However, in both cases, performance requirements must be defined and met.
- *Usability, user interface, user experience.* If the user interacts with the product, it must be sufficiently usable and easy to learn how to use. Care must be taken to determine the target user and how they intend to use the product. For instance, a user interface designed primarily for novice users will frustrate advanced users.
- *Architecture and tools.* The product might need to fit into an existing infrastructure or technology. The product might be an add-on to an existing product and so need

to work in that infrastructure. Or the product might be intended for use with an existing deployed environment, such as a particular mobile phone operating system. A common architectural requirement is a need for software portability if the product is intended to be in multiple environments. Many products have requirements to interoperate with other systems.

- *Compliance.* The product might be required to comply with various regulations or industrial standards concerning safety, privacy, reporting, etc.
- *Commercial.* There are various types of commercial requirements. Often there is a schedule requirement to meet either a market window or a customer commitment. Almost always there is a need for the product to be profitable or fit within a predefined budget.
- *Compatibility.* There are often requirements for backward compatibility with previous releases when adding new features to an existing product. This could be a requirement to work with features such as previously released file formats or existing hardware.

When developing software requirements, all of the above types of requirements should be considered. The resulting set of requirements must be prioritized and pared down to the subset that is necessary to have a viable product.

It can be very useful to also document nonrequirements. This is especially helpful in a couple of different cases. First, there could be controversial requirements for which the ultimate decision is to not include them in the product or release. Documenting that these are not to be implemented can save many hours of recrimination and ill feelings later. The other main reason to document nonrequirements is where there is the possibility of an incorrect presumption of implicit requirements. For example, if there is a specific decision not to support backward compatibility when previous releases did support it. Documenting certain nonrequirements can prevent unnecessary work from being performed or suboptimal design decisions.

12.2   Documenting Requirements

Requirements can be documented in numerous ways. The most common methods are use cases, user scenarios, user stories, and Unified Modeling Language (UML) use case diagrams.

Use cases describe a sequence of steps taken by a set of "actors" to accomplish a task. Actors can be people, system software or hardware components. The sequence of steps is referred to as a "flow." More than one flow can be included in the use case to describe alternate steps. Use cases also include a description of the task and pre- and post- conditions. This kind of requirements documentation is often found in waterfall-type life cycles. Figure 12.1 shows simple example of a use case with requirements.

In Agile processes, requirements are much less formal than in waterfall life cycles. This does not mean that requirements are somehow less effective in Agile. Agile requirements are expressed as user stories with acceptance criteria. User stories and acceptance criteria are intended to be kept short. Traditionally, they were documented on index cards. Today, the user stories and backlog are generally managed with a tool; however, the intent is to keep them short and to the point as if they were still on index cards. If

UC1: Robot avoids colliding with objects
This use case describes the movement of the robot to avoid collisions with objects.

Actors
      Robot

Preconditions
PRE1. The robot is moving
PRE2. The robot has detected an obstacle

Postconditions
POST1. The robot has changed direction or stopped moving

Basic Flow
B1.   The robot is moving.
B2.   The robot's front sensor has detected an obstacle within 15 centimeters in the direction in which the robot is heading.
B3.   The robot detects no obstacle on its left side sensor.
B4.   The robot changes direction by 270°

Alternate Flow – obstacles in front and left
A1.1   At step B3 the robot detects an obstacle to the left.
A1.2   The robot detects no obstacle on its right side sensor.
A1.3   The robot changes direction by 90°.

Alternate Flow – obstacles in front, left and right
A2.1   At step A1.2 the robot detects an obstacle to the right.
A2.2   The robot detects no obstacle on its rear side sensor.
A2.3   The robot reverses direction.

Alternate Flow – obstacles all around
A3.1   At step A2.2 the robot detects an obstacle to the rear.
A3.2   The robot ceases movement and sounds its buzzer.

Alternate Flow – sensor error
A4.1   At step B3 the robot fails to read its left side sensor.
A4.2   The robot ceases movement and lights its error LED.

Requirements

R1.   The robot shall have four proximity sensors located at 0°, 90°, 180°, and 270°.
R2.   The proximity sensors shall detect objects that are <= 15 centimeters from the robot.
R3.   The robot shall change direction to the left if it detects an object in its path.
R4.   The robot shall change direction to the right if it detects objects to the front and left.
R5.   The robot shall reverse direction if it detects objects to the front, left and right.
R6.   The robot shall cease movement if it detects objects at 0°, 90°, 180°, and 270°.
R7.   The robot shall sound its buzzer if it is unable to move in any direction.

Figure 12.1 Sample requirements.

the set of requirements in this example were documented for Agile as user stories with acceptance criteria, they might look like Figure 12.2.

Note that Agile user stories tend to be shorter and less detailed than traditional use cases and requirements written for waterfall-type life cycles.

Once the requirements are collected and documented, there are criteria that can be used to determine if the requirements as documented are of good quality. As

As a user
I need the robot to avoid colliding with anything in its environment
So that the robot is not damaged, nor is anything in its vicinity damaged or harmed.

Acceptance Criteria
1. The robot will change direction when the robot approaches to <= 15 centimeters of an object that blocks its path.
2. The robot will attempt to move in alternate directions if there are obstacles in more than 1 direction.

As a user
I need the robot to notify me if it is unable to move due to obstacles
So that I can fix the situation when robot cannot accomplish its tasks.

Acceptance Criteria
1. The robot will stop moving and sound its buzzer if all directions are blocked by obstacles.

As a user
I need the robot to notify me if it encounters an internal hardware error
So that I can fix the robot.

Acceptance Criteria
1. The robot will light its red error indication LED when it cannot read a sensor.

Figure 12.2 Sample user stories.

documented in ISO/IEC/IEEE 29148 [2], each good requirement must meet the following criteria:

- *Correct.* Each requirement accurately describes an attribute or capability of the system.
- *Unambiguous.* Each requirement has only one interpretation shared by all readers.
- *Implementable.* There is a cost-effective way to implement each requirement.
- *Verifiable.* The requirement is written such that one would be able to determine whether the implemented system satisfies the requirement. Furthermore, there is a cost-effective way to verify that the requirement has been met.
- *Consistent.* No requirement conflicts with any other current requirement for the product.
- *Complete.* The original user need is fully satisfied by the set of requirements.

In addition, the requirements must be prioritized to allow project managers and developers to choose the most important ones to include in the scope of any project. Typically, requirement priorities are limited to a few categories such as "must" (or "shall"), "should," and "want."

- *Must.* The highest-priority requirement that is necessary for the success of the project.
- *Should.* Middle-priority requirement that ought to be done if time allows. A "should" requirement adds significant value to the product, but will not prevent release of the software if not included.
- *Want.* Lowest-priority requirement category. Requirements in the "want" category should only be included if all "must" and "should" requirements have been implemented.

Different development organizations may use different terminology to identify the necessity of each requirement. For instance, some organizations may use "shall" instead of "must." To avoid confusion, the development team should standardize specific terminology and educate the group to ensure that everyone understands the usage.

12.3   Documentation

As with metrics, documentation has its detractors among software industry pundits. Some people will complain that "the documents or comments are out of date as soon as they are written" and that "the code is the vault of truth." Of course, there is some kernel of truth in what they say. However, the benefits of creating good documentation and code comments greatly outweigh the fact that such documentation is not perfect.

There are numerous benefits of creating accurate documentation. Documentation enables reviews; without proper documentation, there is nothing to review. Documentation, especially design documentation, provides a record of how things work and why choices were made. Even when the documentation is outdated, it can still provide a jump start in understanding the software. Documentation provides a mechanism by which different groups can effectively communicate. This is especially important when development groups are not co-located or when a product is developed by interdisciplinary teams, such as when a product consists of hardware and software.

Software industry literature covers all the benefits of good documentation that have been previously discussed. However, one extremely important reason for documentation is often overlooked. The act of articulating a description about a design or code in a way that is understandable by others forces the developers to fully understand their own work.

Early in my career, I had an experience that taught me this lesson. I could not decide which of two approaches to take to resolve a technical issue with my program. In desperation, I asked a colleague for help. My colleague watched patiently while I white-boarded each approach and only occasionally spoke to say, "Yes, go on" or, "Then what?" By the time I had finished explaining all the pros and cons of each approach, I knew exactly which one was better.

This is such a common experience that it has become known as *rubber duck debugging*, as described by Hunt and Thomas in their book *The Pragmatic Programmer* [3]. The act of having to explain the problem to someone else caused me to clearly think through the options, which in turn made the right choice obvious. No one can adequately explain something he doesn't understand. Forcing developers to write documentation and hold reviews is a mechanism to ensure that the developers actually understand what it is they propose to do.

Good documentation aids in communicating across groups and cementing agreement between development teams or with customers. Often, different teams will have different visions of what is being built. Effective documentation is one tool to help combat this. Well-documented requirements provide a mechanism for managing inevitable scope creep coming from customers. They enable the development team to clearly document that changes are being requested and thus describe how those required changes will result in schedule and/or cost changes.

Similarly, adequate design documentation, especially regarding interface definitions, lessens the probability that different development teams will deliver components that do not interoperate as intended. Integrations will go much more smoothly if the development teams cooperatively document and review the designs for components that need to integrate.

Documentation should, at a minimum, include requirements, architecture/design, and test plans. In later sections of this book, we will discuss each of these three types of documentation. Documentation does not have to be particularly elaborate, but it does need to be clearly written. Some organizations have very formal specifications for how each of these types of documents should be written. It might be helpful to have standards in order to facilitate the development of each type of document and to make it easier to review the various documents. However, the important thing is to create the documents and review them. Reviews are critical at every step in the development life cycle: requirements, design, code, test, etc.

In projects that include both hardware and software, it is especially critical for hardware designs to be reviewed by the software team, and vice versa. It is very common for each of the engineering teams to work in isolation. Often, design decisions made either by the hardware or the software teams alone cause the other to be unnecessarily complicated. This leads to overly complex products that are more costly to develop and more prone to quality and reliability issues. One key way to avoid this is to have software engineers involved in reviewing hardware design decisions before designs are concrete and for hardware engineers to reciprocate early in the software design process. The problem of having hardware and software designs get out of sync should be addressed organizationally by creating cross-functional teams that oversee the product as a whole, rather than as individual technology silos.

For example, two hardware designs may be equivalent in complexity and development cost from a hardware-only perspective. However, one of those designs might make the software design significantly less complex. Having software engineers involved in the design reviews for hardware designs can help ensure better overall product design decisions.

No matter how smart or talented an engineer is, she is never as smart as a group of engineers. When an engineer creates a set of requirements, or a design, or writes some code, or a test plan, it is important to have at least one other engineer, preferably a few, review that work product. The reviewers should include someone with domain expertise in the technology as well as at least one person familiar with the end user's use cases.

Documents don't have to be fancy, just clear and concise. They could be text, PowerPoint slides, white-board sessions, etc. The important thing is to get it documented and reviewed.

12.4   Code Comments

It is common for software engineers to modify code initially written by others, perhaps to fix a bug or add a feature. Comments in code shorten the learning curve for a developer other than the original author to work on that code. Properly commented code allows subsequent engineers to more easily understand how the code works and locate the appropriate place in the code to modify. Good code comments are even useful to remind the original author how the code works if it has to be revised after some time has passed.

Many software products have publicly exposed application programmatic interfaces (APIs). Some of these APIs might be invoked by engineers developing other modules in the system. Proper code documentation will facilitate the use of the APIs by external developers and they are much less likely to make mistakes when calling the APIs. Many products have APIs included in software development kits (SDKs) that are meant to be used by customers. Typically, the documentation for these SDKs are written by a technical writer. Even if the technical writer is a software developer, the documentation process is far easier and more productive if the code has adequate comments.

Comment blocks should be included at the top of a source code file and for each method or function in the file. Comments should also be included to describe variables, structures, or meaningful code flows. The comment block at the beginning of the file should include:

- The overall purpose and contents of the code in the file
- A copyright notice if appropriate
- The revision history of the code in the file

Comment blocks at each method or function head should include:

- A description of the method or function (what is does)
- An overview of the usage of the method of function
- A description of each of the parameters
- A description of the output or return code

There are tools that allow comments using specific documentation schemes to be processed into external user documentation. A few of those in common use are Sandcastle, Doxygen, and Natural Docs. The two that are probably the most common are the Microsoft markup format and the Doxygen format. The Microsoft format is based on XML tags. Comments that contain Microsoft XML tags can be easily turned in Microsoft help text or be viewed with IntelliSense (a Microsoft Visual Studio set of features). Doxygen is a widely used tool freely available under the GNU General Public License (GPL). Comments containing Doxygen markup can be extracted into standalone HTML documentation.

Figure 12.3 shows an example of some well-documented pseudocode.

12.5 Reviews and Inspections

Designs, code, and test plans should all be reviewed. The primary benefit of holding reviews is that it is, far and away, the least-expensive method to find errors and defects. Discovering problems early in the project during a review can prevent significant rework or even product failure later. Consider a design flaw that is not discovered until the system testing phase late in the project. Fixing the design flaw at that point could have major schedule impact due to the amount of rework required. Even worse, if the requirements are incorrect and not discovered early in the project, the whole product could be useless.

```
//
// This file contains code to read proximity sensors.
//
// Copyright (C) 2018 Jonathan Rodin. All rights reserved.
//
// Revision History:
// Date                Name                Notes
// 2018 Apr 11         J. Rodin            Refactored after code review
// 2018 Mar 27         J. Rodin            Initial creation
//

// read the ultrasonic proximity sensor to find range to nearest object
//
// Parameters:
//          eSensor - sensor to read
//          range - distance to detected object
//
// Output:
//          sets range to detected range in cm
//          if no object in range, sets range to -1
//
enum ESensor {Front=0, Right, Back, Left};
#define baseTriggerIO    10
#define baseEchoIO       13
#define errorLedIO       100

void proximitySensorCheck( ESensor eSensor, float range )
{
      float duration;

            // pulse sensor for 10 useconds
            digitalWrite(baseTriggerIO+eSensor, HIGH);
            delayuSecs(10);
            digitalWrite(baseTriggerIO+eSensor, LOW);

            // get the echo time
            duration = digitalRead(baseEchoIO+eSensor, HIGH);

            // returns -1 if nothing is in range
            range = convertToRange(duration)
}
```

Figure 12.3 Code comments example.

Another major advantage of reviews is that they facilitate knowledge transfer amongst the technical staff. If more engineers understand how something works, then it is easier to support that component later when maintenance is required.

No single person is as smart as a carefully selected group. Nor is any single designer or programmer familiar with every possible design pattern, platform, technology, etc. The group always has a knowledge advantage and sees things that elude the author. A group of reviewers are going to catch issues that the author has missed. Not to mention that it is extremely hard for an author or coder to recognize mistakes in their own work

[4]; it is easier for a third-party reader to find mistakes than it is for the author. Another good reason for reviews, as previously noted, is that having to explain a topic to other people forces the author to think through the plan.

The review process for each of the technical work products have some common characteristics. Reviews should include representatives from all of the stakeholders, and reviews should be held as frequently and as early as possible. Stakeholders for technical reviews should be representative of a few different areas of expertise. They should include engineers who are responsible for any other components that interact with the component under review. This would include hardware engineers if the component under review interacts with hardware. Conversely, software engineers who will be writing code for specific hardware components should attend design reviews for those components. Reviews should also include subject matter experts on the technology or functionality being reviewed. Test plan reviews, especially for system test plans, should include one or more customer representatives. Depending on the organization and the component being tested, this might include product marketing personnel, product owners, software quality assurance (SQA) personnel, or even actual customers. There is no standard for how many reviewers should be involved in a review.

To maximize productivity, reviews should be iterative, started as early as possible, and broken into manageable small chunks. It is best to start with informal reviews or inspections prior to a formal review where signoff might be required. For instance, early design reviews could be simply white-board sessions where feedback can be gathered and consensus built on a design concept. The least-productive design reviews occur when an engineer spends significant time developing a design document and then finds out during a review at the end that the design is flawed. It is far better to hold iterative informal reviews to get course correction on the way than it is to not do so and get to the wrong destination.

It is especially beneficial to create a narrow scope for code reviews such that small amounts of code are reviewed during any single review. Code reviews should start as soon as possible. It is very difficult and inefficient to wait until the end of a project and ask reviewers to read, comprehend, and give feedback on very large amounts of code. Not only is this inefficient, but asking a reviewer to inspect a large body of code late in the project will almost guarantee that the review will be substandard, inasmuch as the reviewers will most likely be rushed and probably not giving the inspection their full attention.

Reviewers should focus on the functional aspects of the work product. If it is a design review, then the focus should be on whether the design covers all the use cases and meets the requirements, including implicit requirements such as performance goals. Code reviewers should focus on determining if the code implements the design, if the code covers all the error and exception conditions, and if the code is correctly implemented. There is a natural tendency for reviewers to spend too much time correcting typos, formatting errors, and coding standards violations. While it is a good thing to remediate formatting issues and standards violations, it is more important for inspections to verify that the requirements, design, and code is actually functionally correct. Even though code comments are not part of executable code, reviewers should ensure that sufficient comments exist and that those comments adequately describe the code.

It may also be useful to also hold a security review of the code and the design. If the product is connected to a network or it is otherwise remotely accessible (for instance just running on a computer that is on a network), it is at risk for being tampered with or used as a vector for further attack. Since software security tends to be a specialized area of knowledge, it is generally useful to have a specific review or set of reviews that focuses on the security of the product. Security is an area where it may be useful to have an outside consultant participate in the reviews or do security testing.

The manner and count of code inspections can vary, depending on the circumstances. At minimum, at least one senior reviewer should review the code. However, larger, more detailed code inspections would be appropriate in the following cases:

- The programmer is new or inexperienced.
- The technology is particularly complex or is new to the coder or the organization.
- The component has many external dependencies or many other components have dependencies on it.
- There is a need to perform knowledge transfer.

In these cases, the code review team should be as large as necessary. Also, the developer holding the review might hold a code walkthrough prior to the actual review of the code. During a code walkthrough, the developer describes chunks of code as individual units and discusses their purpose and dependencies. The actual code inspection is held after the code walkthrough.

Reviewers should be on the lookout for common types of errors. Often checklists can be a useful tool to make sure reviews are complete and nothing is overlooked. The checklists should list the common types of errors appropriate for each work product (i.e. requirements, design, code, and test plans). Whether in checklists or not, work products should be checked for the following common potential errors during reviews:

1. Requirements reviews
   - Have all the types of requirements (functionality, security, performance, usability, architecture/tools, compliance, and commercial) been listed and prioritized?
   - Are the requirements correct, unambiguous, implementable, verifiable, consistent, and complete?
2. Design reviews
   - Does the design fulfill all of the requirements?
   - Does the design emphasize technology instead of the user? For example, has the designer chosen a technology because it is familiar or the hot new tool, rather than one that best meets the needs of the project?
   - Does the user interface design focus on looks rather than usability?
   - Does the design cover error and exception handling?
   - Is there too much interdependency between components? Is the design overly complex?
   - Is there clear separation of responsibility between the components?
   - Does the design make incorrect assumptions about the hardware? For instance, is the design targeting specific hardware features when it needs to be portable across hardware?
   - Does the design lock the software into a specific platform when it needs to be portable? For instance, does the design require specific operating system features when it needs to run on multiple operating systems?
   - Has analysis been done to determine which parts of the system require performance? Have those parts of the system been designed with performance in mind?

- Does the design take into account multithreaded or parallel operations to prevent race conditions, data corruption, and deadlock situations?
- Can the design scale with more hardware, users, and/or data?
- Are interoperability standards, regulatory compliance, and compliance with industry standards accounted for?
- Is some part of the design high risk? If so, perhaps plans could be put in place to mitigate the risk.

3. Code reviews
   - Are all of the design elements coded?
   - Are there any typos in the code?
   - Is the code well documented?
   - Are boundary conditions met with data structure access, for instance reading or writing off the end of an array?
   - Are there any off-by-one errors?
   - Are there any potential divide-by-zero errors?
   - Can the code use up all available memory?
   - Are all locks freed? Semaphores cleared? Flags set and unset?
   - Are there any overflow and underflow conditions?
   - Are there wrong order of operators or wrong formulas?
   - Are all variables correctly initialized? Do any data structures or memory blocks get reused without appropriate reinitialization?
   - Are there any memory leaks?
   - Are there endless loops?
   - Is there optimization occurring in the wrong or unnecessary code?
   - Is there improper use of global variables?
   - Has assignment accidentally been used instead of comparison (i.e. "=" instead of "==")?

4. Security reviews
   - Are users authenticated? Are external components and services authenticated?
   - Are passwords managed securely?
   - Is access authorized after authentication?
   - Is data storage and access secure?
   - Are operating system security features used correctly?
   - Is encryption used in communications or data?
   - Are there any hardcoded keys, passwords, or unsecure backdoors in the product?
   - Are unnecessary network ports left open on the system?
   - Can third-party code be run without appropriate authorization?
   - Is there any potential for cross site scripting?
   - Can the code be tampered with or modified by third parties or customers?
   - Are the code libraries signed?

5. Test plan reviews
   - Are there test cases for every requirement?
   - Are there test cases for edge/boundary conditions?
   - Are there test cases for error and exception conditions? Are there mechanisms to create or emulate error conditions?
   - Is there performance testing?

- Are there test cases for scalability of data, users, and hardware?
- Is there penetration testing?
- Is there usability testing?

12.6   Traceability

Traceability is a means of verifying that no requirement, use case, feature, or test case has been omitted from the development plan execution. It is a critical step in assuring that the product is delivered as expected. Traceability verifies the linkage from requirements to design to implantation to quality assurance.

Software work products form a hierarchical tree. At the top are requirements. The next level down is architecture and design. Below that is code. At the bottom are test cases. The traceability task makes sure that for each requirement, there exists a design, code, and test cases. This ensures that no requirement go undesigned, unimplemented, or untested.

Traceability is usually both its own process step and integrated into the review process for each work product. Generally, a traceability matrix is created that lists each requirement with its derivative designs, code, and test cases. During each review the matrix is updated to list the associated work products. When all the other reviews are complete, a final review is held of the traceability matrix itself to verify that there are no missing linkages.

Traceability reviews may seem like a tedious bureaucratic process step; I felt that way when I first was introduced to them. However, I learned otherwise when I was surprised to find during a traceability review that one of my projects had missed several requirements.

12.7   Defect Tracking

It almost goes without saying that the quality of the software cannot be known unless the defects found in the software are tracked. If a record of each defect is not kept, then there is no way to know how many defects were found in the software nor if all the important defects were fixed. Defect tracking is critical to ensuring that the software quality meets its goals.

Lack of defect tracking, then, leads to unknown software quality. Moreover, none of the statistics mentioned in this book can be created without some sort of defect tracking. The number of escapes, the escape rate, and the defect density of the code are all unknowable unless each defect is recorded and monitored to resolution.

A defect tracking system (DTS), also known as a bug tracking system, is an application that records and monitors defect (bug) status. There are many open-source and commercial DTS available to facilitate defect tracking. All good DTS should have the following functionality:

- Defects should have status such that the user can tell if the defect is open, fixed, or verified. Other defect states could be supported to best support the software development and verification processes.

- Defects should be assignable to engineers. Any given defect might be assigned to a different engineer depending on its state.
- Defects should be associated with specific products and components of those products.
- The DTS should support a workflow such that the state of the defect and the assignment of the defect should automatically transition when action is taken on the defect. For example, when a defect is opened, it might be automatically assigned to a specific development engineer; when it is fixed, it might automatically be assigned to the appropriate SQA engineer.
- The DTS should provide notifications to the assigned engineers when a defect is opened or transitions state. Additionally, the DTS should allow notifications to additional interested parties, such as relevant managers.
- The DTS should allow a user to add comments and attachments (such as logs and screenshots) to the bug report.
- The DTS should provide a rich set of reports to be used for determining defect counts, defect open and closure rates, and root cause analysis.

Deploying and using a DTS will improve productivity by making sure each engineer knows there are defects to fix in a timely fashion. Additionally, information can be added to the bug report to facilitate the repair of the defect. For example, when a defect is opened, the report could specify the steps required to duplicate the defect. This will make it easier to reproduce and thus fix and test it.

A good DTS is required to track the quality of a product. Once a DTS is put in place, it must be required that engineers enter bug reports in the DTS and use the DTS to track the resolution of each defect.

12.8   Software and Hardware Integration

One of the best ways to improve productivity is to create reusable software. Common functionality can be coded in such a way that it can be used in project after project without making changes. Not only does this result in effort savings for subsequent projects, but it also improves product quality by relying on tested proven code. New code will have new bugs and must be thoroughly tested before release. Reused code has already been tested and shown to work in production. Additional SQA can be limited to merely regression testing.

Code reusability can be applied to hardware and software integration. Typically, hardware exposes a register map for software to use to interact with the hardware. However, registers often change from one hardware implementation to the next. This poses a challenge for reuse; either the code has to change with each register map change or the hardware designers are constrained by leaving registers unchanged. The solution to this problem is to create a thin layer of software called a hardware abstraction layer (HAL). HALs expose a functional interface for the hardware at the function or method level which can be called by higher level software. Typically, the software and the hardware teams jointly design that functional interface. HALs hide the hardware implementation details from the upper level code, so that the higher-level application code can be reused from project to project without worrying about the register assignments of the actual

```
// bool UartRead() reads data from the UART.
//    returns TRUE if successful, FALSE if read fails.
//    int dataLength:              Size of read buffer.
//    out String receiveBuffer:    The data read from the UART.
//    out int returnLength:        The length of the returned data.
//    out String errorMessage:     Error message if call fails. Empty if successful.
//
bool UartRead(int dataLength, out String receiveBuffer, out returnLength, out String
errorMessage);

// bool UartWrite() writes data to the UART.
//      returns TRUE if successful, FALSE if write fails.
//      int dataLength:            Size of data to write.
//      out String writeBuffer:    The data to write to the UART.
//      out String errorMessage:   Error message if call fails.  Empty if successful.
//
bool UartWrite(int dataLength, out String writeBuffer, out String errorMessage);
```

Figure 12.4 Sample UART HAL code.

hardware. The only code that has to change when the hardware implementation changes is the HAL code itself and generally the changes to this code are easy to test. In fact, creating and maintaining HALs also makes it more efficient to perform design verification (DV) on the hardware itself by allowing the DV code to be reused as well as the application code. Thus, the hardware team will also be able to more efficiently provide more stable hardware to the software team when the hardware and software integration tasks begin.

Generally, the hardware teams create and maintain the HALs. Since the hardware developers may be changing the register maps from project to project, or even during the course of developing a particular component, it is most efficient for them to also change the implementation of the HALs in conjunction with the hardware changes.

For maximum payback, HALs should be created for functions that are common to many hardware implementations. This would include generic common functionality such as buses and communications interfaces like USB, JTAG, UART, and SPI. For functionality common to your hardware types, HALs would also be highly beneficial as well. For example, developing HALs to control motors, perform pulse width modulation (PWM), control power supplies, and so on would probably all deliver good ROIs.

A sample HAL interface to read from or write to a UART is shown in Figure 12.4.

Here are three guidelines to follow to create good HALs:

1. HAL interfaces should be named clearly with explicit names so that they are easy to understand and for the higher-level software to consume. If HALs' function names are cryptic, they can confuse the consumer of the HALs.
2. Most HALs should support parameters for initialization and or to pass values between the software and the hardware.
3. HALs should be kept at the highest level of abstraction of abstraction possible without incorporating any application code level logic.

Sometimes specific registers need to be read or written by higher-level code. Often, this is the case with DV code where a test is performed and then registers are read to confirm they are correct. A HAL can be provided to generically read or write a register. Even though code to read or write a register is easy to write, repetitively implementing such code is an opportunity to introduce defects.

Reusing code is one of the best ways to improve software development efficiency and software quality. One of the most effective techniques for providing reuse of software that interface with hardware is to create HALs to isolate the software from the hardware. The benefits of using HALs include less effort to port software between similar hardware, less effort to test implementations of similar hardware, and better-quality software.

References

1 Carroll, L. (1865). Alice's Adventures in Wonderland, first printed. *New York Millennium* 2014.
2 ISO/IEC/IEEE 29148 Systems and Software Engineering – Life Cycle Processes – Requirements Engineering, 2011.
3 Hunt, A. and Thomas, D. (2000). *The Pragmatic Programmer: From Journeyman to Master*. Addison Wesley Longman, Inc.
4 Stockton, N. (2014). What's Up with That?: Why It's so Hard to Catch Your Own Typos. *Wired*.

Further Reading

Cockburn, A. (2001). *Writing Effective Use Cases*. Addison-Wesley.
Doxygen home page, http://www.stack.nl/~dimitri/doxygen
Freedman, D.P. and Weinberg, G.M. (1990). *Handbook of Walkthroughs, Inspections, and Technical Reviews: Evaluating Programs, Projects, and Products*. Dorset House.
Microsoft Inc., How to: Insert XML comments for documentation generation, https://docs.microsoft.com/en-us/visualstudio/ide/reference/generate-xml-documentation-comments
Microsoft, Inc., Recommended Tags for Documentation Comments (C# Programming Guide), https://docs.microsoft.com/en-us/dotnet/csharp/programming-guide/xmldoc/recommended-tags-for-documentation-comments
Wiegers, K. and Beatty, J. (2013). *Software Requirements*, 3e. Microsoft Press.

13

Why Hardware Reliability and Software Quality Improvement Efforts Fail

After blending the reliability processes and tools into your system, you can still fail, even with the best intentions. There are other problems that you will encounter that can stifle or seriously block your effort. In Chapter 2, we discussed the barriers to implementing the reliability process. Here, we consider how poor execution or poor follow-up can cause the reliability effort to break down.

13.1 Lack of Commitment to the Reliability Process

Commitment to a task doesn't guarantee success, but the lack of commitment is certainly a guarantee of failure. Commitment to a reliability program must come from the top management. But commitment by itself still will not guarantee success. Top management must understand what it is that they're tasking their managers to do. It is a high-level understanding of the elements of the reliability process, the cost, the requirements, the time it will take to fully implement, and what to expect from the effort. The implementers of the reliability effort must truly believe that top management has resources committed to their success. Management's everyday actions, such as signing purchase orders for equipment and materials, give believability to managers' commitment. They must recognize that the costs to implement reliability are easily calculated; yet the short-term results of all these actions are much more difficult to measure.

The shortsighted view of commitment to reliability is to redouble efforts toward correcting product failures by focusing on field failure analysis and corrective actions. Certainly this is part of the reliability effort, but the main effort must be to reinvent the process. The commitment must be to change the process so that failures are caught and corrected before the product is ever shipped. Management must commit to developing the know-how to change the process. At first, this know-how will come from reliability engineers, specialists, and consultants. These few individuals will impart their knowledge to the rest of the workforce. After a time, the processes will be well established, understood, and in place as part of the day-to-day ongoing activities of the company.

Reinventing the process will be a team effort. A football team has top management, a head coach, assistant coaches, many support individuals, a wide array of resources, and of course, the players. The players are the ones who have to implement top management's and the coach's plays to be successful. The players must believe in the game plan (process). To be a winner, the coaches know that they have to have a strong running game, a deceptive passing game, and a versatile kicking game.

Improving Product Reliability and Software Quality: Strategies, Tools, Process and Implementation, Second Edition. Mark A. Levin, Ted T. Kalal and Jonathan Rodin.
© 2019 John Wiley & Sons Ltd. Published 2019 by John Wiley & Sons Ltd.

If the line coach sees weaknesses in the right side of the defensive line, he will study the plays and the players to learn their weaknesses. If the offensive coach has a quarterback who can throw the ball into an opening that was created by deceptive running back and hits his receiver perfectly, and yet the ball falls incomplete to the ground, he knows he has to improve performance. Through observations, the coach may find that the receiver is taking his eyes off the ball. As a result, the player's hands aren't ready to clasp the ball at the precise moment. One last detail, one seemingly trivial task needs to be controlled for completion of the pass. Follow-through by everyone who is part of the process is absolutely required to be successful to win the game.

The failure modes and effects analysis (FMEA) process is not unlike the pass play in a football game. It is completed by a group of people who gather to identify weaknesses in a design. In a typical FMEA, the team may identify a small resistor, which, if it were to open, would cause power supply voltage to double, thus destroying the surrounding components. As part of the FMEA process, the group readily determines that because resistors are extremely reliable this failure is an unlikely outcome. Upon further investigation, one member of the team points out that the resistor is to be located near a corner-mounting hole. He points out that when printed circuit boards are installed and removed, there is a good deal of flexure of the circuit board at and near the mounting holes. Resistors placed in close proximity to significant board flexing will cause the solder connections at the resistor to flex, possibly enough, to cause a crack. This failure may occur at the first time of flexure or over time. An open resistor or open connection to the resistor will, in this case, cause power supply overvoltage and much damage. A probable outcome of this observation will be to make a recommendation to ensure that the resistor is mounted where little flexure will take place. This means that one of the team members will be assigned that task with a date for completion. The person assigned this task must be certain that the information is given accurately to the printed circuit board designers and that they understand where acceptable resistor locations might be. Then, after the printed circuit board is fabricated, this FMEA team member must verify that the resistor is in an acceptable location. This is closure. This is follow-through. This is reliability.

In football, a lack of follow-through may range from an incomplete pass, to a missed block, to running in the wrong direction. Too many of these mistakes will lead to a lost game and a lost season. Knowing what you are supposed to do and executing every detail leads to success. Follow-through to closure when practiced in football, or in business, will ensure reliability of the outcome.

Follow-through in Highly Accelerated Life Test (HALT), is no different. In the FMEA process, the findings are theoretical and probabilistic. In HALT the findings are real. Remember, that the failures discovered in HALT will bear a strong correlation to the failures that may be found in the field. Correcting them before shipment is the intent of the process. After the failure is encountered during HALT, the first step is to find the actual failure. Then you must investigate further to determine the root cause and the actual physics that led to the failure. At this point in the process, you are half done. You must still identify what action is needed to prevent this failure from reoccurring. It will very likely require a design change. So one of the outcomes of the HALT process is a list of recommendations driven by failure and root cause analysis that need to be implemented. And you're still not done. You must be certain that the recommended changes have been implemented and retested to ensure that the changes perform correctly. Again, it is follow-through to closure. Without complete closure, the HALT process will not yield any improvement in reliability.

No matter how many items you find that need to be corrected in a product, your reliability efforts will fail if you disregard follow-through to closure. Finding the problems is only part of the task.

13.2   Inability to Embrace and Mitigate Technologies Risk Issues

To lead the competition, companies are hard-pressed to become proficient in new technologies. This can be risky. Oftentimes, new technologies haven't been time-tested. As a result, the company risks poor return for its effort if it hasn't taken steps to mitigate this risk. For example, as electronic components become more complex, the need for connectors having a very high number of connecting pins continues to increase. If a company is planning on using a new high-density, high-pin-count connector just for its design, hoping that it will suffice, it will, most assuredly, lead to disaster. First and foremost, this new high-tech connector must be recognized as a potential high-risk component. Simply because it's new is reason enough for it to be classified as a high-risk component. Later, the connector needs to be investigated, its physical characteristics defined, how it will be installed in the manufacturing process, how it will perform in the product, and how it will perform in the various field locations. Only after identifying the parameters of the connector that make it a high-risk component and taking steps to mitigate the risks (i.e. environmental stress screening (ESS) testing) can the connector be deemed acceptable for use in new products. Companies that overlook the risks of any part of their new product development will suffer from low reliability.

Sometimes, a single component can be a product's Achilles' heel. It's often caused by selecting a component that has not been used in the company before, and by not identifying how this component may cause problems. Usually, a team is formed to identify all the risk items on an assembly. There will be a range of risks. Some risks are higher than others. The team must identify tests that every risk item must successfully pass before the component can be an acceptable part of future products. Obviously, just selecting some tests is not adequate. Using internal and external resources, the team must identify the right tests and test environments to ensure success. Virtually every component in an assembly carries its own risk. Many are low-risk and can be set aside so you can spend more time on the higher-risk items. Sometimes risks are weight, flammability, rapid wearout, or operating temperature range. The list is endless. Each risk must be identified and mitigated to the satisfaction of the risk mitigation team. Again, each identified risk item must be tracked to mitigation closure.

Using a connector as an example, the risk mitigation team may determine that the end user will use a connector 100 times in the 20-year life of the product, and that the connector manufacturer specifies that the connector be designed to withstand 100 insertion/removal cycles with acceptable reliability. This connector will be used on a printed circuit board. During production and testing, 20 insertions and removals will be consumed, leaving 80. If the end user clearly needs all 100 insertions in the 20-year life of the product, then this oversight could cause undesired failures at the end of the product's life. The risk mitigation team must either find a more acceptable connector or develop a means to produce and test the product, without unnecessarily consuming needed insertion counts. Defining the success requirements is absolutely necessary when qualifying new technologies. Not doing so is another way in which companies fail in implementing reliability.

13.3 Choosing the Wrong People for the Job

Many of the very best companies promote from within. They may take a design engineer and cross-train him/her as a manufacturing engineer. They may take a system architect and groom that person into marketing. Financial analysts can be trained to be program managers. When a company has an individual who has been performing very well, this person can be moved into an area that will continue to challenge the individual and benefit the company. This will keep individuals interested and will increase employee retention. This is a good idea if there is someone in the company who can train this individual into his/her new area. Without proper training, however, the promoted employee will probably start slowly and may never grow to full competency. Digital electronic engineers can be trained as programmers, as there is a considerable job similarity. Also, they will come up to speed more quickly if they work with other programmers. But training is the key. This is especially true for reliability engineering. If you do not have someone to train these talented people, they will have difficulty in delivering what is expected of them. If they are asked to go off on their own, they may not deliver what is really needed.

Reliability engineering is one of those areas that easily fall into this category. Companies that do not have a reliability program will often identify several of their best engineers from manufacturing, test, or design to become a reliability engineer or be the reliability manager. Even though these individuals are hardworking, talented, and respected by their peers, they don't have the tools to identify the reliability weaknesses and recommend process changes. This is one of the downsides of installing reliability in a company. Simply put, you will get better results if you hire an experienced reliability engineer.

The reliability engineer must have a firm understanding of the processes and concepts needed to develop and enhance product reliability. This person must have the drive and initiative to install processes in a company, even though there may be some resistance to a methodology new to everyone. This is a difficult task, and it requires dedication and perseverance of the highest order. He/she must have a personality that adjusts to the personalities around him/her. This person must know that he/she has the full backing of the management. And finally, the reliability engineer must be a teacher. A smaller company can probably only afford one reliability engineer, and yet needs someone with all the skills. One engineer cannot do all the tasks, but must be able to impart the reliability knowledge to everyone. This process may take several years but, when done correctly, it will have trained other employees in the reliability process. Companies that try to install reliability on their own without outside help will probably fail.

13.4 Inadequate Funding

When a company chooses to implement reliability as part of its new product development, it must consider the up-front funding needed to be successful. Even before that reliability person is on board, the company must spend resources to identify and find reliability talent. Management must commit to this increase in salary expense as a minimum to get started. Very soon, the new reliability engineering hire will submit budgets to management with a timetable for implementation. Some of the budget items will be reliability laboratory space, tools and test equipment, test chamber costs (either internal or external), electrical and mechanical fixtures and training, to name just a few top items. The timing of the expenditures within the budget must be funded by the ongoing operations of the company. Financial planning must include provisions to meet these needs. A major portion of resources must be brought to bear on creating this reliability capability with the understanding that the return on investment will not be realized until after the product's release. Truly, this takes commitment and the understanding that returns are not immediate.

Early in the commitment phase, management will have high hopes for the results. As time progresses, management sees a lot of effort, many reports on product improvements, increasing development costs, and estimates of increased reliability. At this point, all they see is money going out and none coming back. This is the part of commitment where companies often fail. At this phase of the process, management has reports on field failures on previously developed products and financial reports as to what this is costing in terms of warranty dollars. Bookkeeper's ledgers are constantly adding up the cost of reliability, yet, no improvement in reliability is seen. Even though management initially understood that the return would not come until after new products were released, over time they are easily persuaded that this expenditure was a bad idea. Management must understand that they have to become true believers in the process. If they do not keep their commitments, they will certainly not be successful in their effort to initiate reliability.

One of management's major misconceptions is that they can measure increased product development time and cost. This new reliability process is delaying delivery to the customer. Upon first inspection, this is clearly true. But by accepting this delay, product development will go through fewer redesigns, which were causing the delays of the past. The new reliability process significantly reduces expenditures for multiple circuit board redesigns and software revisions, and that's just the beginning. This delay caused by the new reliability process happens only once. Because the new product will be much more reliable, engineers will not be required to develop corrections for field failures as they have in the past. This is like money in the bank. For the next new product, the same engineers will be able to apply much more of their time to new product development and much less time to fixing problems that exist in older products. Management must be patient and wait for the completion of the full product development cycle.

Companies that initiate reliability programs without follow-through will probably still fail, just a little later than if they had done nothing. Put simply, if a company does not deliver high-quality reliable products to its customers, the competition will. The marketplace will move toward manufacturers of high-quality reliable products. If there is not an effective software quality and product reliability process in place, you'll probably go out of business. If the reliability and quality process is halfhearted, it may take a little longer and the market for your products will diminish. Without the commitment from the top down, any effort to improve product reliability and software quality is destined to fail. Putting lots of money and engineering resources to improve software quality and product reliability will not bear fruit without a top level down commitment and the management follow through to make sure the effort is successful. A weak commitment to this is even worse than no commitment. Making a commitment and failing to stay the course is a major reason companies fail.

HALT consumes a lot of hardware dollars. In terms of circuit board count, depending on the cost of the board and other resources, anywhere from three to six circuit boards are needed to perform HALT properly. In early product development, the engineering designers need the very first prototypes to learn how well their designs perform. After investigation, these circuit boards will usually undergo some revisions. At this point, the circuit board development is no different from what was done in the past. After learning what is needed from the prototype evaluation, several engineering changes are usually incorporated. At this point, the design team, very often, believes the reliability of the product is complete. What makes matters worse is that given tight budgets, management may decide that the reliability team must HALT just the prototypes. This decision is disastrous.

Sometimes, after the prototype fixes are incorporated and products are manufactured to the latest revision, they are parceled out to other product developers, programmers, test engineers, manufacturing engineers, and so on. The reliability engineers are not provided the latest and possibly final revision so HALT can be perform. They must wait until these secondary developers have finished their activities. Here's where commitment to HALT funding is critical. Dollars must be set aside for the HALT process even though prior to HALT it is understood that the design is not complete. This is the development point in time where the designers feel the product is nearly complete. This is where the prototype discoveries were implemented, and this also includes all the FMEA findings. The only thing that is not included in the new design are those failures that will be precipitated by stress testing of the product and determining the Highly Accelerated Stress Screens (HASSs) profile for manufacturing. The reliability process cannot tolerate this expenditure failure. Don't fall short at this critical juncture. Failure to do so will be a failure of the reliability process.

When the HALT process is complete and all the changes are implemented in the new design, all the reliability work done to this point, essentially, ensures that the best product that can be designed will be manufactured using the existing process. Your field failure data probably indicates that a significant part of your field failure causes are directly related to manufacturing errors. The commitment to manufacturing reliability includes HASS. At this point, engineering management feels that they have a very good product, and they do. If the manufacturing process is flawless, there is no need for HASS. (Accept the likelihood that this reliability screening process is needed because the manufacturing process probably is not perfect.)

Adding HASS to the production process adds cost and production time. These resources can be significantly reduced through proper planning. Commitment to HASS means early planning. Environmental chambers will need to be purchased and installed near the end of the production line. The mechanical fixtures required that support the product to be tested to the chamber is, in itself, a significant design task. Instrumentation and test software must also be developed as part of the process because the product will be stress-screened dynamically, while it is in operation. Adding the HASS process slows production; it's an added step. Without strong commitment, management may rationalize that HASS is not needed. (The authors do agree that if the production line processes are well controlled and the product design is reliable, HASS may not be necessary. This may seem contradictory at first glance, but a review of the field failure data may show that the manufacturing process is in control.)

Remember, that the HASS process has more than one purpose. Besides it being a production screen, it can also be used as a field failure screen. After boards are repaired, they can be sent through HASS to ensure that they meet the screening standards of the production process. The HASS process identifies weaknesses in the process. It will also find weaknesses in the repair process. Whatever is fixed in repair, HASS screening will verify. When HASS is reduced to an audit process, because the process is in control, a properly scheduled Highly Accelerated Stress Audit (HASA) will ensure that there are no quality escapes. If the production line is adding contributors to field failures, the reliability process will not meet its original expectations. Companies that skimp on HASS and HASA may well fail in their reliability efforts.

There is a great disconnect between new product designers and field reliability data. Most engineers only know if their designs work for a relatively short time, typically, a year or so, then they move on to other design tasks. They do not know what the manufacturing and field failure Pareto breakdown is over time. Even when field failure reports are presented to them, they find it difficult to attach their design effort to the actual field failure data. In fact, most companies do not provide field failure data to design engineers. They will often have a department that specializes in fixing problems in the field. This disconnect is actually the broken link that allows inadequate designs to continue to propagate. Companies must communicate field failure information to the new product designers so they can evolve. Engineering management must provide this information to their designers (using failure reporting, analysis, and corrective action system [FRACAS]) to help spread the knowledge of what doesn't work.

This leads management to require the design staff to design for reliability (DFR). Most engineers believe they are already doing it. When designers do not receive feedback of field failure information, there is no reason for them to believe that their designs are not reliable. Designing for reliability is not well understood and DFR information certainly is not readily available. Engineering teams that have, over many years, developed DFR tools, do not publish the information because this knowhow is hard-won. The reliability engineer must provide this information in training classes in order to make designers aware of things that can go wrong with what they believe to be good designs. Collecting the field failure information and presenting it in an understandable fashion to designers will greatly help them in eliminating faulty designs from new products.

When a design is complete, it is usually tested to ensure that it performs to specification. This may not be enough. Product design validation often does not include testing to identify design margins. This can be done on the test bench and/or as part of the HALT process. During testing, it is learned that under normal operating conditions a product will operate well. But if the product is not tested at the design margins, it may never be known that it is precariously close to falling out of specification, or even failure. Design changes need to be made to widen the margins for product reliability. Failures can often be attributed to designs that operate the product too close to a limit or margin. Investing the time and effort to learn the margins lead to higher reliability and fewer field failures. A lack of commitment to this effort will lead to reliability failures.

Designing too close to the operating margins is often a source for field failures that cannot be reproduced at the factory repair center. These are often referred to as *no fault found (NFF)*; meaning that the customer sent the product back for service and the factory could not duplicate the field failure. This unit may well be returned to the field only to repeatedly fail and be returned for service, to the consternation of the customer. The actual environment in the field may be just outside the environment that existed when the product was bench-tested as acceptable, and no trouble was found. Products that fail in the field may work well after they are returned to the factory because the factory test environments and conditions do not represent the customer use environment and conditions. If this "no fault found" product is returned to the customer, it may well fail over and over again, until someone decides to scrap this troublesome unit. Incidentally, FRACAS can capture this repeat field failure unit and offers the opportunity to focus on why it keeps failing repeatedly.

When management first embarks on improving product reliability and software quality, they are usually driven by their awareness of excessive costs and high levels of customer dissatisfaction. Earlier in this book we pointed out that warranty costs could be significant. Setting a realistic product reliability and software quality goal is very helpful in determining trade-offs that meet the needs of the business. Complex designs that use redundancy to enhance reliability add cost to the product. Yet, this initial cost may not be a long-term cost. The cost/warranty/reliability/quality/redundancy analysis trade-offs should be performed and understood as part of the product development process.

Companies that don't know their actual warranty costs do not know how much money is being lost that could be returned through improved reliability (see Table I.1 in the preface, 1st edition). This dollar figure is actually the source of funds from where the reliability budget can be funded. One of the most important things to do initially (as management works toward their commitment toward reliability) is to put in place a warranty metric that can be tracked as the reliability process develops. Clearly, the initial reliability development funding must come from sources other than the lost warranty dollars. These dollars are not returned until after the reliability improvements have been installed. But, as reliability improvements are made, this metric will indicate how much money is no longer being lost in terms of warranty dollars. The warranty dollar measurement is a strong indicator of reliability program success. Without using this metric, a company may fail because it may well have installed reliability practices that are yielding little or no results.

There is a logical place for reliability activities. The reliability budget and estimates should be made early in product development. Design FMEAs should be scheduled near or at the end of product development but usually before any production analysis review. After corrections and improvements have been made to prototypes, the HALT process should begin. These are just a few of some of the major steps in a well-defined process. Making sure that all the reliability steps are included and in their proper order establishes a well-defined reliability process. Overdoing or underdoing reliability by not having a well-defined process will lead to failure.

Part of a well-defined reliability process is establishing reliability budgets and reliability estimates in phase 2 and phase 3 of product development respectively. The reliability budget refers to the mean time between failures (MTBF) goal set for the system, subsystem, module, and possibly key components. There are other reliability budgets that need to be decided in Phase 2 of product development (i.e. staffing, material costs for testing, project timeline for planned activities like FMEA, HALT, etc.). However, we are referring to the reliability MTBF budget, which is a breakdown of all the major parts of the product in reliability MTBF budget terms. The reliability budget must support the high-level system MTBF goal set for the project. The reliability estimates come from an analysis of the reliability of every component that makes up the product. If an assembly consists of parts and materials very similar to parts and materials of a previous generation assembly in the field, the new product reliability estimates will be similar to what was experienced in the field for the previous generation product. The reliability of the previous generation product is a good metric for setting expectations if you do nothing new and for measuring reliability improvement. These reliability estimates are based on customer field experience that is track a database. In-house data is the best information for preparing reliability estimates. There is another way to model reliability estimates that is much more unreliable.

Reliability estimates can be made using dated practices and military standards that have long since become impracticable. Many companies still use them, and some purchase requirements specify that these standards be used to make reliability predictions an additional requirement of the purchase specification. The authors believe that reliability estimates do little to improve a product's reliability. Reliability estimates are only as good as the data and judgment used to derive the estimates. Manufacturing and field failure data (FRACAS) that is accurately collected in your own business and in your own industry is the most accurate reliability information available. Estimates made (using your supplier data) is better suited for determining which components have the highest failure rate. Then you can determine if the failure rate is acceptable, if the component(s) can be designed out, or the impact reduced by using fewer components. Companies that use those obsolete standards will find that it would be better to focus on reliability activities like FMEA, HALT, accelerated life testing (ALT), and accelerated reliability growth (ARG) testing. Wasting resources is never a formula for success. Also, there have been several studies that have shown that reliability estimates can vary by a factor of 0.5 to 5 times the estimated value. With such a large variation between reliability estimates and observed MTBF, it is hard to see their benefit.

13.5   Inadequate Resources

Regarding staffing, inadequate resources falls into two different categories. The first category is too few resources. The second is the wrong resources. In the case of too few staff assigned to the project, the result will be a trade-off between scope, quality, and schedule. Applying the wrong resources to a project can likely result in quality loss that can be extremely costly to rectify in terms of cost and schedule.

If too few personnel are assigned to a software development project and all of the quality processes are followed, the project will simply take longer to complete. However, the project may have committed milestones to meet. Even if milestones are not inflexible, there is generally a tendency by management to drive to meet a desired schedule. The most common course of action is to take shortcuts when project staffing is too thin to support the required work. This will almost certainly result in poor-quality software since the shortcuts are generally to skip required software quality activities such as reviews or unit tests. Another common detrimental response to a late project is to perform all of the software development activities, but deliver the software to software quality assurance (SQA) late and give inadequate time to thoroughly verify software functionality. This also results in a software release with poor quality. If insufficient number of resources are available to execute the project and project schedule is critical, an safer alternative is to cut scope from the project. It is better to release a software version with less functionality than it is to release a poor-quality product. Bear in mind that the cut functionality can be released later in a subsequent software version.

The impact to software quality is more severe when the software staff does not include personnel with adequate skills. There is a strong probability that the software will not achieve its goals, if staff does not include members with the proper domain expertise with the technology and/or the intended usage. The kinds of quality issues that arise from inadequate expertise applied during design and implementation might result in the software release being unable to adequately perform expected tasks. No amount of software testing and bug fixing will remediate such errors. Usually, the result of applying inadequate expertise is the need to step back and redesign and reimplement portions of the product. This can cause major schedule delays as well as add significant project cost to cover the rework.

When a release gets late, there is often a desire to throw bodies at the project to recover lost time. This is an understandable reaction; however, it will probably be even more detrimental to the project. Certainly, most projects can achieve some schedule benefit from applying small increases in staffing. This is especially true if additional developers are assigned to the project relatively early on, as soon as it is determined that the schedule is at risk. However, applying many staff to a late project, particularly later in the project lifecycle, will almost certainly result in making the project even later. This problem is described by Fred Brooks, where he succinctly summarizes the problem as "nine women can't make a baby in one month." [1] What has become known as "Brooks's law" states, "Adding human resources to a late software project makes it later." [1] There is a threefold impact from adding staff to a late project. First, the new staff will generally have a learning curve, so they will not be capable of contributing to the project right away. Second, in order for the new staff to ramp up that learning curve, they will require assistance from the developers already on the project. This has the effect of reducing the productivity of the current project staff. Finally, most projects have a dependency sequence where some work has to be completed prior to starting subsequent tasks. This dependency sequence dictates a natural maximum number of staff that can effectively work on any given project. Exceeding that limit will result in wasted effort.

In summary, the best course of action is to staff the project with the right number of people with the right expertise required to achieve the desired schedule. If that is not done, the best remediation plans should be to extend the schedule or to reduce the scope of the project so as to maintain quality. Omitting quality process steps and/or adding large numbers of staff late in a project will almost certainly cause the project to be even later or to miss its quality goals.

13.6 MIL-HDBK 217 – Why It Is Obsolete

When the US government started purchasing manufactured assemblies, among the specifications that manufacturers were required to meet was a reliability metric. This reliability metric was essentially a measure of the time the product would function to the specification without failure. At the time it was initiated, there were no acceptable means to determine product life, so they had to be defined.

It was understood that the more sophisticated the assembly, the more likely it was to fail sooner, at least when compared to a less complicated assembly. It was also understood that the failure rate of individual components could be combined in such a way as to determine a reasonable figure for the failure rate of the finished assembly. Systems with more components were supposed to have higher failure rates and lower life expectancy. There was a problem, however. At that time, there was no established database of component failure rate for all the components that went into a typical electronic assembly. So, the government went about collecting data from their sources for generate failure rate information.

From its many repair facilities, including field repair stations, mobile repair stations, and depot maintenance locations all over the world, the military collected failure data for component failures – how they were used, when they failed, in what environment, and other parameters. They had categories for benign environments like an office; and high stress environments like shipboard, tank, helicopters, canon, and missile. Based on regression of the data collected, empirical models were developed and used to back out pi factors for various categories of environmental conditions. The environmental conditions included stress factors such as temperature, humidity, shock, applied voltage, and more. This collection of stressors eventually grew into a uniform document now known as *Military Handbook 217* (US MIL-HDBK-217). It is titled the *Reliability Prediction of Electronic Equipment.*

In the title is the word *prediction,* implying that the set of guidelines, set forth in tables and formulas, can be used to combine many thousands of component supplier variations and individual life expectancy figures into a lump sum called a failure rate prediction. There are many problems with this method.

For the most part, the data was gathered from military personnel who were trained to repair things as rapidly as possible. Oftentimes, many components were replaced in the assembly before it was determined to be fixed. Nonetheless, all the components collected were added to the collective database of failed parts. First-line repairmen generally completed the repairs (the author was one of these technicians in the US Air Force from 1961 to 1965). These personnel were discharged in a few short years after their technical training. This means that relatively inexperienced technicians did repairs. Their mission was to get the assembly fixed as soon as possible. The number of unnecessary parts that went into the final result didn't matter at the specific location, just the speed. This tended to generate erroneous data. There were many more failed components reported than the real number of failed components.

The component failure rate data for the initial release of MIL-HDBK-217A was for single point component failures that was collected in the 1960s and released on December 1965 by the Navy. Then starting with MIL-HDBK-217B (which was a major overhaul), environmental factors were included for each component. There were six revisions in total between 1965 and 1975. The component failure rate and reliability prediction method did get significantly revised with the release of MIL-HDBK-217F, which was released in December 1991. MIL-HDBK-217F had two additional revisions, Notice 1 released in July 1992 and Notice 2 released in February 1995. One of the problems even with the updated component library list is that there can be new components not covered in the document, so there will be no failure rate information available. The standard has been periodically updated to reflect on technology improvements and include components not covered, but for the most part, it is always lagging. Component technology will always move far faster than industry standards to cover the new technology.

The standard had been highly acclaimed as the foundation of reliability predictions. The prediction method consisted of two pieces, one for the part count and one for part stress. The part count method provides an overly conservative failure rate estimate for the system. The part failure rate can be adjusted by considering stress factors like part complexity, maturity of the part technology and manufacturing quality, stress applied, environmental stress factors, and temperature. The reliability estimate after correcting for stress factors often was still very conservative, predicting low reliability for a product that could be highly reliable. There have been a few studies that showed the estimates to be too conservative as well. In general, it is very likely that the reliability estimates will be very wrong.

MIL-HDBK-217 began to lose acceptance, especially in the 1990s, because many of the consumer products that used the standard to determine a prediction were much more reliable than what the standard predicted. Some who used the standard applied multipliers from 1.4 to as high as 10 to increase the final calculated predictions because they learned, over time, that the predictions were simply wrong. The MIL-HDBK-217 reliability predictions were completely inconsistent with reliability performance. The Pareto of failures predicted by MIL-HDBK-217 had no correlation to actual field failures. This is extremely problematic when the reliability predictions are being used to plan for repair spare inventory. It can be extremely expensive when overestimating the number of parts needed in inventory to support repair and the parts in inventory don't correlate to the actual component that are failing. The standard doesn't take into account human factors, where a part can be designed for use in a way wasn't intended or outside the manufacturer's recommended operating specifications. Finally, correction factors for temperature may have no correlation to component failure rate and may be driven by factors not considered like humidity and mechanical shock to name two. Even so, the myth of the standard remains.

The manufacturing quality control processes used today produce components that are orders of magnitude better than the components manufactured in the 1960s when this document was first released, and even better than the latest update (Notice 2) released in 1995. The improvement in component reliability has been occurring at a rapid pace. The standard, essentially, has fallen behind these changes, and what doesn't work is tossed aside for what does.

There are other industry standards that are being used that may be slightly better than MIL-HDBK-217F, but they have many of the same shortcomings.

Telcordia SR–332 (formally Bellcore) is a commonly used standard, particularly in Europe. Telcordia SR–332 was originally developed at AT&T Bell Labs and is based on MIL-HDBK-217F. The Telcordia document allows reliability predictions based on three different methods. The first is based on the standard method of using the generic failure rates provided by Telcordia SR–332, which are less pessimistic than the generic standards based on MIL-HDBK-217F. The second method allows for the combination of lab test data with the generic failure rates. The third method allows the use of field failure rate data in combination with the generic failure rates. Finally, the standard provides guidance on how to compensate for the higher first-year failure rate (infant mortality) and allows for a credit when burn-in strategies are applied to weed out infant mortality failures.

In addition, there is IEC 62380 TR edition 1 (formerly RDF2000 and UTEC 80810), PRISM, China 299B (GJB/z 299B), Siemens SN29500, 217Plus, and Nippon NTT procedure.

All of these document standards addressed electronic and electromechanical components, but they do not cover mechanical assemblies like pumps, seals, springs, compressors, electric motors, valves, bearings, or belts. There is a nonelectronic component standards database NSWC-06/LE10 (based on NSWC-98) and developed by US Naval Surface Warfare Center that covers these components. This standard has many of the same shortcomings. For example, it can be hard to predict when a spring will fail based on a standard. There is also the "Non-electronic Parts Reliability Data" NPRD-95 released by RAC, which has been updated by Quanterion as NPRD 2016.

There are software programs that use these reliability estimate standards and their generic failure rate data to run reliability predictions. Many of these programs allow you to choose between different reliability predictions methods, like MIL-HDBK-217F or SR-322 to run reliability estimates. Most of these reliability estimate programs allow you to also use your own reliability failure rate data instead of the generic failure rates. These programs are still popular and are also used by reliability consultants who specialize in reliability predictions. These too are failing to provide reasonably accurate reliability predictions. Using incorrect, unreliable information when making business and engineering decisions is not the way to be successful when implementing a reliability program.

13.7   Finding But Not Fixing Problems

HALT, HASS, HASA, and FMEA will reveal problems that need attention. If not addressed, all these problems can lead to low reliability. Each issue must be carefully studied for the root cause and recommendations must be made to mitigate these problems. Each recommended corrective action must be tracked to final closure and audited by reliability engineering to verify completeness. Here, often, is where the reliability effort fails.

In the rush to ship the product, the time it takes to correct a problem and make design or process changes can seriously delay delivery of the product to the customer. These delays are very visible to the bottom line, and no one wants to be blamed for causing delays in shipments. Many times, the needed corrective measures are skipped, just to make shipments. This can be a reliability disaster.

Less visible are the unaddressed reliability problems that can lead to early failures in the field. The drive to ship as soon as possible, to beat competition and capture early market entry dollars can be wiped out by low reliability and poor customer satisfaction. The money gained by early market delivery can be lost due to excessive warranty claims. If the failures are serious enough to require design changes, the cost to do the design changes are considerably higher now since there are many units in the field. Fixing the problem(s) early in the development stage is the least expensive and the fastest way to make corrections. All the reliability efforts in the world will be completely wasted if the issues that need to be fixed are not addressed.

13.8   Nondynamic Testing

Product reliability testing has evolved over the years. There has been temperature testing, vibration testing, shock testing, and so on. Much of that testing is done on the product when it is operational. Nonoperational testing rarely reveals failures because the failure mode often goes away when the stress is removed. If you are going to invest the time and resources to reliability test a product, it should be done when the product is operational. Field failures occur when the system is operating, so if you want to precipitate field failures, you must operate the system under a stress test.

13.9   Vibration Testing Too Difficult to Implement

Vibration testing is even more difficult. There is usually a mechanical apparatus or fixture that has to be designed to affix the product in test to the chamber for vibration testing. This means that the time cost to install the product to the vibration test fixture, run the test, and remove the product from the test fixture may seem to be prohibitive. The vibration test fixture has to mechanically couple the product to the vibration table to ensure that the forces are actually working on the product. This means that fewer units can be placed into the chamber at a time. This reduces test flow-through.

Operating the product during stress testing will require test equipment and may require test software. The test equipment adds cost. Developing the test software adds to the cost of the test process and consumes programming resources that can be used elsewhere.

Avoiding these costs will generally save money up front. The missed reliability problems will probably cost much more. Being thorough in the stress testing will return more reliability discoveries and more warranty dollars. Lack of dynamic testing is where the success of the reliability process can be lost.

13.10   The Impact of Late Hardware or Late Software Delivery

Software needed for the stress testing is on a critical path. If it is late, the tests cannot be done dynamically. Improperly done stress testing will result in poor reliability. Planning to ensure that the test software is ready when the stress testing is scheduled is critical for success. Poorly planned test software will be a major cause of the failure of the reliability effort.

13.11   Supplier Reliability

When you begin transforming your company, you may well be doing the same with your suppliers. An added function of the purchasing group is to ensure that suppliers are closing in on all the reliability issues.

Transforming the product development process to achieve higher reliability and improved customer satisfaction requires the implementation of many strategies. Doing them only half way will not lead to success.

Reference

1 Brooks, F. (1995). *The Mythical Man-Month*, 2e. Addison-Wesley, First published Department of Computer Science, University of North Carolina, Chapel Hill, 1974.

Further Reading

Peter, A., Das, D., and Pecht, M. (2015). Appendix D: Critique of MIL-HDBK-217 National Research Council. In: *Reliability Growth: Enhancing Defense System Reliability*. Washington, DC: The National Academies Press https://doi.org/10.17226/18987.

Relyence Corporation, A Guide to MIL-HDBK-217, Telcordia SR-332, and Other Reliability Prediction Methods, Relyence.com July 16, 2018, https://www.relyence.com/2018/07/16/guide-reliability-prediction-methods/.

Weibull Reliability Engineering Resources, Military Directives, Handbook and Standards Related to Reliability, Weibull.com, Copyright © 1992–2018 HBM Prenscia Inc. All Rights Reserved. https://www.weibull.com/knowledge/milhdbk.htm.

14

Supplier Management

14.1   Purchasing Interface

One of the many factors that influence the bottom line is supplier quality. The ability to receive purchased materials on time, to specification, at the quantity and quality specified, is critical to business. Many companies have a purchasing department, but what they really have are buyers and expediters. There is a vast difference between these two material procurement methods. Buying materials for production purposes looks easy. One picks up the phone, calls the supplier's order desk, places an order, uses a credit card, check, or purchase order, and expects on-time delivery. If your supplier has the specified material in the quantity needed, it is reasonable to expect prompt delivery. But what happens if your supplier is out of stock?

You thank your first supplier very much and call another. You continue to do this until the needed material is found. You may get the material you want in the quantity and even at the price you need. Then, you do it all over again for the next needed item. You repeat this cycle as you buy the material that goes into your product. When the bill of materials needed for production has finally been ordered, you can't be comfortable because things can still go wrong.

You need to know the following: Will all the purchased materials arrive on time for the planned production run? Will you get complete or partial shipments? Will your supplier fill the order exactly as specified? Will some other components be substituted, at the discretion of your supplier, because it was out of stock? Will the price of the purchased materials be within what is needed for you to stay within your cost margins? Chances are that these, and other problems, will occur that will negatively impact the production run, all of which will drive up your costs and lower your overall quality and reliability.

When the orders start coming in and you realize that there are discrepancies, the buyers are converted into expediters. Now, the buyer stops everything and scrambles to get the needed materials that were short-shipped. What would have been productive buying time has turned into a state of panic. Even if you are fortunate enough to get the needed material with the follow-up expedition, the mix of components that will now go into your product may cause problems that will eventually drive up costs through in-process rework, scrap, and accumulating warranty costs. All these problems would have been solved through better materials planning.

Purchased materials planning is the establishment of goals, policies, and procedures that work together to create a continuous flow of quality materials that are on time and at a price that meets the needs of your business. This is the difference between buying and

Improving Product Reliability and Software Quality: Strategies, Tools, Process and Implementation, Second Edition. Mark A. Levin, Ted T. Kalal and Jonathan Rodin.
© 2019 John Wiley & Sons Ltd. Published 2019 by John Wiley & Sons Ltd.

purchasing. There are many variances that impact production. These variances need to be identified so that resources can be brought to bear toward minimizing their impact on your business.

An obvious planning variance would be sales volume. Is your business cyclical, seasonal, or growing at a continuous rate? Do you have some products that are declining in sales while there are others that are increasing? Do you have some products that have just completed development and which you are ramping up in production to fill anticipated orders from your marketing efforts? Do you have committed purchase orders from some of your customers, with some others straddling the fence? Sales variances are major drivers in determining the need for materials. When the sales and marketing departments can generate accurate sales forecasts, the production levels can be met. From the production requirements, the materials and their quantities that go into your product can be known. The longer the range and accuracy of your sales forecasts, the better will be your ability to more accurately plan materials purchases. This gives you great leverage in materials planning.

14.2   Identifying Your Critical Suppliers

It is important to identify those parts and material that are critical, where early planning reduces risk. Based on experience, using consultants, referencing journals, and industry reports, you can be aware of supplier risk issues associated with long lead-time items, new components not in production, preliminary/unreleased datasheet, small startups with little to no production history, and suppliers with known quality and reliability issues. For suppliers with unusually long lead items, this will allow you to place orders for those critical items well in advance, before they become critical to your engineering development or manufacturing process. Of course, there will be a range of criticality. Many of the small components that go into your product still require planning but usually have shorter order lead times. When business is booming for you, it is probably booming for many businesses. This usually increases the lead times for items for which your suppliers have huge demand from many of their customers. Remember, your suppliers have variances, too. You may be in a business in which you need a component that is specifically designed by you and made for you by one company. The planning that is needed to ensure that this critical part will be on time is crucial to your business.

Depending on the size of your business, you will be buying materials from either manufacturers or their distributors. Establishing a good relationship with your suppliers is an absolute must. Without a doubt, it is obvious that when you receive materials that are exactly what you ordered, it is very important to make full payments on the invoice to maintain a good supplier relationship. In effect, there is a "business handshake;" you get the materials you want and your supplier gets paid on time. This is important in maintaining a good supplier relationship. Well before that, however, selecting a supplier that will satisfy your needs for the present and the future is critical and very time-consuming.

14.3   Develop a Thorough Supplier Audit Process

As part of purchased materials planning, it is imperative to know what you are looking for in a supplier before you even begin the selection process. Create a supplier audit list that identifies the important parameters you need in a supplier. These lists can be found in magazines covering the topic of purchasing, in articles and pamphlets offered by the American Society of Quality (based in Milwaukee, Wisconsin), and the knowledge about your specific business and its needs. From a combination of inputs, a supplier audit list can be constructed that will become a general template for most or all of your supplier selections.

A major part of the supplier selection is the process itself. Uniformity of the process is important when you are visiting several suppliers who may supply one type of item. Later, after auditing several suppliers, you can fairly and critically compare them against a uniform standard – your audit list. The results of the supplier audit can be used to identify strengths and weaknesses in your supplier. There is a possibility that suppliers may be unaware of their weaknesses; they might not even know that there is something they need to provide as part of their product that their customer wants. This is where you, the customer, can work toward "partnering" with your supplier.

Partnering is a concept that began in the 1980s with the total quality management (TQM) boom. In its simplest form, the idea is to use the strengths of two companies to identify and improve on the weaknesses of the other. For example, the company doing the purchasing may make sporadic purchases that are difficult to fill by the supplier. The purchasing company might not view this as a problem. The supplier, on the other hand, cannot satisfy small orders and then big orders without accumulating large inventories and accepting risks that might not be in the supplier's best business interests. In this case, the supplier might be well suited to work with the purchasing company to improve its materials purchasing planning. In another case, the supplier may be delivering product that does not always meet the quality standards needed by the purchasing company. Here, the purchasing company may be able to send a quality engineer to the supplier to help identify and improve its output quality. The partnering concept can be applied to virtually all segments of the business. As purchasers and suppliers work more closely together, they can minimize business risks and establish and maintain low costs with high quality.

14.4 Develop Rapid Nonconformance Feedback

Even with the best business relationship, on occasion, purchasers will receive nonconforming material from their suppliers. The identification of nonconforming material and the speed with which it is identified will help hold down quality costs. The sooner nonconforming material is identified in the process (failure reporting, analysis, and corrective action system [FRACAS]), the lower is the cost of recovery. Also, the time to recover from discrepant material is reduced. When nonconforming material is identified, it is to be gathered and placed in an area that is controlled so the material does not get mixed with forward production.

The discrepancy is to be identified using some formal process. Typically, a meeting with purchasing, manufacturing, engineering, and any others stakeholders is held daily to discuss the discrepant material. This reviewing group is referred to as a *material review board* (MRB). Often, the location where the discrepant material is held is referred to as the MRB crib. It may turn out that the identification of the discrepant material was incorrect, and if so, the MRB can place it back into inventory for forward production.

If the material is unacceptable and inexpensive, the best disposition may be to scrap it. Here is where supplier partnering is valuable, because either the supplier or the purchaser has to pay for the scrap. If there is a good partnering relationship with the supplier by the purchasing company, the purchaser might be allowed to scrap the material at the supplier's cost, as long as the supplier can review the material at a later date when visiting without driving up the cost of shipping the material back to the supplier where they may scrap it themselves. There are several other dispositions that nonconforming material can have. It might be slightly nonconforming but still used "as is." It might be used with a small amount of rework that can either be expensed by the purchasing company or billed back to the supplier. Here again, partnering helps to smooth out difficulties in these situations. If there is enough time, the material can be sent back to the supplier for corrective action. This, too, can be a touchy situation if there is not a strong working relationship between the two companies.

14.5 Develop a Materials Review Board (MRB)

In any event, what is most important is that the MRB process quickly identifies the unacceptable material and works to reverse the situation, very often, with the cooperation of the supplier. Here too, the supplier is very interested in identifying unacceptable material. With an early warning, they may be able to stop current manufacturing on the very same product that is unacceptable to the purchasing company, until the matter is corrected. A rapid means of identification and feedback to the supplier via the MRB process is important. There are many software applications that are currently available that help speed this information to the supplier. Very often, both parties can share the burden of cost of the software applications. The size and complexity of these software applications vary, depending on size and needs of the businesses. Depending on the software application, the feedback may take the form of a document that is automatically faxed, sent over by a modem, or through the internet to the supplier.

There is much more to the supplier management role that cannot be fully addressed in this reliability text. But one of the most important parts of the purchaser/supplier mix is building a partnership that understands the real needs of each party and working continuously to adapt to the needs of both businesses.

14.6 Counterfeit Parts and Materials

Counterfeit electronic components is a significant growing problem in the electronics industry. There may not be a consensus on the magnitude of the counterfeit problem in the electronics industry, but there is general agreement that it is wide-reaching problem that nobody is immune from. An electronic component is considered to be counterfeit if it is an unlawful reproduction, misrepresented substitute, or misidentified (e.g. manufacturer, part number, date code or lot code). Legitimate manufacturers of electronic components and products can fall victim if counterfeit material slips into the manufacturing process or if they accept returned material and do not verify its authenticity. Some counterfeit parts may be visually detectable because there are imperfections in the company logo, spelling errors, missing information, mismarked, or other visually detectable signs. However, not all counterfeit parts can be visually detected or inspection.

There are several sources for counterfeit parts. The parts may be manufactured to look like the original parts. The part may be repackaged or relabeled – for instance, labeled for higher performance than it was designed or manufactured to. The parts may be refurbished and sold as new. Finally, recyclers take scrap boards and salvage parts that are resold as new.

To prevent counterfeit parts from getting into your production stream requires a proactive approach that includes controls everywhere in the material stream, from raw materials to finished products. Counterfeit parts can be a significant risk when using brokers and independent distributors because the source of origin may not be known or may be misrepresented. When counterfeit material or parts are detected, they should be reported and quarantined to prevent the reintroduction of the material back into the production stream.

It requires supply chain management and a process for material authentication and traceability to prevent unauthorized sources from getting into the production stream. There needs to be a way to verify the origin of the material through trace-ability documentation (i.e. acquisition traceability or conformance certification) and

Table 14.1 Industry standards for managing counterfeit material risk.

AIR6273	Terms and Definitions – Fraudulent/Counterfeit Electronic Parts
APR6178	Counterfeit Electronic Parts; Tool for Risk Assessment of Distributors
AS5553	Counterfeit Electronic Parts; Avoidance, Detection, Mitigation, and Disposition
AS6081	Counterfeit Electronic Parts Avoidance – Independent Distributors
AS6171	Test Methods Standard; Counterfeit Electronic Parts
AS6174	Counterfeit Materiel; Assuring Acquisition of Authentic and Conforming Materiel
AS6462	Verification Criteria for Certification against AS5553
AS6496	Authorized Distributor Counterfeit Mitigation
AS9100	Quality Management Systems – Requirements for Aviation, Space, and Defense Organizations
IDEA-STD-1010	Acceptability of Electronic Components Distributed in the Open Market
IDEA-STD-1010	Acceptability of Electronic Components
IEC/TS 62668-1	Process Management for Avionics – Counterfeit prevention – Avoiding the use of counterfeit, fraudulent, and recycled electronic components
IEC/TS 62668-2	Process Management for Avionics – Counterfeit prevention – Managing electronic components from non-franchised sources
UK MOD Def Stand 05-135	Avoidance of Counterfeit Materiel
AR 42 – IECQ	Counterfeit Avoidance Program
AR 36	Accreditation Program for Avoidance of Counterfeit Electronic Parts Management Systems

authenticity verification (i.e. visual or X-ray inspection, material analysis or testing). Some manufacturers are taking creative steps to help them quickly identify counterfeit material through the use of imbedded markers and anti-tampering mechanisms. It is advisable to have a process in place that monitors for counterfeit parts reported from external sources and to keep informed of the latest counterfeiting information and trends. Finally, there must be training to bring an awareness of the magnitude of problem and about the tools and processes in place to prevent the introduction of counterfeit parts.

There are industry standards that provide guidance on how to manage the risk of counterfeit material and components from getting into their manufacturing pipeline. Table 14.1 provides a good starting place.

Part III

Steps to Successful Implementation

15

Establishing a Reliability Lab

Installing a reliability lab at the company without proper planning and an understanding of the cost considerations can be very expensive. To begin with, the total company sales dollars and the associated warranty costs dictate the magnitude of any plan. This will guide the planner as to the number of personnel involved in the reliability process on a day-to-day, full-time basis. This salary expense is the long-term driver because the returns on investment, in terms of recovered warranty dollars, will take several years to recover. The current salary budget must be able to absorb these expenses for this time period. This is a minimum.

Then there are the other major expenses:

- Equipment costs
- Reliability lab space
- Lab benches, desks and files, and so on
- Support tools and equipment
- Test equipment
- Mechanical fixturing (between the device under test [DUT] and the chamber)
- Dynamic test devices (to operate the DUT during environmental stress)
- Consumables (power, materials, liquid nitrogen (LN))
- Maintenance overhead

15.1 Staffing for Reliability

The reliability lab will not, in and of itself, deliver all the savings to the bottom line, but it will be a substantial part.

To start with, one person must lead the activity. This person must have either the qualifications based on experience or be willing to hire the expertise needed to establish the reliability program. The latter will take a lot longer, but it is recommended that you seek someone from outside the organization with reliability expertise to build the reliability program, establish the reliability lab and implement the reliability process changes. The ideal candidate will have the following characteristic skill sets and experience:

- A reliability engineering background
- Highly Accelerated Life Test (HALT)/Highly Accelerated Stress Screens (HASS) and environmental stress screening (ESS)
- Shock and vibration testing

Improving Product Reliability and Software Quality: Strategies, Tools, Process and Implementation, Second Edition. Mark A. Levin, Ted T. Kalal and Jonathan Rodin.
© 2019 John Wiley & Sons Ltd. Published 2019 by John Wiley & Sons Ltd.

- Statistical analysis
- Project budgeting/estimating
- Failure analysis
- Conducting reliability training
- Persuasiveness in implementing new concepts
- A bachelor's degree in engineering

The salary can vary, depending on experience, qualifications, area of the country, and so on. There are not many recruiters with experience in hiring reliability engineers and managers, so it may require spending time with the recruiter to explain the skills, education, and experience needed for the position. (A website for this information can be found at www.salary.com.)

15.2 The Reliability Lab

The remainder of this chapter will discuss what needs to be considered in establishing the reliability lab. Tables 15.1 and 15.2 are matrixes providing suggestions for the best choices for each issue, based on small, medium, and large companies.

The lab space is not trivial. Besides the space needed for the HALT chamber, there will be requirements for the following:

- Liquid nitrogen tanks (if a large external tank is out of budget). Remember, that there will be some full tanks, some empty tanks, and perhaps one or two partially filled tanks to contend with. The nitrogen tanks are about 30 in. in diameter and 6 ft high and weigh several hundred pounds, when full.
- Lab benches for failure analysis, repairs, and other test equipment.
- Desk space for intranet/internet communications, general report preparation, and so on.
- Room to maneuver test rigs and cherry pickers in and out of the HALT chamber.
- Storage space for other equipment when not in use.
- Space for engineering support staff who will be taking part in HALT activities.
- Chairs for everyone, some at lab bench height and some at desk height.
- Tool cabinet(s), preferably on casters.
- Storage cabinets.
- Wall space will be needed for high-power sources, that is, shop air, coolant water sources, and so on.

Support tools and equipment will be needed. Some necessary tools are:

- *General hand tools.* Tools are needed for soldering, tightening, cutting, holding, and so on.

Table 15.1 Annual sales dollars relative to typical warranty costs.

	Small ($)	Medium ($)	Large ($)
Annual sales	1 000 000–5 000 000	10 000 000–50 000 000	100 000 000 and up
Annual warranty cost	100 000–500 000	1 000 000–5 000 000	10 000 000 and up

Table 15.2 HALT facility decision guide.

		Company size		
		Small	Medium	Large
HALT machine	Rent/lease	X	X	
	Buy		?	X
HALT machine operator	Rent	X	?	
	Hire operator		?	X
Nitrogen tank	Dewars	X		
	Storage tank		X	X
Concrete tank mounting slab			X	X
Multiple Dewar manifold		X		
Safety mats		N/A	X	X
Training	External	X	X	X
	Internal		X	X
Mechanical fixtures		M	M	M
Test instrumentation		?	?	?
Turnkey solution		N/A	?	X
Room exhaust fans		N/A	N/A	X
Room oxygen monitors		N/A	N/A	X
Lab facilities		N/A	N/A	X
Travel costs		X		

X = likely best choice
? = depends on circumstances
R = rent
M = make

- *Hoist.* A hoist will be required for large DUT units that cannot be carried by one individual.
- *Thermal instrumentation.* These include thermocouples, thermocouple welders, extra accelerometers (because they fail too), and so on.
- *Test equipment.* Digital volt meters, digital thermometers, recorders, clamp-on ammeters, portable oxygen sniffers (for leaks), oscilloscope, function generators, RF generators, power supplies, variable frequency strobe light, and whatever your special needs may be.

Also needed will be some sort of universal mechanical fixturing (between the DUT and the chamber) that will mechanically hold the DUT to the chamber table for vibration testing, such as the following:

- *Hardware.* This includes drill rod and cross bars with locking nuts, as well as extra bolts of various lengths, nuts, and flat and split ring washers to attach holding devices to the HALT table (usually, standard 3/8″ thread found in the hardware store).
- *Mechanical hold-downs.* This may need special design for your specific needs.
- *Towing rope or cable.* These are needed to hoist heavy DUTs.

Some of these custom items may take some weeks to design and to fabricate, so planning here is important.

Without a doubt, the dynamic test devices or instrumentation can be a high-cost item. In HALT, the DUT must operate dynamically so that failures are detect when they happen. The instrumentation that will accomplish this may be as simple as a voltmeter and oscilloscope or it can be as complicated as a special hardware assembly with special software designed solely for these tests. (Often, this special gear can be a part of the test gear planned for the HASS in the manufacturing process later.) Here, planning is paramount. Even the length of the cables that go in and out of the HALT chamber are considered. Plan on spares for everything that will see stress during HALT; it will serve both for debugging and replacing anything damaged or degraded during testing.

15.3 Facility Requirements

There are consumables that need to be planned for:

- *Power and light.* Power can be substantial when large temperature excursions are applied during HALT or when thermal cycling large thermal mass assemblies.
- *Liquid nitrogen, sized either by the portable Dewar or large external tank.* This can range into thousands of dollars per month, even with the smallest HALT systems. (The wider and faster the temperature extremes, the more it will cost in consumables.)
- *Maintenance overhead.* This includes replenishment of lab supplies, among others.

15.4 Liquid Nitrogen Requirements

The cost for liquid nitrogen will vary, depending on the volume used and delivery frequency. Fifty-gallon Dewars provide the lowest-cost solution if your usage is small. You can store multiple tanks and roll one out when it is empty. If you are planning to use a lot of liquid nitrogen, then a larger external tank might be the best solution. The placement of the tank can be surprisingly costly. Zoning codes and local community planner preferences can significantly impact the installation cost. In industrial locations, a simple concrete slab may be all that is needed. This can typically cost \$15 000–\$40 000, depending on the size of the tank needed. If the community requires more aesthetics to have the storage tank blend into the environment, this can reach over \$100 000, particularly when earthquake protection is part of the slab design specification. Things like lattice panels to cover the tank, street lighting for evening service and tank filling, and special jacketed piping from a driveway located port where the nitrogen truck connects the filler hose to the tank will all be necessary; the list can be extensive.

The large tank will be beneficial in that the liquid nitrogen cost can be substantially lower by buying in bulk. There will be no need for personnel to manhandle the Dewars, because the external tanks can be set up with a phone line/modem to facilitate automatic refilling. This is a great time saver, especially in the manufacturing process. Consider, as part of your big picture planning, having one tank that supports both the HALT and the HASS process. This may cost more due to the added insulated piping needed, but through careful planning, even these costs can be controlled. The tank can be larger and the volume usage costs will be lower. The number of refills will be less as well, which adds to the cost reduction.

Losses occur in the piping from the tank to the HALT chamber. This can be almost eliminated by using insulated, jacketed piping. Some nitrogen tank suppliers can provide this as part of their complete cryogenic, turnkey services. They custom the piping design as part of the whole system. This is important because where the HALT chamber and the liquid nitrogen tank are placed affects the cost of this piping. It is typically $300 per running foot. Regular piping is less expensive, but the nitrogen losses will soon add up. There will be frost buildup every time the HALT chamber is used if regular piping is selected. This can add to other problems. When the frozen humidity finally warms, the resultant water may cause damage and safety problems.

All liquid nitrogen tanks are not equal. Some do a better job at minimizing nitrogen losses. Check with your supplier. It is recommended that tanks come with an insulated output flow valve. Tanks without this valve will frost up and create a frost bubble that can be 1–2 ft across at the valve. This means that the valve can't be turned until the frost melts. A failure that occurs past the valve might result in the inability to stop the flow until the tank empties. Insulated valves can be shut off because they do not frost up to where they cannot be operated.

Every time HALT commences, the chamber has room air inside that has humidity. It will freeze during subfreezing temperature excursions and later condense when the chamber is heated. This condensation may damage the DUT. It is recommended that a vaporizer be added to the nitrogen tank to convert a small portion of the liquid nitrogen to gas. This gas can be dispensed into the chamber to flush out the humid air from the chamber itself at the start of every test. For those who select Dewars, dry nitrogen bottles can be used for this purpose. Remember that this is another tank to manhandle, reorder, and have available.

Budget the liquid nitrogen cost based on planned usage. A typical HALT will consume from 250 to 1000 gallons per week. This depends on the temperature cycle rate and the temperature swing levels. The liquid nitrogen cost can vary widely from $0.80 to several dollars per liter. The cost of LN will be dependent on factors like leasing an LN storage equipment vs. owning the storage tank and the planned annual volume of LN usage. Many of the liquid nitrogen delivery contracts are ever-green, which means they automatically renew and have set requirements and conditions for termination. If the chamber will see high utilization, the monthly liquid nitrogen cost will be in the thousands of dollars.

15.5   Air Compressor Requirements

Compressors can reach $30,000 for larger HALT chambers. It is recommended that the compressor be placed outside of the HALT lab, outside of the building is preferred, so that the compressor noise does not interfere with personnel. Some compressors can be very loud; check the decibel sound level for the compressor. If the compressor will be in a work area, the sound level should be kept to 65–70 dB. Shopping around will help you discover quiet units. Typically, the best units have full-power running-noise levels at 62 dBA similar to the noise level in a relatively quiet office. This means for a quieter compressor, plan on purchasing a unit that does not need to be operated at its maximum output level. A size larger than the chamber manufacturer specifies will still do the job, work more efficiently, and be quieter. When possible, place the compressor on the roof or outside the building. A rooftop unit requires an automatic restart feature on the compressor so that no one has to climb onto the roof to restart the unit after power failures.

Compressors use outside air for their source, assume that the air will contain water vapor (humidity) and oils (from pollution), which will now be in the airline. The water vapor needs to be removed from the airline and can be removed using a dryer/filter or a dehumidifier. Depending on the method used, the trapped water will need to be drained if it is being collected in an airtight container. The HALT chambers use pneumatic piston hammers to generate the table vibration. The hammers will corrode if there is water/moisture in the airline that was not filtered out. The HALT chamber manufacturers may have a filter system at the inlets of the chamber, but they will rapidly become ineffective if the facility is located in an area of high humidity. Having a water filtering mechanism as part of the air compressor will help lengthen the life of the pneumatic hammers. If you use the compressed air that is already in your facility, this problem may already be eliminated. However, it may not be a good idea to use existing compressed air from your facility in the HALT chamber.

If the facility's compressed air already services other manufacturing processes, like delicate pick-and-place component equipment, then the periodic on and off of the HALT chamber with its relatively high usage, may create problems with other processes or equipment. This could create problems that will be almost impossible to diagnose. Therefore, it is best to have a local air compressor for the HALT and HASS process.

15.6   Selecting a Reliability Lab Location

Next, where to put the lab? The closer it is to the product development lab, the better. You want to make the necessity of the design engineers to walk over to the HALT lab as easy as possible. Do not place the lab in another building; this often tends to engineers being reluctant to make the journey and the loss is yours. It is desirable to have windows in the reliability lab so that passersby can see observe HALT and other reliability testing activities. This is especially beneficial when you have stockholders, customers, and other management heads tour the facilities. The reliability lab and reliability testing to make products reliable is part of the organizations core competency and a source of pride and distinction. It also communicates the importance of this activity.

The lab has to be large enough to contain the HALT chamber and all the other equipment. The HALT lab planning needs to take into account how product will be moved/wheeled in and how easily it can be placed into the chamber. Typically a 20′ × 28′ floor plan is needed. Anticipate growth; there should be sufficient space for a second HALT machine to be added in the future. It can be very costly if you have to move the HALT lab to a different area to make room for more HALT chambers.

The lab requires sufficient amperage for the chamber, typically 480 VAC, three-phase, at 200 A. This is for the larger chambers. Less amperage is required for the smaller chambers. This specification is available from the manufacturers. Compressed air for the product and test methods is also frequently used.

The lab can be completely up and running in 90 days from start to finish. This can be faster if sign-offs and community construction permits are expedited. In some communities, the zoning and construction permits can be exasperating in the amount of time they seem to waste. Be prepared for these delays. Get the city or town inspectors and zoning people involved early. This can help expedite the building process.

Often, companies begin with an outside service to start HALT, and later transition it in-house HALT facility. This might improve the planning of in-house facilities as the team becomes more experience in HALT. Outside services usually have a two- to six-week waiting period, because they also need to plan their facilities relative to typical five-day HALT exercises. HALT service will cost typically $5000 per week for just the rental. Other costs related to the actual testing described earlier should be included when considering total expenses.

If HALT will be done at an outside lab, additional planning will be needed for mechanically fixturing, cabling, supply and monitoring equipment, since the DUT needs to be operated dynamically during HALT. This will take time and resources and the HALT facility can provide guidance and support and may have some of the equipment and fixturing needed. Depending on how far away the outside HALT facility is, these costs can quickly add up. If you have to travel out of town for a week, the cost of the hotel, rental car, meals, and so on will have to be considered. A partial list of HALT facilities is including in the appendix.

15.7 Selecting a Halt Test Chamber

In choosing a HALT chamber, there are important items that need to be taken into consideration.

First, there is the cost of the HALT chamber itself. A HALT chamber can cost from $75 000 to $500 000, depending on the size needed and the manufacturer. Custom chambers can be even more expensive. We have found that, as a practical matter, the published costs are relatively competitive. A good negotiator can reduce the chamber costs, particularly if you know you will be growing and require additional HALT chambers in the future. If the plan is to do HALT *and* HASS, determine if one chamber will be adequate. Typically, the HALT chamber is located in an engineering environment and the HASS chamber is in a manufacturing/operations environment. If manufacturing material needs to be segregated from product development, two similar chambers will be required. Buying two or more chambers can help in price negotiation, even if you are not planning to purchase them at the same time.

Before any dollars are committed for HALT equipment, determine the magnitude of the dollars lost to cover product warranty. A significant portion of this warranty cost could be eliminated with HALT. Quantify the potential saving from reduced warranty cost due to the investment of a HALT chamber. The savings from removing product design failures before product release is often significant enough to show an return on investment (ROI) in less than a year after product release. Table 15.1 sizes your reliability budget relative to your warranty cost. (Remember that typically the warranty cost in a company that has little or no reliability processes in place usually falls in the range of 10% of the total sales dollar.)

The warranty figures may appear extremely large at first glance, but these are typical to a variety of industries. It is not unrealistic to achieve a 1–2% warranty cost after implementing an effective reliability and software quality processes, as outlined in the book. The savings can be substantial and provide an appealing ROI. Focus on all the costs that are subtracted from the sales dollars due to returned goods, rework, scrap, field service, costly design engineering changes, manufacturing process changes, hand versus automated processes, outsourcing costs, supplier relationships, inventory losses, and so on. These total costs are quite significant. Conduct a special audit, and use a consultant who specializes in identifying the cost of quality (reliability too), if needed, to objectively determine the true warranty cost.

Once at a division of a substantial Fortune 500 company, I was trying to establish the cost per warranty repair. I came to a figure of about $3,000 per fix. The management didn't believe it. I was asked to work on this figure with the senior financial manager to get the "right" number. After a time, we agreed that, depending on the specific product, the figure ranged from $2,700 to $3,100. I also told them that the warranty cost as a percentage of the sales figure would be in the 10–12% range. The top manager was in disbelief. He looked over his financial data and quickly came to the realization that it was 11.5%. He signed the purchase order for the HALT machine the very next day.

A lot of the warranty costs go unmeasured and are unknown to a manufacturer. These can be in areas such as increased personnel costs related to customer interactions to "iron out" a problem. In essence, any activity that the seller performs to make the customer happy with the delivered product can be warranty cost. Most importantly, what additional sales were lost due to poor quality and reliability? This is a difficult figure to determine, but it is very small when there is good quality and high reliability.

15.7.1 Chamber Size

The HALT chamber cost will usually be determined by the physical size of your product. If you manufacture items that are the size of a satellite receiver, then the HALT chamber can be less expensive. However, the HASS chamber might still have to be the larger-sized unit, so several units can through production HASS at a time in meet manufacturing through put requirements. Specific planning will determine the best mix.

The internal chamber size is, first and foremost, important. If it's too small, the ability to HALT parts of the product or the entire product may be lost. The tables in the chamber are about 4 in. smaller than the chamber wall-to-wall dimensions in the x- and y-directions. This is so that the table can move in vibration in these directions. The table has a grid of 3/8 in. tapped holes that will facilitate screws and threaded rods to hold the DUT securely to the table. Most chambers have internal lights or lamps that can be swiveled for maximum adaptability. These lights take up space at the top of the chamber, so be sure that the chamber height is tall enough, even with the lighting fixtures at the top of the chamber. (The height of the chamber is really smaller by the amount of space the lamps take up.)

15.7.2 Machine Overall Height

The overall size of the HALT machine is important as it has to fit into the intended space (Figure 15.1). Size considerations often come into play when determining how the

Figure 15.1 ESPEC/Qualmark HALT chamber.

chamber will be moved from the loading dock to the lab. Make sure the ceilings and door sizes allow for this move. Obviously, where the HALT machine is going to be placed is important, but how you will get it there is often not a simple matter.

One installation location for the HALT machine was near a laboratory on the second floor of an engineering facility. This particular HALT machine was one of the larger units. A junior engineer was assigned the task of checking the manufacturer's dimensions of the HALT machine and making sure the unit would fit into the HALT lab. The young engineer reported the unit would fit, just barely. He even checked to ensure that the HALT machine would fit into the freight elevator that would take it from the first floor to the second. The day the machine arrived, everything was ready to move the unit from the receiving dock to the HALT lab. Three very strong and burly equipment movers were contracted to do the heavy lifting. Early in the morning, as planned, the truck arrived with the new HALT machine.

It was immediately apparent that a tiny issue was overlooked. The large wooden shipping crate that contained the HALT machine would not fit into the large doors on the receiving dock, so it was temporarily set down on the parking lot. The wooden crate material was removed. Then a rented, industrial forklift was used to place the machine back on the receiving dock. Two, smaller lifts were used to move the chamber to the lab. But it didn't fit onto the elevator. It was too tall, by 2 in. The manufacturer's drawings were in error. Nonetheless, the three movers were undaunted.

With small blocks and pallet jacks the movers raised the unit high enough to remove the four metal legs from the machine, thus giving back 4 in., just enough. They lowered the machine on five, 1-in.-diameter electrical conduit pipes, and rolled the machine onto the elevator. They reversed the process getting it off the elevator. They checked to make sure that the machine would still fit into the lab, and when satisfied, reattached the legs and placed the unit in the lab.

The lab manager had earlier hired a professional engineering firm to certify that the building was strong enough to support this 5,000-lb machine so that the floor didn't cave in on office workers below. As an extra precaution, he placed a 3/4-in.-thick, aluminum plate under the HALT machine to help distribute the weight in the lab. This precaution is usually not necessary on first-floor installations when a unit is placed on a concrete foundation. Some floors are actually a grid work raised above the foundation floor to provide room for cables and wiring and so on. It would still be wise to check the floor strength, if the unit will be placed on a tile floor such as this.

The ceiling height in the hallways from the dock to the HALT lab was higher than 8 ft, but not much. The movers managed to nearly rip off an exit sign and a water sprinkler attached to the ceiling. The machine has two air filters mounted low on the unit. They were low enough for both to be damaged by careless forklift operators. The manufacturer of the HALT machine was very understanding and replaced the filters at no charge.

Needless to say, a lot of things must be considered when installing a HALT machine into a lab facility. Take care to see that the details are considered.

15.7.3   Power Required and Consumption

The power requirements to operate the HALT chamber could be 480 VAC, three-phase at 200 A. Ensure during the planning phase that the appropriate power will be available at the planned location for the HALT and HASS chambers. Plan on providing power to meet other needs logistic needs as well, i.e. powering the DUT, test, and monitoring equipment. Standard 110 VAC, single-phase at 20–30 A may be needed throughout the lab for instrumentation and so on. Compressors often require three-phase power as well.

15.7.4   Acceptable Operational Noise Levels

Years ago, the noise levels from the HALT chambers were so excessive that the machines had to be placed in external buildings. Today, they have much lower noise levels, measured in dBA. Typically a 65- to 75-dBA noise level is acceptable when the machine is operating at its highest vibration levels. Ear protection for extended high-level tests may well be needed with some machines.

15.7.5   Door Swing

The larger units have two doors on each side. This makes the swing space required smaller. This can help lower the size of your HALT lab. Smaller machines have one door and often require the same or even more door swing space. This should be considered during planning.

15.7.6   Ease of Operation

The operation of the machines is essentially the same, but how you operate them varies widely. Some controls are hard to understand, while others are as simple as a cookbook.

15.7.7   Profile Creation, Editing, and Storage

The HALT test profile has to be developed by the HALT operator. It is essentially the single and combined temperature and vibration stress stimuli and how the operator chooses to mix them. Some software systems allow for easy copying and editing of previous developed profiles. Seeing how the machine operates first hand will allow you to determine how easy it will be to modify a test profile. A little due diligence will save a lot of frustration when in the lab creating test profiles. The HASS process uses profiles that are preprogrammed for production. Usually, the HALT process is so empirical that an automatic system is impracticable. HALT is more like an audible; the stress profile and test process is highly dependent on how the DUT performs under stress testing. HALT by its nature, cannot be an automated test. HALT is not a pass/fail test. The buyer should determine the importance of automatic profile capability.

15.7.8   Temperature Rates of Change

The rate at which the chamber temperature changes is a major selling point made by all chamber manufacturers. Some can reach 80–100 °C min^{-1} (i.e. for the smaller DUTs that have a small thermal mass). It is important to note that Dr. Gregg Hobbs has written that he has never discovered a flaw as a function of the temperature ramp rate. [1] So, maybe this "must have" feature is not that important. The ramp rate capability is important, however, in the manufacturing process, to speed throughput. This is where the ramp rate pays dividends.

15.7.9   Built-In Test Instrumentation

Some HALT chamber manufacturers only make HALT chambers. Others make a variety of environmental chambers and provide a wide variety of test instrumentation, as well. Often, this instrumentation is integrated into the HALT software so that they are very compatible. This can be an important feature.

15.7.10   Safety

HALT chamber manufacturers provide second source oxygen sensors and alarms that will alert a user, in the event of a nitrogen spill, where the oxygen levels might become depleted. They are usually from another manufacturer who specializes in gas detection. To save some money, these can be purchased separately.

15.7.11   Time from Order to Delivery

From personal experience, the HALT machines are not purchased off the shelf. They have to be built to order, and this time frame is usually 10–12 weeks from receipt of the purchase order.

15.7.12   Warranty

Every manufacturer has a warranty. The typical standard is two years for parts and labor. Some offer two to three preventative maintenance visits, at their costs, to ensure that the chamber is operating to specification. During these visits, they may discover that other repairs are needed or offer software updates. Customers will probably have to bear some of these costs, if they discover problems that are out of warranty. Other manufacturers offer 90-day service and one-year parts, which will have to cover their logistics as well. The range of warranties vary widely, so review the warranty policies of chamber suppliers carefully.

15.7.13   Technical/Service Support

Technical/service support is important. This may mean that there is a person who you can contact for help. It may mean that they have field service personnel who can rush to the facility after a system failure. Sometimes, it means that they will expedite the delivery of a part so that repairs can be made internally. Make sure there is a clear understanding of the service package offered. It is important to diligently call other users of different manufacturer's chambers to check on their support experiences. We have found that in some cases, especially when the manufacturer is not local to the user, the user soon becomes expert in the repair and maintenance of their chamber.

15.7.14   Compressed Air Requirements

As described earlier, be sure that your in-house air system can accommodate the HALT chamber needs. If not, be safe and install your own dedicated air compressor.

15.7.15   Lighting

Lab lighting is important. Make sure there is adequate lighting. It can be dark inside the HALT chamber unless there is chamber lighting. Be sure it is part of the chamber package.

15.7.16   Customization

Customers frequently have special needs. If the chamber manufacturer will make custom machines, this can be a great asset. It may mean that a temperature-only machine can be purchases and add the vibration section later. The authors believe strongly that temperature and vibration are the two stresses that are required at a minimum to precipitate failures. Your HALT team provides all the other stresses, voltage margining, time margining, and so on. Temperature alone is not enough.

A matrix is provided so that the reader can best decide what to do on the basis of the recommendations in the matrix (Table 15.2).

A selection matrix is provided, so that you can identify those items you deem critical in your selection of machine and manufacturer (Table 15.3).

Table 15.3 HALT machine decision matrix.

Item #	Attributes Quantitative	Example	Base #	Weighted	Company A	Base #	Weighted	Company B	Base #	Weighted	Sorted by weighting factors
	Model #	Model X1			Model			Model			
1	Vibration level max (in Grms)	50	3	15							
2	Rate of temp change (in °C min⁻¹)	60	3	15							
3	Factory compatibility	Can use in factory	3	15							
4	Built in test instrumentation	Yes (not designed in)	1	15							
5	Machine reliability	Needs repairs every 2 months	3	5							
6	Noise level, in dBA	75	3	15							
7	Safety	Door has safety lock	3	15							
8	Machine size H″	102	3	9							
9	Machine size W‴	54	3	9							
10	Machine size L″	52	5	15							
11	Chamber size H″	52	3	9							
12	Chamber size W‴	52	3	9							
13	Chamber size L″	50	3	9							
14	Table size L″	42	3	9							
15	Table size W‴	42	5	15							
16	Delivery in weeks	5 wk	5	15							
17	Warranty (yr)	2 yr P & L	3	9							
18	Cost (base price)	$145 000									
	instrument, etc. options	$25 000	3	9							
19	Compressed air req.	120 SCFM @ 90 PSIG	3	9							
20	Max static load	800	3	9							
21	Number of nitrogen gas ducts	2	3	9							
22	Accessibility (table height to floor)	20	3	3							
23	Cable ports	4	3	9							
24	Window size (L″ × W‴)	18 × 18, Qty 2	3	3							

(Continued)

Table 15.3 (Continued)

Item #	Attributes Quantitative / Model #	Example / Model X1	Base #	Weighted	Company A / Model Base #	Weighted	Company B / Model Base #	Weighted	Base #	Weighted	Sorted by weighting factors
25	Weight (in lbs)	6500	3	3							
26	Multiple Dewar hook-ups	3 total	3	9							
27	Safety pads available	1 per door	5	15							
28											
29											
30											
Qualitative											
1	Software	Easy to use	1	5							
2	Technical support	Yes, local	5	25							
3	Graphical user interface (GUI)	8	3	9							
4	Total of all HALT machines produced	125	5	15							
5	Years doing HALT chambers	3	5	15							
6	Years doing other chambers	None	3	9							
7	Customization	Yes	3	9							
8	Ergonomics	7	3	9							
9											
10											
11											
12											
	Scores =		115	377							

Reference

1 Hobbs, G.K. (2005). Reflection on HALT and HASS. *Evaluation Engineering* (December).

16

Hiring and Staffing the Right People

16.1 Staffing for Reliability

The reliability lab will not, in and of itself, deliver all the savings to the bottom line, but it will be a substantial part of it.

First of all, one person is needed to lead the activity. This person must have either the qualifications from other experiences or be willing to transfer current career ambitions toward reliability engineering. The latter will take a lot longer to grow. It is recommended that you seek someone from outside your present staff with the experience to build your lab and install the processes that will be utilized. His/her main characteristic skill set listed in Table 16.1 will include the following:

- A reliability engineering background
- Highly Accelerated Life Test (HALT)/Highly Accelerated Stress Screens (HASSs) and Environmental Stress Screening (ESS)
- Shock and vibration testing
- Statistical analysis
- Failure budgeting/estimating
- Failure analysis
- Conducting reliability training
- Persuasiveness in implementing new concepts
- A degree in engineering and/or physics

16.1.1 A Reliability Engineering Background

Look for a person with reliability engineering experience that has 5–10 or more years installing reliability tools such as HALT, failure modes and effects analysis (FMEA), component derating guidelines, and so on. Analyze if the engineer has trained others and, if so, how many have been trained, over what time period, and if this mentoring developed other reliability engineers.

16.1.2 HALT/HASS and ESS

Has this candidate used or operated, or better yet, installed and used stress test chambers (HALT/ESS)? There are many things to consider, which are covered in this chapter. Can the candidate produce past stress test reports without compromising any

Improving Product Reliability and Software Quality: Strategies, Tools, Process and Implementation,
Second Edition. Mark A. Levin, Ted T. Kalal and Jonathan Rodin.
© 2019 John Wiley & Sons Ltd. Published 2019 by John Wiley & Sons Ltd.

Table 16.1 Reliability skill set for various positions.

	Reliability skills	Consultant or reliability manager	Reliability engineer	Reliability technician
1	HALT/HASS	A	B	B
2	HALT chamber experience	A	C	C
3	HALT chamber installation	B	C	C
4	ESS	A	C	C
5	Shock and vibration	A	C	C
6	Chamber experience	B	C	C
7	Hired outside test facilities	A	B	C
8	Failure analysis	A	A	C
9	Statistics skills	A	C	C
10	Reliability budgeting	A	C	C
11	Reliability estimating	A	C	C
12	Training experience	A	A	B
13	Mentoring	A	B	C
14	Held seminars	B	C	C
15	FMEA	A	C	C
16	Has done FMEAs	A	C	C
17	Has trained others in FMEA	A	C	C
18	Component derating	A	B	C
19	Persuasive	A	C	C
20	High energy	A	B	B
21	Can show success examples	A	B	B
22	Engineering degree	A	A	C
23	Electronics (EE)	B	A	C
24	Mechanics (ME)	B	A	C
25	Physics	B	A	C
26	Business	C	C	C
27	Advanced degree	B	B	C
28	Electronics (EE)	B/C	B	C
29	Mechanics (ME)	B/C	B	C
30	Physics	C	B	C
31	Business (MBA)	A	C	C
32	Associate degree	C	C	A
33	Electronics	C	C	A
34	Mechanics	C	C	B
35	Drafting	C	C	C
36	Continued studies	B	B	B
37	Classes	B	B	B

Table 16.1 (Continued)

	Reliability skills	Consultant or reliability manager	Reliability engineer	Reliability technician
38	Seminars	B	B	B
39	Publications	B	C	C
40	Books	B	C	C
41	Magazines and journals	B	C	C
42	Well respected by peers	A	A	A
43	Very believable	A	A	A
KEY	A = must have			
	B = nice to have			
	C = least important to have			

confidentiality agreements? Is there evidence of reliability improvements, and how much and over what time period? Was the testing done in or out of house? Learn what this candidate actually did to organize the team of engineers who were involved in the testing effort. Had this skill been passed to others?

16.1.3 Shock and Vibration Testing

Some stress testing is done to ensure that the product can be successfully shipped. Usually, the need for this testing is not continuous, but is done on an on-and-off basis. This usually means that the test is done at a test house, where a variety of environmental and shipping tests are available on an as-needed basis. See if the candidate has this experience. Discover what was learned and what was done when design shortfalls were revealed by these tests. Has the candidate participated on shipping packaging design teams? Skill in this area can be very valuable.

16.1.4 Statistical Analysis

Design engineers have a great deal of training in mathematics, but unfortunately, statistics is usually not part of their toolset. See if the candidate has had formal statistics training; two or more college-level semesters would be good. A statistical tool that has gained great acceptance is Weibull analysis. This is used to help identify field failure patterns. Of course, degreed engineers can learn to use this tool (the math is relatively straightforward), but having skills in statistics helps to get them going much quicker. See if they have experience with Pareto analysis. This will help them to quickly take action to correct the most important things. Statistics can be used to demonstrate the reliability of the final product. This is useful for management to ensure that the reliability goals have been met.

16.1.5 Failure Budgeting/Estimating

Not knowing the reliability of a new product until after it is produced can be a financial disaster for a company. The ability to budget the several segments or assemblies that make up a system will help engineers to identify and focus on those parts of the assemblies that will probably be the weakest link in the system. See if the reliability engineer candidate has this in his/her background. Reliability budgeting allows early identification of high reliability risks.

This can offer opportunities to make alternative choices in a design solution before it is too late for change because the product "has to be shipped." Comparing reliability estimating (i.e. derived from reliability estimation tools or in-house data) with budgeting closes the loop in that it is a sanity check of the budgets. Anyone can determine (guess) reliability budgets. But how well do these figures align with actual data? Look for examples of reliability budgeting and estimating. Learn how the candidate does it.

16.1.6 Failure Analysis

Whenever anything fails, there is a reason, or in engineering terms, a *root cause*. When an engineer is investigating a failure, it is common to "jump to the cause." More often than not, the cause found this way is incorrect. See if the candidate has failure analysis skills: ask for examples; ask how long it took. If the candidate has solved problems, very often, a short-term solution is implemented to fix the problem right away and then a long-term solution is added to the system as a design change. See if the candidate has been involved in these activities. Evaluate how effective and timely the corrective actions were.

16.1.7 Conducting Reliability Training

Seek someone who can be the seed that grows a reliability capability in your company. Discover what this person has done to pass on this knowledge to others. Have they mentored individuals – how often and how many? Also learn if there is a continuing effort by the candidate to learn more so that they stay in touch with progressing reliability technology. Have the reliability tasks been taken over by these newly qualified individuals? Has the candidate performed formal training classes, seminars, given papers, and so on? If so, what were the topics? Have they been published in journals, magazines, or books? Are they aware of new software tools that will make their job easier and more efficient?

16.1.8 Persuasive in Implementing New Concepts

The authors consider the ability to recommend and persuade others to use unfamiliar tools as the "secret ingredient" to success. The evolution from little or no reliability capability to a strong reliability capability will take several years. This takes a reliability engineer who possesses great persistence and an equal amount of support from top management. Some consider the message from the reliability group a "broken record" and it is, to some degree. The message has to be delivered on a continuous basis and the messenger has to be able to do it in a way that is persuasive, yet won't upset the rest of the staff. To sell anything, the person delivering the message has to be liked or the effort will be unfruitful. (Few of us have ever bought anything from a salesperson that we didn't like.) Study the candidate; see if this is part of their personal makeup.

16.1.9 A Degree in Engineering and/or Physics

Reliability engineers interface with engineers of all kinds. No one reliability engineer can possess the skill of all disciplines, yet the person must be believed and respected by the engineering staff, as well as management. If they hold an engineering or physics degree, they will at least have a sound set of tools by which they can communicate with other engineers. See that the candidate has this tool. There are reliability engineers who have several degrees, and many have advanced degrees. Some, however, have technical backgrounds and hold other nontechnical degrees. This can be good, too. *It's what they know that counts*; but having a degree is a good base on which to begin.

Some reliability engineers are members of or attend meetings of various reliability-engineering societies, several of which are the following:

- IEEE Reliability Society (http://www.ewh.ieee.org/soc/rs) with groups all over the world.
- The Reliability and Maintainability Symposium (http://www.rams.org) where many reliability engineers and others meet to discuss the subject, present papers, and where the American Society for Quality offers the Certified Reliability Engineering Exam.
- The Society of Reliability Engineers (SRE) (http://sre.org) and others.

Using these recommendations, select a reliability expert. If one reliability person fills the bill, then you're done. Now the task is for this new person to pass on the reliability knowledge to others. If it is clear that the new reliability engineer will need support, from where can this support person be found?

We recommend that you select a well-respected engineer from the existing staff. One who has a proven track record of successful designs, who has run high productivity production lines, or has a quality engineering background. Have the new reliability engineer take a strong role in selection of other internal engineers and technicians from your staff. This will help ensure compatibility and loyalty to the lead reliability engineer. Give the added staff engineer time and tools to come up to speed.

If outside hiring is considered, hire a temporary consultant or a contract reliability engineer. Often, these individuals have a wide range of experience and can get things started quickly and most effectively. If this is a permanent position, then a reliability engineering recruiter may well serve your needs.

Salaries can vary considerably, depending on experience, qualifications, area of the country, and so on. A source for current salary information can be found at www.salary.com; there is a fee for the service. Five levels of reliability engineering salaries can be found there.

16.2 Staffing for Software Engineers

For the software quality assurance (SQA) to be fully effective, there is some specialization required among the members of the staff. The SQA specialization is along two different vectors. The first category of differentiation is infrastructure engineers versus testers. The second specialization dimension is by technological or use model domain expertise.

Automated build and testing is a critical capability necessary for attaining efficient and repeatable software quality execution. Developing and maintaining that automation requires software development staff (infrastructure engineers) to create the tools. These infrastructure engineers will set up automated build environments, write scripts and code to execute test suites triggered by new builds, and develop tools to track test execution and notify managers and software developers of test outcomes.

There are many build-and-test automation tools available. The infrastructure engineer's job will be to create an integrated infrastructure from a set of off-the-shelf tools and custom scripts or programs. The infrastructure engineers are primarily software integrators and developers. The languages required by the infrastructure engineers will be dictated by the scripting or programming languages that are used by the selected set of build-and-test automation tools. The number of infrastructure engineers necessary for an organization to carry is dependent on the amount of work required to develop and maintain the build-and-test environment. The requirement may range from a single part-time infrastructure engineer to several full-time engineers. It may be necessary to have one or more full-time infrastructure engineers to support very large software projects, multiple concurrent software projects or for a particularly complex environment. Smaller projects and teams could assign a part-time engineer to do the automation integration work.

The actual test plan development and test execution will be executed by test engineers. The test engineers should have domain expertise in the use of the product. Some of the test engineers will almost certainly need to have domain expertise in specific types of testing or technologies. Depending on the product and type of testing, the following broad domain areas should be staffed:

- *End user use model expertise.* Knowledge of expected customer use of the product is required for at least a subset of the test engineers. These engineers will be responsible for testing the product the way the customer will use the product. The engineers with use model domain knowledge should also be required reviewers of other test plans.
- *Usability engineers.* Usability engineers will provide significant benefit if the product has a substantial user interface (UI). They should be involved in reviewing user interface designs (or even creating the user interface designs) long before those user interfaces are implemented. User interface testers will work with the user model experts to develop test plans specifically to test the UIs to make sure that they are intuitive, logical, and easy to navigate.
- *Application programmatic interface (API) testers.* Many software products expose APIs for use by the product customers. Testing APIs requires writing code to exercise the APIs in an appropriate manner to check that each performs its required function as well as fails gracefully when invoked incorrectly by testing ranges of legitimate and illegal invocations. API testers are software developers who will have knowledge of the customer programming use model. They will write programs that use the APIs in the manner expected by customers.
- *Security testers.* Testing product security is a very specialized expertise. It is probably best performed by outside consultants who have the specialized tools and experience to perform vulnerability and penetration testing.

16.3   Choosing the Wrong People for the Job

Many of the very best companies hire from within. A design engineer can be cross-trained as a manufacturing engineer. A system architect can be groomed for a position in marketing. Financial analysts can be trained to be program managers. When a company has an individual who has been performing very well, it is often a good idea to move this person into a new department. Sometimes, but not necessarily, this will keep this individual interested and will increase employee retention in the company. This is a good idea if there is someone in the company who can train this individual into the new area. Perhaps, external training will meet the needs of the company. Without training, however, the promoted employee will get a slow start and may never grow to full competency. Digital electronic engineers can be trained to be programmers. There is a lot of job similarity. They will come up to speed more quickly if they work with other programmers. If there is no one to train these talented people, then they will have difficulty moving into the new areas. If they are asked to go off on their own, they may not deliver what is really needed. This is especially true of reliability engineering.

Reliability engineering is one of those areas that easily fall into this dilemma. Companies that do not have a reliability program will often identify some of their best engineers from manufacturing, test, and design to become reliability engineers. Even though these individuals are hardworking, talented, and respected by their peers, they don't have the tools to identify the reliability weaknesses and recommend process changes. This is one of the downsides of installing reliability in a company. Simply put, you'll probably have to hire from outside.

The reliability engineer must have a firm understanding of the processes and concepts needed to develop and enhance product reliability. This person must have the drive and initiative to install processes in a company where there will be a natural resistance to a methodology new to everyone. This is a difficult task, and it requires dedication and perseverance of the highest order. This person must have a personality that adjusts to the personalities around him/her. This person must know they have the backing of management all the way to the top. And finally, the reliability engineer must be a teacher. The smaller company can only afford one reliability engineer but needs all the processes. One engineer cannot do all the tasks. This engineer must be able to impart the reliability knowledge to everyone. This process may take several years, but when done correctly, other employees will also be trained in the reliability process. Companies that try to install reliability on their own without outside help will probably fail.

17

Implementing the Reliability Process

Consumers demand products that are reliable, bug free, safe, and secure are changing the way companies develop future products. We saw this happen in the 1970s when US consumers demanded better-quality autos. Consumer discontent was expressed by an increase in Japanese auto sales at the expense of the big three US auto manufacturers. When the American auto industry realized that its market share was decreasing due to inferior quality, it slowly began implementing quality programs. Change was a matter of survival. For the next several decades there was a continuous evolution of new quality programs many of which were short-lived. Today, quality is a significant part of most businesses. In fact, it is widely accepted that "Quality is everyone's job." Experience has shown that it takes many years to fully implement an effective quality program. The same can be said about implementing an effective reliability program. Plan on it taking several years to reach full implementation and effectiveness. We have seen more recently in reliability issues with lithium batteries that can combust, automobiles where control can be taken over through network security vulnerabilities.

17.1 Reliability Is Everyone's Job

The similarities between the need for improved product quality brought on in the 1970s and the need for improved product reliability 50 years later is undeniable. The challenge is in how fast the organization can transform into designing and producing more reliable products. Business success will be based on the ability of the organization to transform into taking a shared ownership for product reliability, "Reliability is everyone's job."

The companies that have been most successful in achieving a reputation for highly reliable products did so by having everyone participate in the reliability process. If a reliability program fails to transform the organization into taking a shared responsibility for product reliability, the results will be marginal at best. The added cost to transform the organization is small, but the cost of not doing so is large.

The reliability process can be applied to any organization and implemented at any time in a product development cycle. Implementing the process late in the development program can delay a product's release date because the process exposes design weaknesses that will likely require a redesign to fix the problems uncovered. These changes probably would have been uncovered without the reliability process when customers complained about them. It may seem to be a difficult decision to implement the reliability process late in a program because of its impact on time to market and profitability, but those profits

Improving Product Reliability and Software Quality: Strategies, Tools, Process and Implementation,
Second Edition. Mark A. Levin, Ted T. Kalal and Jonathan Rodin.
© 2019 John Wiley & Sons Ltd. Published 2019 by John Wiley & Sons Ltd.

can quickly disintegrate into significant losses from product recalls, liability suits, and high warranty costs.

17.2   Formalizing the Reliability Process

An integral part of every reliability program is the plan detailing the activities that will take place in order to ensure success. The reliability plan must be defined and agreed to prior to implementation. The reliability process is formalized into a document that outlines the activities for each phase of the product life cycle. A documented reliability process is a crucial ingredient for success.

The reliability plan describes the reliability activities and defines the expected deliverables for each phase of the product life cycle. After the reliability plan is developed and formalized into a document, the next step is to create awareness within the organization for this new approach to achieve improved product reliability. The entire organization must understand the new process, know what their involvement is, and be aware of any budgetary cost and schedule impacts.

The reliability activities must be incorporated into the product development schedule with adequate time allotted for each reliability activity. Schedule the reliability activities at the beginning of the program so that there are no surprises about the requirements, resources required, and impact on product delivery. At the completion of each phase of the product life cycle, review the reliability process to identify ways to improve its effectiveness for future programs. The reliability process should be continuously improved through feedback from the participants, by periodically reviewing best practices and implementing new tools/techniques to simplify and streamline the process.

In Part IV of the book, we will present a detailed process identifying the reliability activities that take place in each phase of the product life cycle. The reliability plan has been proven to be successful and includes the necessary reliability activities for a company to design and manufacture reliable products. To be successful, the organization must transform its philosophy into "doing the right things well" in order to achieve product reliability and software quality goals. The reliability plan detailed in Part IV represents "the right things" that need to be done in order to achieve product reliability. The implementation of these reliability and software quality activities may be different, based on the specific type of business and business environment. Because not all companies are alike and corporate cultures vary, the way the process is implemented will vary as well. If a process of continuous improvement is implemented, the process can be tailored to best fit any specific business needs. By doing this, a company will not only be "doing the right things," but also "doing the right things well."

Any company implementing a reliability program or wanting to improve its product reliability program can use the reliability process presented here. These activities represent the minimum steps necessary for product reliability. By choosing only those activities that can be easily implemented, sacrifices will be made in the reliability of the product. A common complaint made during the introduction of the reliability process by the design team is how this will delay the product launch date and increase the cost of the product. Use this as an opportunity to remind critics of mistakes made in past products and how those mistakes have impacted profitability, design resources, and product launch dates. If products are taking longer to develop than planned, the reliability process is one tool that will help. The reliability process reduces product development time by identifying significant reliability problems early in the development cycle where they are easier and cheaper to fix.

17.3   Implementing the Reliability Process

The reliability process can be applied at any stage of the product development cycle. Ideally, the process should begin at phase one of the product development cycle. Don't wait for the next new design cycle to begin the process. There is no better time than the present to start a reliability program. The greatest return on investment will always be with a reliability program that is implemented at the concept phase but it can be initiated at any stage of the product development life cycle. The goal should be to identify and fix all reliability issues as early as possible, because the cost to fix a reliability problem increases an order of magnitude in each subsequent phase. Taking a proactive approach to identify the reliability issues, early in product development, will result in a better product, with lower development costs, a shorter development time, and a greater return on investment. Often, the reliability improvements made in the development phase result in a reduction in the number of product fixes later.

17.4   Rolling Out the Reliability Process

There are many reliability activities that can be performed to improve product reliability. Some will produce more benefit than others. A list of references that will provide greater insight about these activities can be found at the end of this chapter in the bibliography.

In Chapter 7, we identified the reliability activities that provide the greatest benefit. How these reliability activities fall into the product life cycle is shown in Table 17.1.

How many of these reliability activities is your organization doing? Your present level of reliability involvement is one of the factors that will determine how best to implement the process. (Do you see anything in Table 17.1 that you are doing now?) Other factors (i.e. staffing constraints, organizational size, capital constraints, level of top down management support, product life cycle phase, and time-to-market constraints) are important to consider when developing your implementation plan. *The most important factor in the implementation of the reliability process is early success.* Expect constant resistance by an overwhelming number of highly intelligent individuals who can explain in painstaking detail why the process will not work in their application. If after implementing the process, the way these skeptics view the reliability activities has not at least changed to some small degree, the whole process will suffer an early death.

There will be glitches along the way with the implementation process, especially if this is new to the organization. The reliability process is a cradle-to-grave approach. It uses continuous improvement to fine-tune the process for the organizational culture and business environment.

In order to ensure success, roll out the reliability process strategically. It is more important to achieve early, recognized successes from rolling out only parts of the process than to push the organization through the entire process. Doing too many new things at one time is an almost impossible task. In other words, it is more important to do

Table 17.1 Reliability activities for each phase of the product life cycle.

| Product concept | Concept phase | | Design phase | | Product phase | End-of-life phase |
	Design concept	Production design	Validate design			
Reliability organizational structure	Reliability plan					
Reliability goal	Reliability budgets	Reliability estimate Design FMEA	Reliability growth Design FMEA		Reliability growth Process FMEA	
	Accelerated life testing plans	Accelerated life testing	HALT		HASS	HASA
		ESS	POS		HASA	
	DFx	DFx	DFx		DFx	DFx
			• Set up FRACAS • Reliability growth		• FRACAS/Design issue tracking • SPC • 6-sigma Burn-in accelerated reliability growth (ARG) and early life test (ELT)	• FRACAS/Design issue tracking • SPC • 6-sigma
Risk issues identified/mitigation plan	Risk issues status/mitigation plan revised	Risk issues status/mitigation plan revised	Risk issues closed			
• Continuous improvement • Quality teams • Lessons learned	• Continuous improvement • Quality teams • Lessons learned	• Continuous improvement • Quality teams • Lessons learned	• Continuous improvement • Quality teams • Lessons learned		• Continuous improvement • Quality teams • Lessons learned	• Continuous improvement • Quality teams • Lessons learned

Note: FMEA: Failure Modes and Effects Analysis; HALT: Highly Accelerated Life Test; HASS: Highly Accelerated Stress Screens; HASA: Highly Accelerated Stress Audit; ESS: Environmental Stress Screening; POS: Proof Of Screen; FRACAS: Failure Reporting, Analysis and Corrective Action System.

the right things well, even if it means doing less, than it is to do all the right things to a lesser degree. Of the list of reliability steps, pick one or two and work hard to install them properly. Success in a few areas will help dissuade the skeptics and gain support. A poorly rolled out process will give added fuel to the skeptics who are trying to convince everyone that the process doesn't work. Letting them see that they may have been wrong, just a little, is more persuasive than clobbering them with a longer list of new processes. Dale Carnegie teaches how to give the skeptics a chance to save face. So do it a little at a time. They will come around and become your staunch supporters, if you let them. Some will even have the "aha!" experience.

> An "aha!" example might be, "Why do they have a gooseneck bend in the pipe under the kitchen sink; wouldn't the water go down more easily if the pipe were straight?" The answer is, "Yes; the water would go down easier but the sewer gasses would come up just as easily too." "aha!" The water in the gooseneck acts as a plug to keep the sewer gases where they belong.

The "aha!" experience usually happens when a colleague realizes something about the design he did not know. This can happen through any of the new reliability processes, failure modes and effects analysis (FMEA), for instance. When the FMEA process reveals an overlooked design element, the lead designer usually is surprised and experiences an "aha!" all on his own. Because the revelation comes from the new FMEA process and not an individual, the FMEA process is more readily accepted. With one or two more "findings from the process" this designer will be won over. Designers take pride in their work and want their design to be successful. Because they truly want to do what is right, once they have this "aha!" experience, they will become strong advocates of the process. If you were to seek early support, which individual would be the best one to select?

Junior engineers will be easier to convince because this is all new to them. However, they will not be able to persuade their more experienced counterparts easily. Senior engineers have much more experience and may even have bad experiences at companies where reliability improvement task teams failed. Focus on the experienced contributors, and the rest will follow. The reverse is nearly impossible. Focus on the senior level designers who are skeptical and harder to convince. Once they understand and realize the value of the process, your job will become surprisingly easier.

So, how many of the reliability activities in Table 17.1 is an organization performing? If the answer is none to very little, then take a slower, incremental approach in implementing these new processes. One reason for this is that the reliability activities do not take place in a vacuum. It is a bad idea to hire a group of reliability engineers and tell them to make the product reliable. This strategy will most likely fail. Remember that reliability is everyone's job. Reliability engineers do not design products, the design team does. Many of the reliability activities require the participation of the design team in order to be successful.

An FMEA requires a design team of cross-functional members participating to identify failure modes and safety issues. Team members are then tasked to remove those issues that have a high risk. A Highly Accelerated Life Test (HALT) test can require several months of preparatory work with the design team. They need to develop fixtures, test procedures, test software, and test access. During the HALT testing, which can last one to two weeks, engineering support will be needed to fix precipitated failures and to identify their root cause. Likewise, risk mitigation, failure reporting, analysis, and corrective action system (FRACAS), and DFx (design for manufacturing [DFM], design for service (and maintainability) [DFS], design for test [DFT], design for reliability [DFR], etc.] are all activities that require the participation of other functional groups. All of these activities can bring product development to a standstill if a reliability program is implemented all at once. The famine-to-feast strategy for implementing a reliability program when all these activities are new to the organization will be devastating.

In fact, there are only a few reliability activities that can be done by reliability engineers alone. These activities would include reliability budgets and estimates, component accelerated life and environmental stress testing, reliability growth tracking, and reliability demonstration. In addition, some of these activities (reliability budgets and estimates, reliability growth tracking, reliability demonstration) contribute little or nothing to improve product reliability.

How to decide the best strategy for implementation? Which of the reliability activities are the most important? Which must be implemented in order to achieve product reliability goals? Not surprisingly, there is no one program that will fit all businesses. If a product is produced for space travel, life support, nuclear reactors, and other mission-critical applications, the reliability effort will be significant and comprehensive. If the company produces extremely low cost, short product life, and disposable products, then the other end of the extreme is experienced. However, our experience has found that there are a few reliability activities that everyone should be doing in order to achieve the needed reliability improvements. These activities can be found in Table 17.2.

Table 17.2 Reliability activities – what's required, recommended, and nice to have.

Functional activities	Concept phase		Design phase		Production phase	End-of-life phase
	Product concept	Design concept	Product design	Validate design		
Every business needs		Design FMEA	Design FMEA	Design FMEA	Process FMEA	Continuous improvement
				FRACAS	FRACAS	FRACAS
				HALT	SPC/6-sigma	SPC/6-sigma
				Early production		
				HASS		
Will make life easier	Reliability goals	DFR	DFR	DFR	DFR	HASA
				POS	HASA	
				Reliability growth	ARG ELT	
Nice to have		Reliability budgets	Reliability estimates		100% HASS	
Do if required					Reliability demonstration	

Note: SPC: Statistical Process Control.

The reliability of a product is determined in the design phase of the product development. Once the product is designed, the reliability can be degraded through poor manufacturing, inadequate testing, or troublesome suppliers. Remember the old saying, "You cannot test in quality." The same can be said for reliability. You cannot manufacture or test in reliability. Product reliability is determined by how well the design meets the design specifications for the different-use environments and conditions and for the expected time of use.

The two most powerful product development reliability tools, which will improve the reliability of a product, are FMEA and HALT. If a business lacks a reliability program, then, the first reliability activities that need to be installed are FMEA and HALT. A design FMEA is a powerful tool, which uses the design team's aggregate expertise to create a synergy that results in identifying design problems often overlooked in design and design review. Of course, these issues can be discovered late in the product life cycle, but it will be at a much greater cost to the company. HALT is the best way to take a working design and precipitate the most likely reliability failures that will occur in the field. If there isn't a reliability program in place, then implementing these two reliability tools first into product development will be what is most needed.

Once there is a successful implementation of these tools, the next step is to develop DFR guidelines. There does not appear to be any DFR guidelines to purchase, so it is necessary to develop them internally. The design guidelines should be based on lessons learned, focusing on the Pareto of top reliability problems of the past. If capacitors are high on the Pareto chart, then a capacitor selection and use guide should be developed. The DFR guidelines should also include issues such as derating. The best way to develop DFR guidelines is to create a Pareto of past reliability issues and, starting with the biggest issue, develop reliability design rules to eliminate repetition of these problems.

17.5 Developing a Reliability Culture

Product reliability must be everyone's job. To achieve this work philosophy, there will be a need to transform the organization's culture into one where everyone talks about product reliability issues. Getting an organization to this point will take time. At the beginning of implementing a reliability program, the following three processes need to be in place before the program rolls out:

1. Formalize the reliability process in a document.
2. Implement top down training for the new reliability process.
3. Prepare a reliability process implementation plan.

In order for the reliability program to be successful, there needs to be a commitment and support from senior management on down. It is best that senior management kicks off the reliability program effort and communicates to the organization the business need and goals and the benefits from a successful software quality and product reliability program. The first step is to define the reliability process that will be followed. Part IV of the book provides the detailed reliability process for successful product reliability and software quality development.

The second step is to develop training to educate the organization on the new reliability process. The training should be rolled out in a top-down approach. Senior- and middle-level managers need to buy into the process before it is disseminated to all other levels of the organization. If there are issues raised by senior and middle management that are not resolved before rolling out the training to the masses, then buy-in is unlikely for success.

The final step involves developing a credible implementation plan that transforms the organization into a culture that is focused on reliability issues and able to achieve the reliability goals. The implementation plan will be different for different-sized companies. For very large companies, consider using a seven-infrastructure approach as outlined in the book *A New American TQM* by Shoji Shiba, Alan Graham, and David Walden, Productivity Press, 1993, Chapter 11. Use the seven-infrastructure approach to transform the organization into a culture that relies on the new reliability process to ensure product reliability. The organizational infrastructure approach identifies seven activities that need to take place. The seven activities are as follows:

1. Goal setting
2. Organizational setting
3. Training and education
4. Promotion
5. Diffusion of success stories
6. Incentives and awards
7. Diagnosis and monitoring

A New American TQM provides an effective framework that can also be used to implement a reliability program in an organization. Implementing a reliability program is no different from implementing a quality program. Today, most companies have quality programs in place. Do you remember how difficult it was to implement these programs and how many of them died within 6–12 months? In most companies, there was a significant amount of resistance to changing the way they manufactured and developed products. It is difficult to change the way an organization operates. In essence, the culture of the organization is changing. The changes are usually slow to take hold and often take years to fully implement. Therefore, take time to review the organization's effectiveness in past rollout programs like total quality management (TQM), quality circles, and continuous improvement. Identify what worked well and what did not. This way, the organization can learn and benefit from this experience. Then, define an achievable plan for implementing the reliability process.

17.6   Setting Reliability Goals

It's time to set reliability goals. There are two types of goal setting that take place in a reliability program. First, there are the high-level non-program-specific goals. The highest-level goals are the mission and vision statements for the organization. The mission and vision statement addresses the business need for improved product reliability. Before the mission and vision statements are created, determine the business environment driving the need for greater reliability. What is the customer's perception about your product's reliability? Is it different from that which is measured, observed, or perceived? What is the perceived reliability of the market leader? Is improved reliability a strategy to maintain or gain market share? Has there been a problem with highly publicized product recalls? Are product liability lawsuits a problem?

Knowing the answers to these questions can prevent the implementation of a very costly and misdirected reliability program. There are costs associated with improving product reliability. These costs affect the bottom line. When implemented effectively, they will bring significant long-term gains. However, a reliable product that is not cost competitive can have an adverse effect on market share.

If a reliability program is being implementing for the first time, there should be high-level goal setting focused on the implementation of the reliability program. These goals focus on the following:

- Forming the reliability organization
- Installing the reliability lab
- Defining and documenting the reliability process
- Implementing a reliability process into the organization
- Developing reliability design guidelines and checklists
- Implementing FMEAs
- Implementing HALT, HASS, and HASA
- Implementing FRACAS

The second sets of goals are the low-level goals. The low-level goals are program or product specific. The goals are measurable, result oriented, customer focused, time-specific, and support the high-level goals. They can be different for different products. Examples of program goals would include the following:

- Will perform without failure for a specified time and under defined environmental use conditions
- Reduced repair time
- Reduced product development time through fewer design spins
- Reduced product development costs through fewer design changes
- Improved manufacturing first-pass yield (through improved design margins)

The goals that are defined should be measurable and supportable in the business environment.

17.7 Training

The greatest benefits of a reliability program are the design improvements made before the first prototype is ever built. Unfortunately, reliability usually takes a back seat in the early phases of the product development cycle. Design engineers do not like to be told how to do their job. We often assume that the people we hire are experts or at least competent in all facets of their job. Unfortunately, this is not always the case. It is not that they are bad designers; it's often simply that they lack the knowledge and skills required to improve a DFR. Simply providing training to designers on making the right design decisions will not improve reliability. While in the proactive phase of the program, it is important to provide training to the design team, so the decisions they make will lead to reliable product designs.

Focus the training in the areas where the reliability of the product has been a problem. For example, suppose the product has had an occasional problem with catching on fire. The product was designed with safety fuses that were supposed to prevent failures that led to fire. An investigation reveals that the fuse was improperly designed. Because the proper selection and use of fuses are extremely important to the product, it is important and beneficial to provide training to the design team on the proper use of fuses. The training should be offered periodically as new employees' cycle in and out of the organization and the past lessons learned are forgotten. There are several ways to provide training within the organization. Some of the more common approaches are as follows:

1. Develop training internally using in-house expertise.
2. Send employees to symposium classes and conferences.
3. Use outside experts/consultants to teach classes.
4. Encourage higher-level education.
5. Provide a library of books on the subject.

Ideally, there is an individual within the organization who is an expert on the design and the use of fuses. Work with this individual to design a training class on fuses. Providing training using in-house expertise is always the preferred method when the expertise exists. This is a low-cost approach that has the added benefit of communicating where the resident expertise is for a particular subject. A large organization will typically have many experts whose knowledge is often underutilized in solving known problems. Make the training materials, presentations, and so on easily available for those who could not attend the training when offered. Be sure to identify the author of the training, as this person is the in-house resource on the subject.

While individual companies will have unique and specific training needs, the following subjects are needed universally:

1. Capacitor selection and use
2. Redundancy
3. Connector selection and use
4. Derating guidelines and use
5. Mechanical reliability
6. Torque and hardware stack-up
7. Electromagnetic interference (EMI)/ radio-frequency interference (RFI) shielding
8. Electrostatic discharge (ESD) protection and susceptibility
9. Solder reliability
10. Corrosion
11. Cooling techniques
12. Materials selection

17.8   Product Life Cycle Defined

The best thing about a reliability program is that it can be applied successfully at any stage of the product life cycle. The reliability process can be applied to a product revision, derivative product, new platform product, or a leapfrog technology. The timing of the process does impact the level of risk taken, the level of effort required, the resources required, or the time frame necessary to ensure product reliability.

In Part I of the book, we described why product reliability is vital to any business that wants to compete in the current business climate. In Part II, we developed the reliability tools that are needed to improve product reliability. In Part III, the process of forming a reliability team and installing a reliability facility was developed. The only piece missing from the puzzle is the implementation of a reliability process. Therefore, in Part IV, we shall put the pieces together to form the reliability process.

The reliability process is a comprehensive cradle-to-grave approach to improve product reliability. The process should be part of a continuous improvement program that applies lessons learned from past products to continuously improve next-generation products. The product life cycle consists of six phases:

1. Product concept phase
2. Design concept phase
3. Product design phase
4. Validate design phase
5. Production phase
6. End-of-life phase

These are shown graphically in Figure 17.1. Because each company may define the product life cycle phases differently, we briefly describe each phase so you can align them to your unique product development structure.

17.8.1 Concept Phase

In the concept phase, the product is conceptually defined sufficiently for a team to design the product. First, the product concept is defined based on market and business needs. The design concept is developed defining the product architecture, physical features, inputs and outputs, assumptions, and so on. The concept phase defines the design requirements, constraints, features, and limitations that will be used to direct the design team. The concept phase may also include a design priority selection list (i.e. in order of priority: cost, time to market, performance, reliability, and manufacturability) that designers use uniformly to make design trade-offs. A list of the desired outputs from the concept phase is as follows:

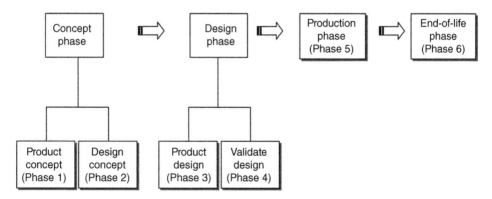

Figure 17.1 The six phases of the product life cycle.

Product Concept Phase Deliverables

- Market-driven product concept
- Product features requirements
- Product functions requirements
- Performance specifications
- Product positioning
- Market/business-driven time-to-market date
- Staffing required to achieve time to market
- Capital required to achieve time to market

Design Concept Phase Deliverables

- System and subsystem design architecture
- Preliminary design concept
- Design specifications
- Define what design needs to do
- Define what design will not do
- Define design decision trade-offs (i.e. in order of priority: cost, performance, time to market, size, weight, etc.)
- Maintenance and serviceability requirements

17.8.2 Design Phase

The next phase of the product life cycle is called the *design phase*. It too, is composed of two separate phases. The design phase begins with the product design phase where design teams create the design details necessary to achieve the concept requirements. It is in the design phase where working prototypes are developed for design validation. The design phase is also where the product documentation package (printed circuit board [PCB] design, schematics, bill of materials, mechanical drawings, etc.) is created.

The product design phase is followed by the design validation phase, where the working prototypes are tested to verify that the design meets the requirements called out in the concept phase. At the end of the design validation phase, the design is verified to be manufacturable, testable, and serviceable. By the end of the design validation phase, the product cost and profit margins are well understood along with strategies to reduce product cost. A list of the desired outputs from the design phase is as follows:

Product Design Phase Deliverables

- Schematics
- Theory of operation
- Bill of materials
- Mechanical drawings
- Product costing
- Working prototypes (hardware, software)
- Supplier selection
- System and subsystem test strategies
- Test fixtures
- Manufacturing fixtures

Validate Design Phase Deliverables

- Verification of design performance to specification
- Verification of adequate design margin
- Verification of production test and fixturing
- Verification of manufacturability of product and fixturing
- Engineering change orders implemented and changes verified
- Shippable product

17.8.3   Production Phase

The production phase begins with the transitioning of the design for production and manufacturing. The engineering effort in the production phase is significantly reduced to a support effort. It is in the production phase that manufacturing begins ramping product to meet customer demand. The production phase activities are focused on supporting product manufacturing, test, and customer support. The activities that take place in the production phase are as follows:

Production Phase Deliverables

- Manufacturing process control
- Volume production tooling
- Supplier management
- Inventory control
- Cost-reduction programs
- Cycle time improvements
- Defect-reduction programs
- Field service and tech support programs

17.8.4   End-of-Life and Obsolescence Phase

The last phase of the product life cycle is called the end-of-life phase. All products have a useful life and eventually reach a point of obsolescence. The end-of-life phase includes all activities associated with the eventual termination of the product. This is the last phase of the product life cycle and the one most often overlooked.

End-of-Life Phase Deliverables

- Eliminating obsolete parts and materials from inventory
- Disposing of manuals, documents, and so on no longer needed for product support
- Introduction of a transition product

17.9   Proactive and Reactive Reliability Activities

The reliability activities in the product life cycle can be considered either proactive or reactive (see Figure 17.2). The proactive activities consist of everything that can be done to improve product reliability and serviceability before the first customer shipment. In essence, we are trying to remove customer failures by identifying what is likely to fail

Hardware Reliability Process

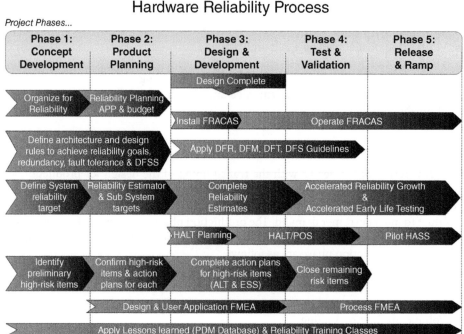

Figure 17.2 The hardware reliability process.

in the field. The reliability activities turn from proactive to reactive once the product is released for manufacturing. Design changes made after this point are more expensive and take longer to implement. In fact, a design change after a product is released receives greater scrutiny and is less likely to be implemented because of its impact on the bottom line. The same change requests made early in the design cycle are likely to be implemented because they are significantly less expensive to implement, do not impact product in the field, and the design team has not moved on to the next project. Design changes represent business decisions, because they impact product and development cost, product release date, and warranty costs. The focus clearly needs to be on optimizing a DFR in the proactive portion of the product life cycle.

The proactive region of the reliability program identifies all potential reliability issues before the products are shipped to customers (Figure 17.3). The proactive phase identifies potential risk and safety issues and resolves all potential reliability problems, which are likely to occur. The proactive reliability activities are as follows:

1. Failure modes and effects analysis (FMEA) before design is complete
2. Applying lessons learned
3. Applying design guidelines:
    (a) Design for reliability (DFR)
    (b) Design for manufacturing (DFM)
    (c) Design for tests (DFT)
    (d) Design for serviceability and maintainability guidelines (DFS).
4. Identify, communicate, and mitigate (ICM) approach to mitigate technology risk

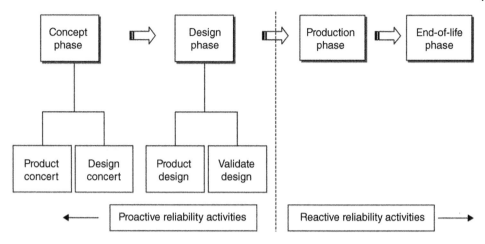

Figure 17.3 Proactive activities in the product life cycle.

5. Complete design simulation and modeling
6. Complete design specs and requirements before design phase
7. System and subsystem reliability budgets and estimates
8. Highly Accelerated Life Tests (HALT)
9. Four corners testing, testing at design margins

These reliability activities will provide the greatest benefit early in the development program. By applying these activities early in the concept and design phase, there can be a reduction in development time, non-recurring engineering (NRE) costs, and the number of design spins. By performing these steps early in the development cycle, the design is likely to be error free the first time.

The reliability activities become reactive once the product is approved for manufacturing. The reactive reliability tools are used after a product is released to manufacturing. It is, then, we find ourselves scrambling to deal with difficult issues regarding product recalls, product alerts, and retrofits. Finding these problems is often costly and time-consuming. When these problems are found late in the development cycle, they usually lead to expensive and time-consuming design changes, which end up delaying the product release date. Some of the reliability activities used in the reactive phase are as follows:

1. Highly Accelerated Stress Screening (HASS)
2. Proof of screen (POS)
3. Reliability growth curve
4. Design maturity testing (DMT)
5. Functionally test at design margins
6. Failure reporting analysis and corrective action system (FRACAS)
7. Root cause failure analysis
8. SPC
9. Six Sigma

Further Reading

Reliability Process

H. Caruso, *An Overview of Environmental Reliability Testing*, 1996 Proceedings Annual Reliability and Maintainability Symposium, pp. 102–109, IEEE (1996).

U. Daya Perara, *Reliability of Mobile Phones*, 1995 Proceedings Annual Reliability and Maintainability Symposium, pp. 33–38, IEEE (1995).

W. F. Ellis, H. L. Kalter, C. H. Stapper, *Design for Reliability, Testability and Manufacturability of Memory Chips*, 1993 Proceedings Annual Reliability and Maintainability Symposium, pp. 311–319, IEEE (1993).

Evans, J.W., Evanss, J.Y., and Kil Yu, B. (1997). Designing and building-in reliability in advanced microelectronic assemblies and structures. *IEEE Transactions on Components, Packaging, and Manufacturing Technology, Part A* 20 (1): 38–45.

S. W. Foo, W. L. Lien, M. Xie, E. van Geest, *Reliability by Design a Tool to Reduce Time-To-Market*, Engineering Management Conference, IEEE 251–256 (1995).

W. Gegen, *Design For Reliability – Methodology and Cost Benefits in Design and Manufacture, The Reliability of Transportation and Distribution Equipment*, pp. 29–31 (March, 1995).

Golomski, W.A. (1995). *Reliability & Quality in Design*, 216–219, IEEE. Chicago: W. A. Golomski & Associates.

R. Green, An Overview of the British Aerospace Airbus Ltd., *Reliability Process, Safety and Reliability Engineering*, British Aerospace Airbus Ltd., IEEE, Savoy place, London WC2R OBL, UK, (1999).

D. R. Hoffman, M. Roush *Risk Mitigation of Reliability-Critical Items*, 1999 Proceedings Annual Reliability and Maintainability Symposium, IEEE, pp. 283–287.

J. Kitchin, Design for Reliability in the Alpha 21164 Microprocessor, Reliability Symposium 1996. Reliability – Investing in the Future. IEEE 34th Annual Spring. 18 April 1996.

Knowles, I. (1999). *Reliability Prediction or Reliability Assessment*. IEEE.

Leech, D.J. (1995). Proof of designed reliability. *Engineering Management Journal* 169–174.

Novacek, G. (2001). Designing for reliability, maintainability and safety. *Circuit Cellar*, January 126: 28.

S. M. Nassar, R. Barnett, *Applications and Results of Reliability and Quality Programs*, 2000 Proceedings Annual Reliability and Maintainability Symposium, IEEE (2000).

Part IV

Reliability and Quality Process for Product Development

18

Product Concept Phase

Product development begins with the concept phase. It consists of two parts, the product concept and design concept (discussed in Chapter 14). In the concept phase, a decision is made to develop a new product. During the concept phase marketing, engineering, operations, and field inputs yield product concept requirements. It does not matter if the product is a new platform or a derivative product; the process is the same. The concept phase is often conducted in a vacuum between senior engineering and marketing management. The decisions made during this time have a dramatic impact on the entire organization. It is in the concept phase that the product is defined on the basis of market needs, customer focus, product features, product cost, business fit, and product architecture. This may seem like a strange time to begin activities regarding product reliability because there's so little known about the actual product itself – after all, it is only a concept. No detailed design effort has started, so there's no work to be done on improving the design. The main reliability objectives in the concept phase are to form the reliability team, define the reliability process, establish product reliability requirements, and a first-pass risk assessment to conduct a Pareto study on previous reliability problems. These top reliability problems become design constraints for the design concept phase. A summary list of the reliability activities performed in the product concept phase along with the expected deliverables is shown in Table 18.1.

18.1 Reliability Activities in the Product Concept Phase

During the concept phase, design decisions are made which may require new technologies, materials, and processes. The decisions made here can impose significant risk to the design, manufacturability and rampability, testability, serviceability, and product reliability. The product concept phase represents the first opportunity to identify significant risk, which can jeopardize the success of a program. The risk issues impact the entire organization. These issues need to be identified, agreed upon, and a risk resolution strategy (identify, communicate, and mitigate [ICM] plan) laid out. If the risk issues can be identified and agreed upon early in the program, the organization will have the needed time to plan and mitigate all significant risk before first customer ship. The product will be more reliable if the risk issues are resolved prior to product release.

Improving Product Reliability and Software Quality: Strategies, Tools, Process and Implementation,
Second Edition. Mark A. Levin, Ted T. Kalal and Jonathan Rodin.
© 2019 John Wiley & Sons Ltd. Published 2019 by John Wiley & Sons Ltd.

Table 18.1 Product concept phase reliability activities.

Participants	Product concept phase	
	Reliability activities	Deliverables
• Marketing • Design engineering • Reliability engineering • Field support/service	1. Form reliability organization and responsibility. 2. Define the reliability process. 3. Define product reliability requirements. 4. Capture and apply external lessons learned. 5. Develop risk mitigation form and have meeting to review each risk issue.	1. Reliability team formed with agreement on the reliability activities. 2. Description of the reliability activities that will be performed. 3. Product level MTBF, mean time to repair (MTTR), availability defined. 4. Pareto top VOC reliability issues and recommendations. 5. Completed risk mitigation form and meeting results in acceptance of risk issues and planned mitigation.

There are five major reliability activities that take place in the product concept phase. The five activities are the following:

1. Establish the reliability organization.
2. Define the reliability process.
3. Define product reliability requirements.
4. Capture and apply lessons learned.
5. Mitigate risk.

18.2   Establish the Reliability Organization

Every reliability effort requires staff to implement the reliability process. The reliability staffing may be small, and in many cases a single reliability engineer will suffice. There is no set rule as to how large the reliability team needs to be for success. Staffing the reliability team at 1% of the design team size is a good starting point. (The topic of staffing for reliability is covered in Chapter 11. Chapter 8 also addresses some of the problems with selecting the wrong individuals.)

In forming the reliability team, selecting an individual who has strong leadership skills is the key. If the reliability program has been established, then selecting an individual with management or engineering skills will suffice. Small organizations may have only a single engineer supporting all the reliability activities. If you have a small organization, then select someone with good management skills.

Finally, the reliability workload is not constant in each phase of the product life cycle. The workload is small in the concept phase and reaches a peak in the design phase, when accelerated life and Highly Accelerated Life Testing (HALT) consume significant resources. In a high-quality organization, the reliability effort can transition to a quality effort in the design validation phase. Having quality and reliability teams engaged jointly during design validation ensures shared ownership, communication, and cooperation between the two functions. Having the two teams report to the same manager will ensure that there are no barriers formed between the two groups.

The reliability design members should have a strong technical background with past experience in product design. The quality team's members should have a strong technical background in manufacturing, process control, and failure reporting, analysis, and corrective action system (FRACAS). Having separate teams is preferred because it is rare to find individuals with strong experience in both design and manufacturing.

18.3   Define the Reliability Process

The reliability process is outlined in Chapter 12 and presented in detail in Part IV of the book. Use the detailed reliability process and tailor it to suit your particular needs. The goal should be to define the process up front, schedule it into product development, and then get everyone to agree to follow it. Once the process is defined, plan on conducting training in the concept phase to educate everyone on the process.

18.4   Define the Product Reliability Requirements

An integral reliability activity in the product concept phase is setting the product reliability requirement. It should be market-driven and focused around the targeted customer requirements. It is usually described in terms of mean time between failure (MTBF), availability, serviceability, and maintainability requirements. In setting the system-level reliability requirements, consider the previous product's reliability performance and how they compare with key competitors (benchmarking).

When defining a system reliability requirement ask:

- Has market share been lost due to unacceptable product reliability?
- What increase in market share is expected if the reliability of the product is increased?
- What improvement in profit margins can be expected with improved product reliability?
- What is the customer expectation for the reliability of this product?
- What is the customer willing to pay for improved product reliability?

Use the answers to better describe reliability goals and objectives.

18.5   Capture and Apply Lessons Learned

It is in the product concept phase that we take the time to reflect on the reliability problems from past programs. It is hard to understand why companies continually repeat the same mistakes. Reliability problems continually reappear, even though it is relatively easy to apply lessons learned to new designs. Large companies with divisions scattered around the globe have difficulty communicating lessons learned. The larger the company, the greater the problem. As a result, the mistakes made in past programs get repeated until they get high visibility, often through extreme customer dissatisfaction or significant financial exposure. By the time this occurs, years have passed, resulting in a logistical nightmare to fix the problem. Without a formalized process to capture, train, and apply lessons learned, past reliability problems will continually be repeated. These lessons learned should be incorporated into a design for reliability (DFR) guideline. A design review checklist is another effective way to verify that lessons learned are getting implemented. The solution needs to be part of a formal process to capture and implement lessons learned.

It is the external lessons learned that are most relevant in the product concept phase. (The internal lessons learned will be captured and applied in the design concept phase of the product development cycle.)

There are four areas to focus on when capturing the external lessons learned. They are as follows:

1. Conduct an external voice of the customer (VOC). The customers in this case are the end users, product support groups, and individuals who service and repair the product.
2. Review the FRACAS reports, then Pareto the field failures.
3. Identify past product recalls and safety warnings.
4. Review the customer complaints file.

Conduct an external VOC regarding reliability issues on previous products. Focus the VOC around product reliability, serviceability, and maintenance issues. This can be done through in-person interviews, phone surveys, mail questionnaires, or the internet. Some methods will be more effective than others, and experimentation may be necessary to determine the best method. It may be a good idea to use a third party for this activity because internal eyes and ears may not be sensitive to every complaint. Whichever method you choose, keep the survey unbiased and the questionnaires the same for all who are surveyed. A weighting mechanism can be used to differentiate the importance of issues to the customer, so a meaningful priority can be associated with each issue. Test the questionnaire in advance (internally and externally) to ensure there's no confusion regarding the language, weighting scale, and intent of each question. Once the data is collected, tabulate the results to determine what reliability issues must be included in the concept phase documentation.

Another source for capturing lessons learned is from the FRACAS database. FRACAS tracks field failures, ranks them by severity of occurrence, documents root cause, and tracks resolution effectiveness. It is important to know the root cause associated with each failure. If the root cause has not been determined, then it may be necessary to launch an activity to determine the root cause prior to completing the design phase. If the root cause cannot be determined, then there is no assurance that the problem will not resurface in the next product.

One big challenge that every organization faces is determining the root cause of a failure. One reason for this is that many organizations lack the technical skills and expertise to link complex failures with the root cause. Getting to the root cause requires an understanding of the physics behind the failure. Once the root cause of a failure is identified, correcting the problem becomes easy. If the root cause isn't identified, the fix can end up correcting a nonexistent problem, and the problem will resurface in the future.

One technique used to determine the root cause of a failure is the "5 Whys" (discussed previously in Chapter 10, Section 10.2). To illustrate this point, consider a standard flashlight that does not illuminate when it is turned on. The problem is traced back to the light bulb not getting the required battery voltage. The problem is further traced to the battery voltage not making good electrical contact with the spring. At this point, it is hard to determine what the root cause of the problem is. The problem could be corrosion on the spring, an incorrect or weak spring, the spring not being installed properly, the spring material is too resistive, contamination on the battery contacts, or an intermittent battery.

So we ask the question a second time, "Why is there poor contact between the battery and the spring?" The answer to this question turns out to be a loose spring. There are many reasons why the spring could be loose: for example, it could be due to a bad spring lot, the wrong spring installed, poor installation of the spring, the clip that holds the spring could be bad, and so on.

So we ask the question the third time, "Why is the spring loose?" This time we find out that the wrong spring was installed. At this point we may seem to be at the root cause of the problem, but we do not know why the wrong spring was installed. Some of the reasons why the wrong spring was installed could be: the bill of materials was wrong, the wrong spring was pulled from the stockroom, the supplier mislabeled the springs, or the wrong springs were ordered. Because there are still unanswered questions, we have not yet reached the root cause of the problem.

So we asked the question a fourth time, "Why was the wrong spring installed?" This time we find out that purchasing ordered the wrong spring. As you continue to peel away each layer of the onion, you get closer to the root cause of the problem. However, until you understand why purchasing ordered the wrong spring, you will not have reached the root cause the problem.

So we asked the question one final time, "Why did purchasing order the wrong spring?" This time you find out that the spring needed for the assembly was unavailable and that a distributor recommended that an equivalent spring be purchased. The spring turned out *not* to be an equivalent. Now that you've reached the root cause of the problem, you can begin to explore solutions that will permanently correct the problem. For example, engineering must approve all part substitutions and manufacturing must test all alternative parts to verify compliance prior to using the alternative component. Since the purchasing agent was told that this spring was an equivalent part, it is possible that engineering could make the same mistake. To prevent this from happening in the future, all substitute parts must be verified in the system before they can be used in the production line. You can also use HALT to evaluate substitute part acceptability. (In tight situations, you may still have to order the alternative component in order to meet production schedules, but the evaluation of the alternative component must be done prior to placing it on the production line.)

Finally, identify past customer complaints, safety warnings, and product recalls along with the root causes associated with each problem. Pareto the list on the basis of severity and occurrence.

Combine all the lessons learned from past problems into a Pareto of severity. This list will be used in the design concept phase to describe how past problems will not be repeated in the new product.

18.6   Mitigate Risk

In Chapter 5, we presented a process to manage the risk issues inherent in product development. Technology, component, process, supplier, and safety are all areas in which there can be significant reliability risks. If these issues are left unresolved, they will probably surface later as reliability problems found in manufacturing, test, or by the customer. The best way to eliminate this problem is through risk mitigation. A risk mitigation program not only removes reliability issues in design, it also prevents delays in product development associated with risk issues that require redesign late in the development cycle.

In Chapter 5, we also showed how to capture and document the risk issues. All of the functional groups are responsible for developing risk mitigation plans. The initial risk mitigation plan is developed at the product concept phase. The mitigation plan is updated in each phase of the product development cycle to reflect the risk mitigation progress and add new risk issues that were discovered. Before the end of the product concept phase, a meeting is scheduled for all the functional groups to present their risk issues. The meeting is a formalized event (i.e. *risk mitigation meeting*) that is part of the product development process.

The risk mitigation meeting is designed to review the significant risk issues and agree on the risk issues, costs, and resources to mitigate these issues; and that there is sufficient time to mitigate before shipping customer units. The development cost at the product concept phase is small and increases significantly in the design phase. The risk mitigation meeting is a useful way for management to determine if a program should proceed with funding to the next phase or be canceled because risk issues are unlikely to be resolved in time to meet the product market window. How many products, from experience, were canceled before being completed? Were they canceled because the product was too far behind schedule, it was too costly, or technical issues were not resolved? Where in the product development cycle were these projects canceled? Usually, it is late in the development program after the development budget has been exceeded and a litany of design issues still need to be resolved. The programs from which there is little hope of resolving critical risk issues are identified early in the program and terminated well before excessive amounts of capital and personnel resources are wasted. If the program is vital to the success of the business, then resources can be allotted to determine the cause of the problem and recommendations made to remedy the situation.

Plan the risk mitigation meeting toward the end of the product concept phase. If there are only two groups involved in product concept, marketing, and engineering, then the meeting will be short. However, as we have shown in Table 18.1, it is a good idea to include as many of the functional groups as possible. In particular, reliability, customer service/customer support, manufacturing, and test should be included in the concept phase risk mitigation meeting. At least a week before the risk mitigation meeting, a risk mitigation package should be put together containing the risk issues identified by the participating functions. The meeting will run smoother if the information is circulated with sufficient time for each organization to review the issues, understand the problems, and discuss them before the mitigation meeting. The package should include all the material that will be presented in the meeting. A sample product concept phase risk mitigation form is shown in Figure 18.1. By starting risk mitigation in the product concept phase, you significantly improve the likelihood that the product will be developed on time and within budget.

Date: 07/01/2001

Product name:

Owner: Mr. Jones

Reliability ICM – product concept phase

Item no.	Identify & analyze risk	Risk severity	Date risk identified	Risk accepted Y/N	High level mitigation plan	Resources required	Comp date	Success metric	Investigate alternative solutions?
	Investigate		Communicate				Mitigate		
1									
2									
3									
4									
5									
6									
7									
8									

Risk mitigation sign-off: _____

Figure 18.1 Product concept phase risk mitigation form.

18.6.1 Filling Out the Risk Mitigation Form

The risk mitigation form has nine parts. We have tried simpler versions of the form but always end up going back to this greater level of detail. Details on filling out the form follow.

18.6.1.1 Identify and Analyze Risk

There are two parts to investigating risk: identifying the risk and analyzing the risk. The first part, identifying the risk issues, is usually the hardest. Determining the risk issues this early in the program can be a real challenge. The product concept phase risk issues will be technology and reliability related. Since the design has not started, the risk issues may be few and not detailed. The following questions can help you identify product concept risk issues:

Places to look for *technology* risk issues:

- Are new technologies required?
- Are there new technologies at the bleeding edge, state of the art, or leading edge of technology?
- Are new processes (manufacturing, rework, and test) required for this technology?
- Is there little or no information published regarding these technologies?
- Is there only one supplier for this new technology?
- Is the new technology not available yet commercially?
- Any issues critical to program success?

Places to look for *reliability* risk issues:

- FRACAS report and Pareto of lessons learned from past programs
- The results from the external VOC
- File of past product recalls and safety warnings

Once the risk issues are identified, they are evaluated and managed through closure. Identifying the risk answers the question, "What are the critical risk issues, and why are they critical?" In analysis, we answer the question, "What needs to be done to eliminate the risk?" Things to consider when analyzing the risks are as follows:

- Are there special skills, equipment, or resources required that either do not exist within the company or are unavailable?
- What testing is required?
- What is the impact to the program?

18.6.1.2 Risk Severity

It is vital to associate with each risk a level of severity. This can help in managing the vital resources to mitigate the most severe and critical risk issues first. To differentiate severity, use a numbered scale, for example, 1–10. Ten represents the most severe risk and one is the least severe risk (see Figure 18.2). The scale is further differentiated into bands of high, medium, and low risk. The scale can be color-coded to give greater visibility to higher risk. High risks are red, medium risks are yellow, and low risks are green.

18.6.1.3 Date Risk Is Identified

Documenting when risk issues are identified is useful while reflecting back to determine how to improve the process. It is important to know if critical risk issues are being identified late in the product development cycle. Keeping track of these dates is also useful in monitoring your success to identify risks earlier in subsequent programs.

18.6.1.4 Risk Accepted

Each risk issue is reviewed at the risk mitigation meeting. A decision is made to accept the risk or reject the risk issue. Use this block to keep track of the risk issues that will be mitigated. A follow-up meeting may be needed if there are risk issues that are not accepted.

18.6.1.5 High-Level Mitigation Plan

Detail the activities that will take place to mitigate the risk. Include any outside contract or service provider work that will take place. The high-level plan should have sufficient detail to show progress.

18.6.1.6 Resources Required

Identify at a high level of detail the resources required to mitigate the risk. The resources should include manpower (i.e. 10 months, 2 people) and capital resources (include equipment, testing and evaluation services, consultants, and other associated R&D costs). Once the risk issues are accepted, the high-level resources are to be inserted into the work breakdown schedule and the departmental capital budget forecast.

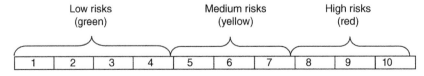

Figure 18.2 Risk severity scale.

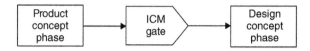

Figure 18.3 ICM sign-off required before proceeding to design concept.

18.6.1.7 Completion Date

Enter the expected completion date to mitigate the risk. The completion date must support the first customer ship date. Note whether this date has slipped, especially if it has slipped several times. This is a strong indication that the risk is high and/or the resources to mitigate the risks are low.

18.6.1.8 Success Metric

The success metric is one of the most important and often overlooked factors in mitigating risk. All too often, risk mitigation activities are launched without clear direction of what results are needed for success. This can often lead to either doing much more than what is needed or doing the wrong activities to mitigate risk. By having clearly defined success metrics, the team that is launched to mitigate the risk will be focused on those activities that support the success metric. By getting the other functional groups to agree on the success metrics up front, there will be no disconnects discovered later in the product development.

18.6.1.9 Investigate Alternative Solutions

For any risk on the critical path that is vital to program success, or if the risk is between a level 8 and 10 (red high level), there should be a contingency plan in place to mitigate the risk. Often, the contingency plan can have a significant impact on the product design or product delivery date. Because of this, it is important to include a date when a decision needs to be made to launch the contingency plan activities.

18.6.2 Risk Mitigation Meeting

The last activity that takes place before proceeding to the design concept phase is the risk mitigation meeting where all the risk issues are reviewed. In the risk mitigation meeting, each group presents its risk issues, with the objective of achieving the following results:

1. Communication of all significant technology risk issues
2. Agreement on the severity of the risk issues
3. Agreement on the strategy to mitigate the risk issues
4. Agreement on the metric for successful mitigation of risk
5. Agreement on the resources required and availability of the resources to mitigate risk
6. Agreement on the timeline required to mitigate risk
7. Agreement on the need for the risk and/or the need to pursue an alternative solution

The risk mitigation meeting requires sign-off on every risk issue. By requiring sign-off on every issue, the ICM process acts like a project gate as to where the program will proceed, only if the risk issues are manageable (Figure 18.3). Project funding in the form of capital and staffing for the development of the next phase is contingent on obtaining sign-off by the senior management on the risk issues.

19

Design Concept Phase

Previously, in the product concept phase, the product requirements are defined on the basis of market-driven product features – cost, forecasted demand, target customers, and business fit. This is where we produce a set of design requirements, and possibly, high-level system architecture. Once these product requirements have been defined, a design concept must be developed to meet these needs. The design concept phase uses the product requirements to develop lower-level design architecture. Upon completion of the design concept phase, the specifications for the outline dimensions, weight, input and output (I/O), power, cooling, and so on is determined.

19.1 Reliability Activities in the Design Concept Phase

Decisions made in the design concept designate what type of components, materials, and technologies are required to design the product. These decisions have a significant impact on the product cost, development time, design complexity, manufacturability, testability, serviceability, and reliability of the product. At the completion of the design concept phase, about half of the product cost is defined (see Figure 19.1). Product cost is a significant factor in profitability. This is where the design concept phase is used to ensure that the product cost goals are obtainable.

Five reliability activities take place in the design concept phase:

1. Set reliability requirements and budgets for subsystems and board assemblies.
2. Define reliability design guidelines.
3. Revise risk mitigation.
4. Schedule reliability activities and capital budgets (the reliability activities are included in program development schedule).
5. Decide risk mitigation sign-off day.

These steps are provided in more detail in Table 19.1. Following these steps, the team should reflect on what worked (and what didn't).

The product concept team is still typically small, consisting of key design person-nel from marketing, engineering, and reliability. The development cost and staffing resources expended to this point are also relatively small. Once the product is out of the concept phase and in the design phase, the staffing and capital resources required increase considerably. Because there is a serious commitment of staff and capital resources required in the design phase, there must be a high level of confidence that the

Improving Product Reliability and Software Quality: Strategies, Tools, Process and Implementation,
Second Edition. Mark A. Levin, Ted T. Kalal and Jonathan Rodin.

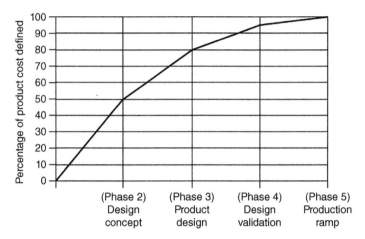

Figure 19.1 Opportunity to affect product cost.

Table 19.1 Design concept phase reliability activities.

	Design concept phase	
Participants	**Reliability activity**	**Deliverable**
• Reliability engineering • Marketing • Design engineering (electrical, mechanical, software, thermal, etc.) • Manufacture engineering • Test engineering • Field service/customer support • Purchasing/supply management • Safety and regulatory personnel	1. Define lower-level reliability design goals. 2. Define reliability design rules. 3. Revise risk mitigation. Include internal reliability VOC (manufacturing and test) and new technology issues. 4. Reliability budget (capital and personnel) and reliability activities are included in program development scheduled. 5. Review status and agree on each risk issue and mitigation plans. Risk mitigation meeting.	1. Subsystem and board-level reliability budgets (MTBF), service and repair requirements (MTTR and availability), and product useful life and use environments. 2. Identify guideline requirements for DFR, DFM, DFT, DFS, etc. 3. Pareto top reliability issues with recommendations. Risk mitigation plan updated with changes. 4. Reliability expenses are budgeted and reliability activities (risk mitigation, FMEA, HALT, HASS) scheduled into project timeline. 5. Risk mitigation meeting and agreement to proceed to next phase.

Note: DFM: design for manufacturing; DFR: design for reliability; DFS: design for service (and maintainability); DFT: design for test; FMEA: failure modes and effects analysis; HALT: Highly Accelerated Life Test; HASS: Highly Accelerated Stress Screens; MTBF: mean time between failures; MTTR: mean time to repair; VOC: voice of customer.

development program will be successful. The reliability activities in the design concept phase primarily focus on capturing the risk issues that can sidetrack a development program. By capturing these issues and developing mitigation plans, an assessment can be made at the end of the product concept phase to either proceed with the product design or stall its progress until key risk issues can be resolved.

19.2 Set Reliability Requirements and Budgets

After the product has transitioned from a product concept to a design concept, guidance will have to be provided by the reliability team detailing the reliability design requirements. These requirements can include the following:

1. Product use environment
2. Product useful life
3. Subsystem and printed circuit board assembly (PCBA) reliability budgets
4. Service and repair

19.2.1 Requirements for Product Use Environment

The customer use environment is defined either for the typical customer or for the extremes of customer use. Setting the requirements for the typical customer will optimize the design for product cost. This can, however, result in higher customer failure rates for those customers operating at the environmental extreme. Designing for the extremes of customer use will result in a more complex and costlier product that will, however, be more reliable (this is evident in high reliability military products). The product may have a larger potential market but will sell for a premium. This higher product price will impact sales. It is best to design the product for this optimum customer base, often referred to as the market "sweet spot." Once the environment is decided, these requirements become the product's environmental specifications.

Environmental requirements include both operational and nonoperational specification. The nonoperational requirements consider storage and transportation. Some shipping and transportation environmental concerns are as follows:

1. If the product is shipped overseas, will it be exposed to salt air, and for how long? For what length of time will the product remain on a loading dock? Will it be exposed to rain, snow, sleet, or dust? Air transportation can avoid many of these problems.
2. What type of vibration shock levels and vibration frequency spectrums will the product be exposed to? Studying the shipping environment to design a proper shipping container/method can avoid the "dead on arrivals" or so-called *out-of-box failure* complaints by customers.
3. What environments will the product be stored in? Will it be shipped to a desert area where cargo temperatures can be excessive? Will it be stored in a humid climate? How long can it be stored in any of these environments?

The operational limits are defined for the environments where the product is expected to normally operate. Operational limits can include operating temperature range, vibration range and frequency spectrum, duty cycles, line voltage/frequency variation, drop

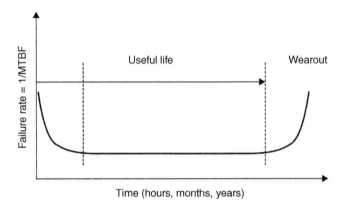

Figure 19.2 The bathtub curve.

or shock test, moisture exposure or water immersion test, corrosive environments, and so on. Once defined, test plans can be developed to validate the design.

19.2.2 Product Useful Life Requirements

It is important to define what the useful life requirements are for the product. The useful life is the period of time when the failure rate turns from a random event to a predictable event based on normal wear (see Figure 19.2). Most product developments proceed without any direction regarding the required useful life of the product. If it is not defined up front, there will be different perceptions about the useful life requirements among the design team. A common example of wear out would be a light bulb with a 2,000-hour life. The 2,000 hours defines the useful life of the light bulb. It can be easily repeated and if a test was performed on a large enough sample, similar results would be achieved each time.

Useful life design requirements are important because they impact design cost, architecture, and complexity. Setting the requirements too stringently will make the product more costly and prolong design development. These requirements are also needed to properly select components in design. The requirements are market-driven, supported by market research, and specific to the product being developed. The useful life requirements can be designed to meet the sweet spot in the customer market, or for the extremes of expected customer use.

The useful life of a product is typically defined by some amount of customer use time. The product development team may define the system operating useful life in years, that is, seven years.

The useful life can be defined in different ways for the same product. Some ways to define useful life requirements are

- Minimum number of failure-free mating cycles (connectors, removable accessories, etc.)
- Minimum number of on/off cycles (relays, hard drives, lubricated bearings, etc.)
- Minimum number of running hours (fans and motors, pumps, seals, filters, etc.)

These other requirements correlate back to the system requirements. There is consistency between the different useful life requirements even though they are defined differently. Consider a product with a seven-year useful life. The product has an accessory that is removable and marketing research expects the accessory to be removed five times a week. The useful life for this accessory can have a minimum number of failure-free mating cycles defined for it. In this case, it would be:

$$\text{Cycles (min)} = (7 \text{ years}) \times (52 \text{ weeks/year}) \times (5 \text{ cycles/week}) = 1{,}820 \text{ cycles}$$

Once these requirements are defined, the next task is to verify down to the component level that every part complies with the useful life requirement. This may appear to be an overwhelming task, but it is not. Many of the components will be grouped and evaluated by their device and component package type. For example, all carbon composite surface mount resistors can be grouped by package type (i.e. 0805, 0603, 0402, etc.). Some component manufacturers will specify the useful life. The useful life of a part is often a function of its environment and use conditions. An electrolytic capacitor manufacturer, for example, will specify the maximum useful life of their capacitor at a particular temperature, that is, 2,000 hours at 105 °C. The temperature that the product operates may be only 55 °C. This will increase the expected useful life of the capacitor. For many electrolytic capacitors, there is a rule of thumb which states that the useful life increases twofold for every 10°C temperature decrease. Therefore, we should expect the capacitor's useful life to extend to 64,000 hours.

Those components that will wear out before the useful life of the product occurs need to be designed for maintenance and service. By setting requirements upfront, there is a process to capture and address those components that do not meet the customer-expected life requirements. Designing the parts that will wear out before the end of the product life to be *easily replaceable* will help reduce service cost and improve customer satisfaction.

19.2.3 Subsystem and Printed Circuit Board Assembly (PCBA) Reliability Budgets

In the previous phase, system reliability requirements were defined. For example, the system needs to have a 10 000 MTBF (mean time between failures). In the design concept phase, the system reliability budget is tiered down to the subsystem and circuit board level. Reliability budgets are not defined to the component level for several reasons. First, in the design concept phase, the components required to design the product are only partially known. Second, and more importantly, one of the purposes of the reliability budget is to provide guidance to the design team as they begin to design the subsystems, circuit boards, and interfaces. The reliability budgets are used to identify reliability risk issues. Because resources and time are limited, the reliability budgets can steer the reliability activity in a direction where it will most benefit the product. The reliability budgets can also provide guidance regarding which component is more appropriate to achieve the reliability goal and for determining the need for redundancy to improve reliability.

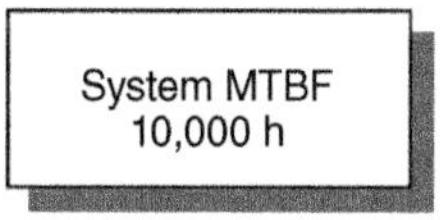

The system reliability requirements were defined in the product concept phase; an example of a system reliability budget is shown in Figure 19.3. The system reliability budget can also include requirements for mean repair time, serviceability, availability, and product life requirements. Once the architecture is

System MTBF
10,000 h

Figure 19.3 System MTBF requirement.

defined, assign reliability budgets for all the subsystems, circuit boards, and mechanical assemblies that make up the system or product. The sum of all the subsystem reliability budgets must be equal to the system reliability defined earlier.

If the reliability is defined in MTBF (in hours), then add up the reciprocal of each MTBF. The MTBF budget numbers for each of the subsystems cannot be added directly. If the reliability for the subsystems is defined in FITs (failures in time, per billion hours), then they can be directly added to determine the equivalent system FIT rate. FIT numbers can be converted to MTBF by taking the reciprocal of the FIT number and multiplying it by one billion to arrive at MTBF numbers in hours.

$$MTBF = \left(\frac{1}{FIT}\right) \times (1 \times 10^9)$$

An example of how the subsystem MTBF budgets can be added together to equal the system reliability budget in shown Figure 19.4. Suppose that you are working on a new product development whose system reliability budget (stated in MTBF) was defined in the product concept phase to be 10,000 hours. In the design concept phase, the system architecture shows the system to consist of five major subsystems. The reliability organization is then tasked to define the MTBF budgets for the five major subsystems. The MTBF budgets represent reliability's best guess at what the MTBF requirements need to be for the different subsystems. If the widget is a completely new product, then the reliability budgets may be nothing more than an educated guess for each of the different subsystems. However, if the new product has many similarities to previous products, there is a database of knowledge regarding the previous product's reliability, broken down into their subsystems to the board level. A failure reporting, analysis, and corrective action system (FRACAS) will contain information regarding the previous product's reliability, which can be broken down into its system and subsystem components. To learn more about MTBF refer to Chapter 6 and Appendix B.

System MTBF

$$= \left(\frac{1}{\left(\frac{1}{MTBF_1}\right) + \left(\frac{1}{MTBF_2}\right) + \left(\frac{1}{MTBF_3}\right) + \left(\frac{1}{MTBF_4}\right) + \left(\frac{1}{MTBF_5}\right)}\right)$$

$$= \left(\frac{1}{\left(\frac{1}{17,000}\right) + \left(\frac{1}{28,000}\right) + \left(\frac{1}{350,000}\right) + \left(\frac{1}{940,000}\right) + \left(\frac{1}{650,000}\right)}\right)$$

$$= 10,000$$

How does one determine the budget numbers? The best way is to review the past performance of similar subassemblies, the key contributors to this performance, and the expected result if these issues were removed in the new product.

The first time you set the reliability budget (for each subsystem) will always be the hardest. With each new program it will get easier. This is because you can leverage the lessons learned from the previous programs and have a better idea as to what the reliability budgets should be for the subsystem and circuit board levels. Appendix B discusses MTBF calculations in more rigorous detail.

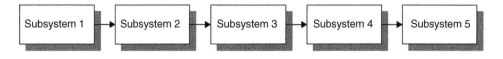

The above system is comprised of five subsystems

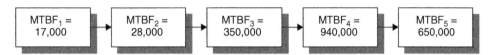

Figure 19.4 Subsystem MTBF requirement.

19.2.4 Service and Repair Requirements

Service and repair requirements can be as simple as the maximum time allowed to replace a faulty device. It can also include statements regarding the modularity of devices that are expected to be replaced during normal use. Can special tooling or equipment be used during replacement? It is a good idea to design out the need for special tooling and equipment that is required for normal service. Finally, will there be any requirements to prevent the incorrect installation or alignment of replacement parts?

19.3 Define Reliability Design Guidelines

The reliability staff is tasked to ensure that the new design will be reliable. However, reliability engineers do not design products. Designers do. Therefore, it is important that the design team have a set of design rules that they can follow for making the right decisions for a reliable product. This can be achieved through a set of design rules for product reliability. These rules are referred to as the design for reliability (DFR) guidelines for product design. Adhering to the DFR guidelines should be one of the requirements agreed to in the design concept phase. The DFR guidelines help the designer to make the right decisions. These guidelines are broad-based and cover all aspects of the product design. They must incorporate all of the reliability lessons learned from previous programs. The DFR guidelines should be part of a design checklist or other similar means used to ensure that the design is ready for manufacturing. The reliability design rules change over time because of new problems, new technologies, and new capabilities. The best way to communicate these changes to the design team is by periodically providing DFR classes to the organization. Provide training sessions as an opportunity to discuss new reliability guidelines, changes to previous guidelines, and to solicit feedback about how these guidelines will impact design engineering. Use the training as an opportunity to find out if there are gaps in the DFR guidelines or if reliability design issues are not being applied.

The DFR guidelines should be easily accessible to the design team. One way to achieve this is by placing the guidelines on the company intranet where access is readily available. Placing the guidelines on the intranet empowers the design team to make the right decisions without reliability involvement on every problem. Each reliability guideline should state the problem clearly and specifically. The DFR guidelines should not be so general that they cannot be applied in design. The guidelines should state the conditions in which reliability is a concern. Each DFR guideline should define the impact it has on product reliability. Finally, each guideline should call out ways to improve the reliability via methods like derating or reducing the operating temperature, and so on. Each DFR guideline should include suggested alternatives. (It is good to provide what not to do; it is better to also provide alternative solutions.)

19.4   Revise Risk Mitigation

The identify, communicate, and mitigate (ICM) process is repeated in each phase of the product development to identify new risk issues and to report status of the risk mitigation progress from the previous phase. There was very little known in the product concept phase regarding the design of the product. Because of that, the risk issues identified were mostly customer-focused. In the design concept phase, details begin to emerge regarding what the structure of this design will look like. As more details begin to develop about the design of the product, new risk issues will emerge. Therefore, the risk mitigation process is repeated in the design concept phase.

These new risk issues, if unresolved, will cause significant delays in product introduction. The delay results because risk issues, if unaddressed, will surface later in the program, which will require resolution. The later in the program the risk issues are addressed, the more significant the impact they will have on the program. Finding major problems in prototype is costly and delays product launch due to the time required redesigning and implementing fixes. Ideally, risk issues are identified and mitigated before design validation.

19.4.1   Identifying Risk Issues

Risk mitigation requires a 180° approach. Risk issues can be found by reflecting back on past problems and looking ahead to see what new issues pose significant risk (Figure 19.5). In the product concept phase, we reflected back on the external lessons learned. In that phase we captured the lessons learned from customer feedback, FRACAS, complaint files, product recalls, and safety warnings. Using a Pareto chart, the top issues were identified and transferred to a risk mitigation plan. These issues were then inserted into a high-level risk mitigation plan and were tracked to closure.

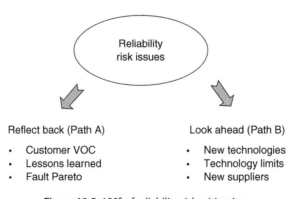

Figure 19.5 180° of reliability risk mitigation.

These risk issues are the lessons learned from the past. The process of capturing these issues, developing plans to alleviate them, and tracking the progress made on these plans is risk mitigation. Incorporating the external lessons learned into the concept requirements will help ensure that they are not repeated.

On reflecting back, the external lessons learned were captured into the product concept. In this phase we capture the internal lessons learned from manufacturing, test, design, and reliability. The internal lessons learned are added to the risk mitigation plan as new risk issues.

19.4.2 Reflecting Back (Capturing Internal Lessons Learned)

The most effective way to reflect back is through a review of the (hopefully) documented lessons learned from the past. If there is no mechanism to capture the lessons from the past, you can expect to continually repeat the same mistakes. This is especially true during a growth phase when a product is ramping up and staff is being added to meet demand. A comprehensive review of the lessons learned from the previous program includes an examination of what went wrong, what didn't work, and what worked well. This is the same activity performed in the product concept phase except that it is focused on identifying internal lessons learned. There are four areas to focus on when capturing the internal lessons learned. The four lessons learned come from:

1. Manufacturing and rework
2. Test
3. Component and supplier issues
4. Reliability

The best way to capture the lessons learned from previous programs is by documenting the issues from the past program one year after it has reached volume production. Each of the functional groups should be responsible for documenting the issues that were significant during product development, early production, and product ramp. Capturing the issues that impacted a program while in production minimizes the effect of people forgetting things over time or those who have left the company. Be specific about the impact of the problems in the documentation. The problem may have caused a low first-pass yield in test; including the yield expected and the actual yield helps to quantify the magnitude of the problem. Some of the common ways to define the yield are the parts per million (PPM) and process control charts. Establishing required PPM rates for test and process control limits for production will aid in identifying problems early in production. If your company has an intranet, then the lessons learned from the previous programs can be posted and made available for everyone.

19.4.3 Looking Forward (Capturing New Risk Issues)

There is also risk in the unknown (Figure 19.5). Looking ahead is the process used to identify new risk issues and is often based on past experience. For example, if application-specific integrated circuits (ASICs) have a history of design-related failures, then ASICs should be a risk mitigation issue. Looking forward requires identifying technology, supplier, manufacturing, and test and design issues, which pose unique and challenging problems difficult to resolve. These risk issues, if not mitigated, will probably manifest themselves later as reliability problems (Figure 19.6).

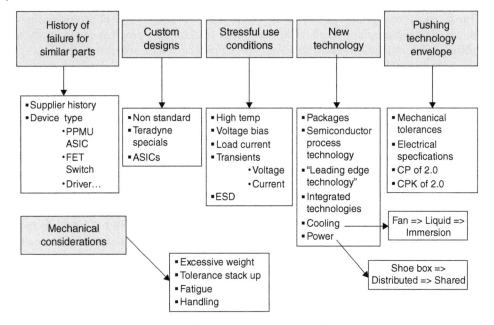

Figure 19.6 Where to look for new reliability risks.

The following questions will help in identifying new risk issues:

Places to Look for Design Risk

- Are new technologies required?
- Are new component packages used?
- Are any aspects of the design approaching or exceeding technology limits?
- Are complex ASICs or hybrids required?
- What parts are custom or nonstandard?
- Will you be approaching or exceeding the stress limits of any component, packages, or designs?
- Are there tight electrical specifications?
- Are there tight mechanical tolerances and tolerance stack-ups?
- Is excessive weight an issue?
- Are there material mismatch or incompatibility issues?
- Are there corrosion or other chemical reaction issues?
- Are there unique thermal issues?
- Are there any premature component wear out concerns?
- Is there a history of problems with similar parts or device types?
- Are there components whose electrostatic discharge (ESD) sensitivity exceeds capability? There are three ESD models to consider (human body model, machine model, and charge device model) on the basis of use conditions.
- Does the component count exceed capabilities?

Places to Look for Manufacturing Risk

- Are new processes required?

- Are new packages required?
- Are there high-pin density parts that exceed present process capabilities?
- Does the new design push the present limits of manufacturing capability?
- Does the component count exceed manufacturing capabilities?
- Does the design require high part density (parts/square inch)?
- Will there be any special handling requirements or concerns?
- Are there new components requiring small package sizes that exceed present capabilities?
- Are there new components with large packages that exceed present capabilities?
- Are there new components with heavy packages that exceed present capabilities?
- Are there new components with high defects per million opportunities (DPMO) packages that exceed present capabilities?

Places to Look for Supplier Risk

- Are there new custom parts that pose special risks?
- Do the new designs require printed circuit board (PCB) technology that exceeds present supplier limits (e.g. size, weight, lines and gaps, number of trace layers, core size, copper weight, material selection, and material compatibility)?
- Is there a history of problems with any of the proposed suppliers?
- Will new designs require supplier qualification?
- Does any of the supplier's small size pose special risk issues (supply capacity, financial stability, etc.)?
- Are there any critical components that cannot be dual-sourced?

Places to Look for Test Risk

- Are there new components or new packages that cannot be tested due to access restrictions?
- Do high-pin count/density component packages pose special test challenges?
- Does the product push the limits of testing capability?
- Is the PCB part count high enough to exceed test node capability?
- Are new component packages too small to probe?
- Are there new component packages that cannot be tested because of access?
- Are there hybrids or other parts that cannot be fully tested?
- Are there parts with high PPM rates that will result in low first-pass yield?

Use the risk mitigation tool to effectively capture risk issues and assign values identifying the magnitude of the risk. Each of the risk issues is assigned a number that represents the degree of risk severity. The risk issues with the highest severity ranking need to be resolved first. All high-risk issues should have plans for an alternative solution if they cannot be resolved before first prototype. These risk issues will have a significant impact on the ability to meet the time-to-market goals. These risk issues, if unresolved, will probably return as reliability problems once the product goes into full production.

Once the risk issues are agreed to by the team as being necessary, risk mitigation plans are put in place with timelines defining when the risk will be mitigated. Finally, the risk mitigation plans identify the metrics that will be used to determine successful mitigation of the risk issue. Defining the metrics for successful risk mitigation is often overlooked.

Too many programs expend significant amounts of time and resources toward resolving problems. The work performed to resolve the risk issues is often very good, but not focused around what is needed to mitigate the risk. In other words, a lot of testing and analysis is performed, but because it was not focused around the problem statement and success metric, the problem is not rectified. By clearly defining what is required to achieve successful risk mitigation, the problem statement is more easily resolved.

One way to ensure that all the risk issues are identified early in the design phase is to have a cross-functional risk mitigation meeting to review the design concept and identify risk concerns. The risk mitigation process begins early in the design concept phase with a kickoff meeting to discuss the design concept and to solicit feedback from the cross-functional teams. It is then the responsibility of each of the functional groups to begin documenting and tracking risk issues. Periodically through the design concept phase, the cross-functional team should meet to discuss the status of the risk issues, identify new issues, and discuss alternative solutions.

19.5 Schedule Reliability Activities and Capital Budgets

An often-overlooked activity in reliability is the upfront planning that needs to take place. Planning is crucial. By having the reliability activities scheduled into the project timeline, common arguments like "*These reliability activities will prevent us from meeting the market window,*" "*I cannot free-up the resources you need because everyone is working on items on the critical path,*" "*It is too late in the program to fix anything we find. It would have been nice to have done this earlier in the project*" are avoided. There is no excuse for not planning these activities into the project timeline. Scheduling ensures that these activities take place in the appropriate stage of the product development cycle, and not when the product is in production. You can be assured, any findings from Highly Accelerated Life Test (HALT) and failure modes and effects analysis (FMEA) after a product is in production will, in all probability, not be implemented.

Scheduling the reliability activities into the project timeline also confirms the commitment of the management to these activities, and the importance of these activities to the program. Participation is not optional. A typical FMEA can take between a half to a full week to complete. This does not include the time required to fix problems found. Scheduling a week for this activity will usually suffice. Complex systems should be broken into smaller parts so that they can be performed in a week. Remember that this is a concurrent activity that will require other functional groups to schedule this activity into their project timeline.

FMEAs should also be planned and scheduled even if an outside supplier will perform them. Subsystems and custom designed parts that are purchased by an outside supplier also require an FMEA. The FMEA should be a joint effort between the supplier and the customer. If the outside supplier is unfamiliar with an FMEA, then additional time must be allotted for training. The training can be performed all at once in the beginning or before each of the three major parts (functional block diagram, fault tree, and FMEA form). If the supplier is a significant distance away, it may not be practical for the participants to all be in the same room for the FMEA. We have had significant success using Microsoft NetMeeting® to tie in different facilities and keep everyone focused. There can be issues with computer firewalls that may need to be worked out, but the process can be performed very effectively with remote sites.

The other significant activity that has to be scheduled is HALT. Allow a week to two weeks for the HALT activity for each subassembly. A second HALT test is usually performed to verify the design fixes and to ensure that no new failure modes were designed into the product as a result of the design improvements. The HALT scheduling is critical because there are many activities that take place before the HALT test is performed. HALT requires a significant amount of up-front preparation. Before the HALT test can be performed the following things are needed:

- HALT test plan
- Stress level limitations
- Mechanical fixturing
- Input/output cabling
- Instrumentation
- Test software ready
- Product test hardware built and debugged
- HALT team identified (the team must be there during the HALT testing as the test evolves for optimum efficiency)

These activities often take place months before the actual HALT test is performed. Scheduling these activities far enough in advance will ensure that when it is time to test hardware, you will be ready. Last minute planning guarantees sputtering starts and lost time due to oversights, and so on. The HALT planning activities are covered in detail in Chapter 15.

Finally, there will be capital expenses associated with the reliability activities that should be budgeted into the program development cost. The majority of the expense is for HALT testing. The HALT test typically requires four working test devices, three of which are destructively tested. The fourth is a golden unit for troubleshooting that can be returned for other uses or sold. However, the material cost for the three boards can be significant depending on the product. In addition, there can be other costs associated with HALT, i.e. fixturing, cabling, test hardware/equipment, electric power, and liquid nitrogen used to run the chamber. If the HALT testing is performed by an outside test service, this too will need to be budgeted. HALT testing may identify a bad part that requires further failure analysis. Thus, it is wise to plan and budget for some amount of outside failure analysis work and possibly accelerated life testing.

By doing a thorough job in planning for the reliability activities, the risk of roadblocks surfacing is minimized. The planning phase will lay out all the activities that need to take place, their estimated duration and time frame when these events need to occur. There will be no surprise of its impact to the program and the ability to meet the marketing window. The resources needed to perform these activities will be budgeted into the program early on so that the funds are available when needed. If there are capital constraints, then adjustments should be made early on so that they do not affect critical assemblies.

19.6 Decide Risk Mitigation Sign-off Day

At the completion of the design concept phase, the entire team understands the program risk issues and should be working cross functionally toward risk mitigation. The majority of the program risk issues should be identified by the completion of the concept phase.

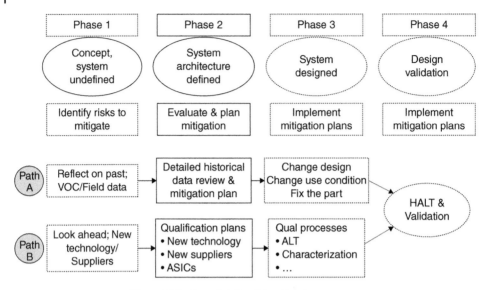

Figure 19.7 The reliability risk mitigation process.

A new synergy will form and the team will work more efficiently to eliminate the shared program risk. The team will better understand and appreciate how the decisions made in the design concept phase will affect product reliability, cost, and time to market.

Remember, toward the end of the design concept there is a planned risk mitigation meeting to review all significant risk issues. The risk mitigation meeting should occur before the design is allowed to proceed to the design concept phase. The risk mitigation meeting acts as a project gate where a decision is made to proceed to the next phase of product development (Figure 19.7). If the risks are not being adequately managed, then the project success can be at risk. By the end of the product concept phase, only a small percentage of the total program development cost has been spent. Proceeding past this point requires significant capital investment and resources. Risk management is an effective tool to determine if the program is ready to proceed further (Figure 19.8).

The structure for the risk mitigation meeting is the same as that used in the product concept phase. Attendance is required at this meeting. Prior to the meeting, a risk mitigation issues package is put together for all participants. The package includes all the material that will be presented in the meeting. The risk mitigation meeting reviews all significant risk issues to determine if they will be mitigated in advance of first customer ship. The risk mitigation meeting is scheduled toward the end of the design concept phase.

Figure 19.8 The ICM is an effective gate to determine if the project should proceed.

19.7 Reflect on What Worked Well

The last activity before the concept phase ends and the design phase begins is reflection. In the reflection step you look back at the concept phase to see what worked and what didn't. Capture the lessons learned early so that they are not forgotten and destined to be repeated. Document the findings and recommendations so that they may be incorporated into future programs. Reflection allows for continuous improvement of the reliability process.

20

Product Design Phase

20.1 Product Design Phase

Now that the concept, requirements, and architecture for the product have been completely defined, the design of the product can commence. In the product design phase, everything required to produce a working prototype is developed. At the end of this phase, there will be a complete product documentation package. This will include the schematic, theory of operation, outline drawing, bill of materials (BOM), software, assembly, and mechanical drawings. There will also be a working prototype suitable for design validation, which will be performed in the next phase of product development. The decisions made by the end of this phase will determine the product cost, design, manufacturing, test, and service complexity and will also determine how difficult it will be to ramp production. Unless the product is completely redesigned, 80–90% of the product cost is determined. Cost down efforts to reduce product cost here are usually limited to reducing material cost because redesign at this point is not cost-effective.

Product cost is emphasized because there is a cost associated with reliability. The cost equation works two ways. First, there is an added material cost associated with using higher reliability components and adding redundancy. The added cost should be evaluated against the improvement expected in reliability and how much the market is willing to pay for increased reliability. There is also a cost saving associated with designing the product right the first time. Typically, early in the product launch, there is an extensive number of engineering design changes to the product (refer to Chapter 1). Design changes at the end of the product development increase the cost to produce the product and can reduce product reliability when there is extensive rework and retrofit required.

The design phase has two parts (Figure 20.1), a product design phase and a design validation phase. The majority of the engineering design activity takes place in the design phase. The product development team has been relatively small up to this point and increases significantly to design the product, produce a working prototype, and create a documentation package. A well-designed and reliable product that is manufacturable, brings the added benefit of ensuring that the design team is released to work on the next project. If there are problems with the design, then engineering support will be needed to fix design problems. Pulling back key design resources to fix past problems negatively impacts resources needed to support future product designs. Some companies get around this by forming a sustaining engineering group. Their focus is to resolve design problems after a product has been released to manufacturing. The sustaining engineering group typically consists of less-experienced designers and engineers to fix

Improving Product Reliability and Software Quality: Strategies, Tools, Process and Implementation, Second Edition. Mark A. Levin, Ted T. Kalal and Jonathan Rodin.

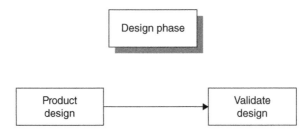

Figure 20.1 The first phase of the product life cycle.

designs in which they were not involved. When there is a sustaining group to fix problems, it is vital that their findings make it back to the design team so that improvements are put in place to prevent the problem being repeated.

There is usually a cost associated with developing a more reliable design. Warranty cost savings will often justify these early-added costs. The capital expense, engineering resources, and time required to fix design problems can be significant. When sustaining engineering activity costs are added, this fuller picture will provide a better indication of what the actual cost is of an unreliable product.

Finally, the design phase is the last opportunity where the product design can be proactively improved. Once a product enters the design validation phase, the costs and schedule impact of design changes increase significantly. The problems found in product validation are more expensive to fix and can impact the product release date. Therefore, significant focus needs to be given in the product design phase to ensure that the product design is reliable.

There are eight reliability activities that take place in the product design phase (Table 20.1).

20.2 Reliability Estimates

As the program proceeds from concept to design, the materials required to produce the product are defined. Once the BOM is known, a reliability estimate can be performed. The reliability estimates are performed for each circuit board and subsystem that makes up the system. The estimate considers each item in the BOM and matches it to a reliability number. This reliability number is often expressed in FITs (failures in time per billion hours). Each BOM item's FIT number is added together to calculate the total FIT. The sum of all the individual FITs provides an estimate of the reliability for that board or subsystem. Reliability engineering is usually responsible for determining the reliability estimates for system, subsystems, and circuit boards. Estimating the reliability is more complicated when redundancy is involved or the components do not have a constant failure rate. The mathematics to deal with this is complex and beyond the scope of the book. Those who wish to gain a better understanding of this subject are referred to the bibliography section at the end of this chapter.

Reliability estimates provide insight into the areas where the product is likely to be least reliable. These estimates also point to those BOM items that are the major contributors to lowering the reliability estimate. There will not be sufficient resources (capital and personnel) to perform all the reliability tests on everything or to life test every

Table 20.1 Reliability activities for the product design phase.

Participants	Product design phase	
	Reliability activities	Deliverables
• Reliability	1. Reliability estimates developed for all lower level assemblies. Identify all items, early wear out items.	1. Reliability estimate spreadsheet and a Pareto by component and subassembly highest failure rate items. Pareto early wearout items for service and maintenance strategies.
• Marketing		
• Design engineering (electrical, mechanical, software, thermal, etc.)	2. Implement risk mitigation plans.	2. Risk mitigation meeting and agreement to proceed to next phase.
• Manufacturing engineer	3. Apply reliability design guidelines (DFM, DFR, DFT, DFS).	3. Checklist or review that design guidelines has been followed and variances are acceptable.
• Test engineer	4. Perform design FMEA.	4. Completed FMEA spreadsheet (Table 7.1) and closure on FMEA action items.
• Field service/customer support	5. Install FRACAS.	5. Structured FRACAS database and user input interface.
• Purchasing/supply management	6. Begin HALT planning.	6. Detailed plan and schedule for HALT.
• Safety & regulation	7. Update lessons learned.	7. Updated lessons learned database, communicate to design team new issues and revise risk mitigation plan if needed.
	8. Review status and agree on each risk issue and mitigation plans. Hold risk mitigation meeting.	8. Risk mitigation meeting and agreement to proceed to next phase.

Note: DFM: design for manufacturing; DFR: design for reliability; DFT: design for test; DFS: design for service (and maintainability); FMEA: failure modes and effects analysis; FRACAS: failure reporting, analysis, and corrective action system; HALT: Highly Accelerated Life Test.

component. By knowing where the reliability problems are likely to reside, the vital resources can be focused on the areas where the product is expected to be the least reliable. The reliability estimates are one technique to ensure that there is significant reliability focus around those areas of the product that are expected to be the least reliable.

The reliability estimate, when sorted from the highest (negative) impact on reliability down to the lowest (least) impact on reliability, provides a Pareto list of components to focus reliability improve effort. The Pareto identifies the components that are most significantly impacting reliability. In these situations, you can evaluate alternative components or design changes (including strategies like redundancy) to evaluate the reliability improvement. Reliability estimates may also help in determining the need to spend more for a more reliable component.

There is significant disagreement regarding the need to do reliability estimates and the value that they provide. First, reliability estimates can be significantly different from the reliability that you realize in the field. There are many factors that contribute to this uncertainty. Reliability estimates can vary significantly from actual observed reliability based on the component or device manufacturer, use conditions, applied derating, and the accuracy of the supplier supplied reliability data. Your ability to account for this variability will lead to a more realistic reliability estimate.

However, reliability estimates take a significant amount of time and resources to calculate. It is important to determine whether value is being received from the reliability estimates. It might be necessary to do a reliability estimate for contractual reasons or because customers have come to expect it. However, if this activity does not lead to reliability improvements, it should either be discontinued or the reason it is not providing value should be determined.

20.3 Implementing Risk Mitigation Plans

The risk issues were first identified in the concept phase and high-level plans for mitigation were developed. The risk mitigation process continues in the product design phase. In this phase, detail plans for mitigating the risk are developed along with implementation. By the end of the design phase, all the details to build and test a working prototype are developed. As the system design architecture, BOM, schematics, and mechanical drawings are being developed, additional risk issues are identified and added to the risk mitigation plan. By the end of the design phase, all the risk issues should be identified with plans for mitigation. This occurs before the end of the design verification phase.

20.3.1 Mitigating Risk Issues Captured Reflecting Back

Using the 180° approach, risk mitigation issues were captured on retrospection and by looking forward. The strategies and plans for mitigating past problems are different from those looking forward. The strategies to mitigate known problems usually fall into one of four categories. The strategies to mitigate past problems are as follows:

1. Design out.
2. Change use conditions.
3. Fix part.
4. Fix process.

20.3.1.1 Design Out (or Use an Alternate Part/Supplier)

Parts that have had a history of being unreliable in design can often be designed out. Look for an alternative design solution that does not require using that component. It may be possible to eliminate the unreliable component altogether; the purpose for the component may be no longer necessary. Designs are often recycled and modified for use in the next-generation design or derivative product. Recycling designs avoids reinventing the wheel. However, with the passing of time, some of the circuitry may no longer be needed, because its function is no longer needed, used, or required. The original designer may have moved on and the new designer copies it for convenience. Design reviews and failure modes and effects analysis (FMEAs) can reveal when part of a design is no longer used or needed. Another potential risk with recycling is part obsolescence. Designing in parts that are soon to be obsolete can cause undue delay in product development. On the other hand, recycling older designs will leverage the intellectual property of the organization. This is highly desirable as long as the designs are reliable.

An example of this is a design that used a three-phase motor for a cooling pump. When the pump motor is incorrectly wired, it may actually operate for a day or two, but soon the motor windings will burn up and the cooling system will fail. Replacing the motor is costly in dollars and lost production time. Over the years, a special relay has been incorporated to circumvent this wiring error possibility. It is connected in such a way as to allow the motor to run only when properly connected. This relay has helped to get new systems up and running without wiring mishaps. It is only needed the first time the system is activated. Once it is clear that the motor operates correctly, the relay has done its job. This solution has been used for many years and is a well-established safety item.

However, in this case the relay was high on the reliability Pareto. The relay contacts and magnetic windings were both prone to failure. If the relay winding fails, the system will not work because the relay needs to be energized before the correct windings are connected. So the relay winding failure may cause unnecessary shutdowns.

With today's new three-phase motor drives, power is connected directly to the drive electronics. The drive electronics is such that it doesn't matter how the three phases are wired because internal diodes redirect the currents correctly for all wiring cases. You can no longer miswire the motor. Still, safety relays were used because of this recycling of intellectual property. Now because of the risk mitigation process, new designs that incorporate the motor drive electronics do not use a safety relay.

Over time, tried-and-true design solutions are incorporated without a thorough analysis as to its continued need. At times, as in the safety relay case, the old solution lowered the overall system's reliability.

Another reason to design out is because there are now better ways to achieve the same function, it may even require less parts. As electronics gets smaller and more integrated, it may be possible to replace the problematic electronics with new technology that has a proven track record for reliability.

20.3.1.2 Change Use Conditions

There may be situations in which a component is being used that has a history of being reliable but has been problematic in a particular design. Often, the cause is related to the use conditions. An example of this was observed with a semiconductor amplifier that the manufacturers specified as being highly reliable. The amplifier was being used in other designs without any problems. However, in this particular application, they failed at a high rate. The problem was related to an amplifier operating above its rated temperature specification. In this example, the part had a rated operating temperature of $85\,°C$, but exceeded this by $45\,°C$. The high operating temperature caused the device to fail early. Simply reducing the device temperature through improved cooling or more amplification stages running at lower power solved the problem. When problems like this are resolved, they should be added to a design review checklist or incorporated into the reliability design guidelines so they do not occur in a different design.

A significant portion of a product's reliability problems is design-related. Often the problem can be eliminated or the frequency of occurrence reduced to an acceptable level by decreasing the stresses that precipitated the problem. Changing the use conditions requires knowledge of the failure's root cause. Sometimes the problem is assumed to be because of a bad component even though the environment or the way it was used accelerated the failure. Getting to the root cause is a time-consuming effort that may require testing and analysis, often by an outside service. However, if the root cause of the failure is not known, an inaccurate diagnosis will not resolve the problem.

20.3.1.3   Fix Part

Some problems just need fixing. A good design concept that is poorly implemented needs to be fixed. This type of reliability problem is best fixed through a smarter design. The design fix can be mechanical, electrical, or both. It usually requires design modeling and later testing to prove out the design fix. The fix is often easy to implement in the design phase. The key was capturing it in risk mitigation so that resources could be allotted for redesign.

An example of this would be a part that has been unreliable and which has no better alternatives. This is common when working at the leading edge of technology. Once again, getting to the root cause of the problem is vital. This often requires working with the component manufacturer to determine if it is a design or process issue. Once the root cause is identified, suppliers will usually work with you to change the part. It is rare to encounter a problem with a component where you are the only one in the industry experiencing it.

20.3.1.4   Fix Process

Fixing the process is often required when the design is sound, but either the manufacturing or test process is insufficient. It can be an assembly, rework, or test process that causes the problem. An example of this may be seen in component lead prepping. Lead prepping can damage a part, either through electrostatic discharge (ESD) exposure, residue left from a machining process, nicks in the lead, or stress cracks to the component body. It may be a moisture-sensitive part in which proper precautions are not being observed.

There can be problems with the test process as well. Examples are: ESD degradation from the testing process, lead damage through handling, and excessive moisture exposure.

20.3.2   Mitigating Risk Issues Captured Looking Forward

Looking forward, risk issues are captured that address what is new and unknown about the design. The risk mitigation activities looking forward answer questions regarding what the impact to product quality, reliability, and performance will be. The risk issues can be for a new technology, process, material, package, supplier, component, printed circuit board (PCB), application specific integrated circuit (ASIC), hybrid, design, manufacturing, or test process, as shown in Figure 20.2. The mitigation strategies resolve questions and concerns that can significantly impact the program if unresolved. Once answers to these questions are known, a decision can be made regarding the acceptability or fit for use in the design.

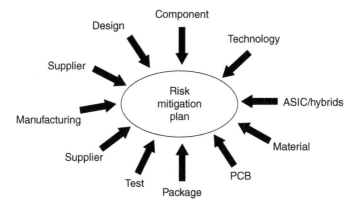

Figure 20.2 Looking forward to identify risk issues.

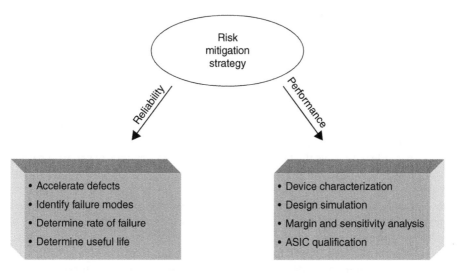

Figure 20.3 Risk mitigation strategies for reliability and performance.

Getting answers to these questions requires testing. The testing to answer these questions will vary. We will consider two types, as shown in Figure 20.3. The testing addresses the reliability- and performance-related questions. The first type of testing (performance testing) simulates various use operating conditions so device characterization, margining, and identifying the limits of use conditions can be found. This type of testing can also be time consuming and tedious. Many choose not to do device characterization. Those who don't will almost certainly regret it later on when design failures surface that could have been identified during this type of testing. The purpose of this type of test is to obtain an understanding of the device's behavior under a wide range of inputs, loads, and use conditions, identify key operating parameters, and understand how design margin may be degraded by environment. Knowing this will help prevent designing in scenarios where the product may fail to operate. This is vital information for the designer, but it can only be learned through testing.

The other type of testing (reliability testing) requires environmental stresses to stimulate defects. Environmental stresses are applied to cause (precipitate) a failure, determine a failure rate, determine the failure modes, and/or estimate useful life. Unfortunately, there is not a single test that will answer these questions. Instead, a test plan is developed on the basis of knowledge of what is expected to fail and why. Knowing this, accelerated test strategies are developed that will answer these questions. There is no universal test. This type of testing can be expensive and time consuming. If you are unsure what testing would work best, then using outside consultants is advisable. Environmental test facilities are a good source of knowledge about which tests are appropriate.

Often products are designed to have a useful life that is significantly longer than the amount of testing time that takes place in the design validation phase. It would be impractical to test the product for five years so you can verify that it will meet its specified useful life. To understand how a product or component will perform (over time) requires acceleration testing. Acceleration testing answers reliability risk issues regarding the following:

- How will it fail?
- When will it fail?
- What are the failures modes?
- When will it wear out?
- Are the failure modes accelerated by stress?

Knowing when the product fails in a more stressful environment and the root cause of the failure helps to answer these questions. Once this knowledge is known, improvements are made to remove the failure mode, reduce the frequency of occurrence, and/or extend the time it takes for the failure to occur. Many types of failure modes that are accelerated through environmental stress testing can be extrapolated to its frequency at a lower stress. The mathematics to determine the useful life based on an accelerated life test is best left to the reliability engineer. The results relate only to that particular failure mode. Most products have numerous failure modes. Therefore, several different accelerated life tests may be required to get a better understanding of the likely failure modes for a particular component, device, or design. The testing that is required to obtain this knowledge is called accelerated life testing.

20.3.2.1 Accelerated Life Testing

Accelerated life testing exposes a device to environmental stresses above what the device would normally experience, in order to shorten the time period required to make it fail. This type of stress testing quickly precipitates failures by compressing the time it takes to fail. The failures are the result of cumulative fatigue at exaggerated stress levels. For example, a paper clip that is bent open $90°$ and then back exaggerates the stress a paper clip will experience. If we repeat this process, we will accelerate the paper clip's useful life.

Accelerated life testing is performed because it is impractical to take a product designed for a 10-year useful life and then test it for 10 years to verify that it conforms to specification. Accelerated life testing compresses the time it takes for failure to occur. The results from accelerated life testing are then used to verify that the product or device will survive for its designed service life.

The mathematics behind accelerated life testing is complex and best left to the reliability engineer to perform. However, the test process is easy to conceptualize. Either mathematically or empirically, an acceleration factor (acceleration rate) is determined for the accelerated life test performed and the intended use environment. By applying the acceleration rate, the useful life can be determined.

A mixed flowing gas test can be used to determine if a device will corrode in the field. This is an example of an empirically performed accelerated life test. In this test, the product is exposed to a mixed flowing gas in a controlled environment. The gas accelerates corrosion. There are different combinations of gas mixtures and concentration levels depending on the environment being simulated. Through empirical testing, it has been determined that, two days of exposure to a mixed flowing gas relates to a year in your use environment. By knowing the product design life, it is easy to accelerate the corrosion that will occur in the device.

Suppose that a design requires a new connector that has many more contacts than anything you have used before. In addition, this connector is attached to the board using small solder balls that are smaller in size than previous designs. A possible risk issue for this connector is that it will not survive its required service life, because the connector may have solder joint failures. An accelerated life test (in this case temperature cycling) will answer this question. To mitigate this risk issue, test boards are built with the connector soldered to the board using the standard manufacturing process. The boards are then placed in a temperature chamber and temperature cycled (i.e. 0–100 °C) to accelerate solder joint failures. By measuring the resistance at the solder joint, we can determine how many temperature cycles it will take to fail. The test is then repeated at a different temperature extreme (i.e. 0–130 °C) to determine the number of cycles to failure. After the two sets of tests are performed, a mathematical acceleration rate can be calculated. The acceleration rate is then used to determine if the connector will have solder joint failures during its service life.

The above two examples illustrate how accelerated testing can be used to evaluate performance over time. There are many different types of accelerated life tests. In addition, there are many different stress levels that can be applied for a particular test. This type of testing takes time and can be costly. However, it can be costly if you design a connector intended for a seven-year service life and it fails after three. First, *expect* every product in the field with this connector to fail around this time period found in the testing. Second, the failure will not be discovered for three years, thus there can be a significant amount of product affected. If the connector failure is a safety issue, then a recall is likely to be costly.

There are many different types of accelerated life test, depending on the failures you wish to precipitate or evaluate (Table 20.3). Some common stresses used in accelerated life test are shown in Table 20.2.

These stresses can be applied singly or in combination depending on the types of failures you wish to precipitate. The key to these tests is to keep the stress levels below the threshold where the physics of the material changes. If the stress that caused this failure is well above the physical limitations of the material(s) in test, the resultant failure(s) will not represent what can actually happen during customer use. In other words, if the stress temperature melts the plastic housing of a component, then you haven't accelerated its time to failure. Instead, you have identified and exceeded the physical limitations

Table 20.2 Common accelerated life test stresses.

Typical HALT stresses	
1	Temperature
2	Vibration
3	Mechanical shock
4	Humidity
5	Pressure
6	Voltage
7	Power cycling

Table 20.3 Environmental stress tests.

Accelerated stress test	Test conditions
High Temperature Operating Life (HTOL)	Temperatures vary as a function of the device in test.
Highly Accelerated Stress Test (HAST)	130 °C, 85% RH for 100 h
Autoclave	121 °C, 100% RH
	103 kPa.
	Between 96 and 500 h
Temperature humidity bias (THB)	Typical: 85 °C &85 RH
	Between 500 and 2000 h
Temperature cycling	Varies
	500–1000 cycles typical −65–0 °C Low temp.
	100–150 °C high temp.
Temperature storage	Varies
	200 °C, 48 h
	150–1000 °C
	125–2000 °C
	175–2000 °C
Operating life	125–150 °C
	1000–2000 h
Thermal shock	500–1000 cycles typical −65–125 °C
	15 min dwells at each temperature
Random vibration	Varies greatly on the basis of user environment and product

of the component. The information is not useful in determining the components useful life, only its upper use limits.

There are many different standard accelerated life tests used in industry (Table 20.2). There can be many variations in these tests. Some of the more common accelerated life tests are described in Table 20.3:

High temperature operating life (HTOL). This is an operational or biased test where the device is kept at an elevated temperature for an extended period of time. The primary purpose of this test is to accelerate failures that are the result of a chemical reaction. Examples are interdiffusion, oxidation, and Kirkendall voiding. Lubricant dry out can also be accelerated at elevated temperature. The acceleration rate can be modeled by the Arrhenius equation. The results of this test can be used to determine useful life at a lower temperature, i.e. 65 °C.

Highly Accelerated Stress Test (HAST). This can be a biased test where the device is kept at an elevated temperature in the presence of a controlled level of humidity for an extended period of time. The test is highly accelerated by using temperatures above the boiling point of water, 100 °C. The test is performed in a pressurized environment, where the pressure can be raised above 1 atmosphere. The primary purpose of this test is to accelerate failures that are temperature- and humidity-related. Humidity can cause material degradation, corrosion of metallization, degradation of lead solderability, wire bond failures, bond pad delamination, intermetallic growth, and popcorning in plastic encapsulated components (moisture absorbed in package rapidly boils during assembly reflow and cracks the case).

Autoclave. This test is commonly referred to as a *pressure cooker* test. The device is placed in a pressurized chamber that has water stored in the bottom. The device is kept at an elevated temperature 121 °C, while suspended in saturated steam and is pressurized to 103 kPa (kilo Pascal). The concentrated steam is achieved by suspending the device at a minimum height of 1 cm above the water in the chamber. The test highly accelerates moisture penetration and galvanic corrosion.

Temperature humidity bias (THB). This is an operational or biased test where the device is kept at an elevated temperature while in the presence of a controlled level of humidity for an extended period of time. The test is not performed in a pressure environment, the temperature is kept at 85 °C and the humidity level is held at 85% relative humidity. The primary purpose of this test is to accelerate failures that are temperature- and humidity-related. Humidity can cause material degradation, degrade lead solderability, and popcorning in plastic encapsulated components (moisture absorbed in package rapidly boils during assembly reflow and cracks component's body). The acceleration rate for the time to failure can be described using the Peck model. The results of this test can be used to determine useful life at a lower temperature and humidity, that is, 65 °C and 45% relative humidity.

Temperature cycling. This is a test where the device may or may not be powered during the test. The device is cycled to a low temperature extreme and dwelled typically for at least 10–15 minutes before it is transitioned to a high temperature and dwelled again for at least 10–15 minutes. The temperature transitions between high and low temperature extremes are continually repeated. The purpose of this test is to accelerate the effects of thermal expansion and contraction to see what fatigues. This is a common test to evaluate solder joint and interconnect reliability. The acceleration rate is modeled by the Coffin–Manson equation. The results of this test can be used to determine useful life knowing that the device will see less severe temperatures and at a lower frequency of occurrence (i.e. 65 °C when in operation and 25 °C when the device is turned off).

High temperature storage. This is a nonoperational test to accelerate temperature-related defects. The primary purpose of this test is to accelerate failures that are temperature- and humidity-related. These failures are the result of a chemical reaction that accelerates at elevated temperature. Examples are interdiffusion, oxidation, and Kirkendall voiding. Lubricant dry out can also be accelerated at elevated temperatures. The acceleration rate is modeled by the Arrhenius equation. The results of this test can be used to determine useful life at a lower temperature, that is, 65 °C.

Thermal shock. This test is similar to thermal cycling except the time of transition between temperature set points is very short. The short transition time is achieved through a dual temperature chamber that can shuttle the product between the two chambers. Thermal shock can accelerate cracking and crazing of seals and encapsulated materials and hermetic package leaks.

20.3.2.2 Risk Mitigation Progress

It is important to track the progress that has been made to resolve risk issues. By the end of the design phase, the majority of the risk issues should be resolved. More importantly, the highest risk issues should be closed or nearing closure. One way to track the progress being made to mitigate risk issues is the risk mitigation growth curve, shown graphically in Figure 20.4. The risk mitigation curve illustrates the progress being made to mitigate risk. The slope of the curve indicates the rate at which new issues are being identified. When no new risk issues are surfacing, the curve will flatten out.

Not all risk issues have the same severity. The risk issues are grouped into three categories, high, medium, and low. Each risk category is plotted separately so critical risk issues can be tracked separately to closure. The high-risk issues are the most significant and priority should be placed on these over lower-risk issues. The progress made against the medium- and low-risk issues is also plotted in a reliability growth curve.

If high-risk issues are not being resolved, then proceeding to the next development phase may result in moving forward with your commitment for a program that is unlikely to succeed. The risk mitigation growth curve illustrates the progress made to mitigate risk issues.

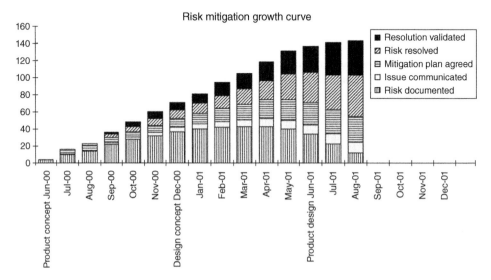

Figure 20.4 Risk growth curve shows the rate at which risk issues are identified and mitigated.

By the end of the design phase, a significant portion of the program development resources has been spent. The resources required, both capital and manpower, to complete product development is significant. By the end of the design phase, the majority of the risk issues should have been resolved. The risk mitigation growth curves indicate the progress made and status of the effort to mitigate risk in the program.

20.4   Design for Reliability Guidelines (DFR)

Product development cycles are continually being compressed. Shortening development cycles places further strains on the design team to develop a reliable product. Making matters worse, products today are designed to be smaller, lighter, faster, and cheaper. Each of these factors impacts product reliability. If the development cycles are too short, there may not be sufficient time to mitigate reliability issues and qualify new technologies, materials, suppliers, and designs before customer ship. As designers struggle to develop products in time to meet tight market windows, it is unlikely they will spend time addressing reliability issues. This problem can be resolved in part through reliability design guidelines that quickly aid designers with guidance on reliability issues. The reliability design guidelines are one of the few tools in the reliability toolbox that proactively improve product reliability. However, there are not many tools that can improve the reliability of a design before building and testing the first prototypes.

The reliability engineer can advise designers on ways to improve a design for reliability (DFR). They can identify components that are traditionally unreliable, suggest alternative design strategies that are more reliable, and evaluate the effect of derating and redundancy. However, unless there is a large reliability team, it is unlikely reliability engineering will be a part of every design decision. There are simply more designers making design decisions every day than there are reliability engineers to review these decisions. In fact, it would be unrealistic to have a reliability review for every aspect of a DFR. Reliability should not be a policing function. It is more effective to develop DFR guidelines focused around past reliability problems and providing training on those guidelines along with updates to the guidelines. Training should be performed periodically to ensure that the design engineers know the DFR guidelines and discuss conflicts that they may have with the guidelines. It is vitally important that the design engineers understand how to apply the DFR guidelines since they are responsible for design decisions. The reliability engineer supports the design team explaining the need behind the guidelines and how they are applied. In situations in which the design guidelines cannot be followed, discussions take place between the designer and the reliability engineer for resolution.

Recall reliability is the responsibility of the designer, not the reliability engineer. The designer is ultimately responsible for the product design, its reliability, manufacturability, and testability. This sounds like an unrealistic requirement. Designers generally do not know what to do to improve the reliability of a design. Having DFR guidelines allows the designer to make the right decisions before the product is ever tested. The reliability engineers are responsible for developing these guidelines and to provide guidance on their implementation. Reliability engineers are also responsible for working with the designer when a reliability requirement cannot be met.

The design guidelines are defined in black and white, but there will be gray areas in their interpretation. Unfortunately, not every application is black and white. These shades of gray represent potential risk areas and sometimes require testing or outside expert opinion for resolution. As the gray areas are resolved, the DFR guidelines are revised to reflect changes. The guidelines will probably not cover every aspect of the design, especially those associated with leading-edge technologies. These issues are better captured in a risk mitigation plan with the results incorporated into the design guidelines. The guidelines are to be continually revised to reflect continuous improvements, and reflect present technology and new reliability issues that surface in the existing product.

Applying reliability guidelines involves evaluating the trade-offs that affect product failure rate, repair cost, safety, image, profit, and time to market. The design decisions for increased reliability can have a negative impact on the product cost, ease of manufacturing, and design complexity. It is best to avoid over design or designing a product that no one is willing to pay for. This requires a marketing understanding of the reliability requirements, design trade-offs, liabilities, and warranty cost and consumer impact. The impact from selling an unreliable product may not be known for years. However, expect a trend of reduced market share as word spreads out about customer dissatisfaction. The cost of an unreliable product also manifests itself in rework, scrap material, and expensive recalls identified late in the product life cycle.

There is an additional cost factor associated with lost business from a dissatisfied customer. Today's consumer is much more knowledgeable about product reliability. With the advent of the internet, dissatisfied customers have a much greater impact on future sales through chat rooms and customer product reviews that are now found at popular sites. This may be the greatest threat. Today's consumer is computer smart and can easily research a product's reliability history. Unfortunately, most of the product reviews on these web pages are the result of dissatisfied customers. To survive, the products developed and produced must be reliable. The best way to achieve this and to meet the demanding time-to-market requirements is by incorporating DFR guidelines in a concurrent engineering effort.

Most companies today have design for manufacturing (DFM) and design for test (DFT) guidelines as part of the product development process. The benefits from DFM and DFT are well understood. Design teams understand how DFM and DFT reduce product development time and reduce the number of engineering changes at product release. The DFM and DFT guidelines are applied in the design phase and usually incorporate a checklist to verify that past mistakes are not repeated. These DFM and DFT guidelines have been developed from lessons learned over time. These lessons are then communicated through a set of guidelines and become part of the checklist required for final design approval and sign-off.

The same techniques and processes that are used in DFM and DFT apply to reliability design guidelines. The process for applying design guidelines is already in place; the problem for most companies is that they do not have reliability design guidelines and do not know how to create them. When it comes to reliability design guidelines, it seems we are back in the Stone Age where the design is thrown over the fence to manufacturing. Then it becomes the responsibility of the quality and manufacturing team or sustaining engineers groups.

So how do you establish DFR guidelines? The DFR guideline format is the same one used for the DFM and DFT guidelines. In fact, you may find that reliability issues are also covered in DFM and DFT guidelines. This is fine as long as the guidelines don't conflict. If there are debates over where the guidelines belong, refer back to the basic definitions for each charter. DFM guidelines focus on issues affecting manufacturability, cost, quality, product ramp, and rework. DFT guidelines focus around the testability, test access, fixturing, and safely operating the device under test (DUT). DFR guidelines focus around product quality over time. If DFR guidelines do not exist, chances are that some of the reliability issues are being covered in the DFM and DFT design guidelines.

The DFR guidelines can take several years to develop, so it is unreasonable for designers to wait that long for guidance on how to DFR. The guidelines must take into account the different users. The guidelines are developed by a team and address the needs of all the design groups. The guidelines are consensus driven. If no guidelines exist, start by creating a Pareto of reliability issues that need guidelines. This is usually derived from field failure data, customer complaints, and manufacturing ramp/test issues. Once created, it then becomes a task of creating, training, and implementing each guideline as it is developed.

The DFR guidelines call out technologies, components, and packages that should be avoided or not used. Be sure, while identifying what not to use, that you offer recommendations as to what is the best alternative.

Reliability design guidelines should be organized for easy access of information. Putting the guidelines into a searchable, electronic database is highly desirable. Over time the guidelines will grow into a significant volume of knowledge (which should be treated as corporate intellectual property). A suggested reliability design guidelines table of contents would be as follows:

1.0 Introduction – the need for reliability
2.0 Component reliability guidelines
3.0 Mechanical reliability guidelines
4.0 System reliability guidelines
5.0 Thermal reliability guidelines
6.0 Material reliability guidelines
7.0 System power reliability guidelines
8.0 Reliability safety guidelines

Each guideline should be defined in a single page if possible. The guideline should address a single thought. In other words, if you were developing reliability design guidelines for capacitors, dedicating single pages for each type of capacitor (i.e. electrolytic, ceramic, tantalum) will make it easier for the end user to apply. Each guideline should address the following:

What is the reliability design requirement?
What is the impact if not followed or the benefit if followed?
What detail is required to properly apply guidelines?

The above three questions should be answered for each reliability design requirement. An example of this is shown in Figure 20.5.

Unfortunately, DFR guidelines cannot be purchased. Most businesses develop their own set of guidelines tailored to their particular business.

Requirements/options:	Reliability impact/benefit:
1. Temperature stress (Rule #1): For every 10 °C increase in temperature the useful life decreases by a factor of 2. 2. Ripple current stress (Rule #2): Stay below 50% of the maximum ripple allowed. 3. Voltage stress (Rule #3): For voltage stress derating above 67%, the expected life increases by the fifth power of the rated voltage (V_r)/applied voltage (V_a).	Derating greatly improves the life of the electrolytic capacitor in addition to ensuring greater protection from spikes that can cause shorted capacitors.

Detail:

The expected life of a capacitor is described as the maximum expected life (Condition a) at rated temperature times the acceleration factors: temperature (T), voltage (V) and ripple current (I).

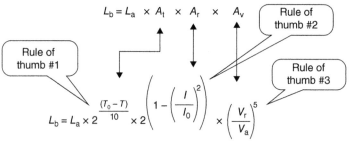

$$L_b = L_a \times A_t \times A_r \times A_v$$

Rule of thumb #2

Rule of thumb #1

Rule of thumb #3

$$L_b = L_a \times 2^{\frac{(T_0 - T)}{10}} \times 2 \left(1 - \left(\frac{I}{I_0}\right)^2\right) \times \left(\frac{V_r}{V_a}\right)^5$$

L_a = Lifetime under condition a,
L_b = Lifetime under condition b,

A_t = Temperature acceleration,
A_v = Voltage acceleration,
A_r = Ripple current acceleration.

Figure 20.5 DFR guideline for electrolytic capacitor usage. Source: Courtesy of Teradyne, Inc.

20.4.1 Derating Guidelines

Derating guidelines are vital to any product development program and should be a part of the DFR guidelines. If there are no derating guidelines in place, chances are good that some of the customer failures are the result of stress levels that exceeded the component specifications. In addition, having sufficient derating in the design will result in increased design margin. Designs with sufficient design margin have lower test parts per million (PPM) failure rates and higher first-pass yields in production.

It is not necessary to develop derating guidelines from scratch. There are several derating guidelines in the industry that can be purchased. The guidelines cover a broad range of users and so they may not be usable by engineers in this form. However, it is a simple task to tailor derating guidelines for your specific application and user environment. Derating guidelines are available from the Reliability Analysis Center. The ordering information is as follows:

Electronic Derating for Optimum Performance
Reliability Analysis Center
201 Mill Street
Rome, NY 13440-6916
http://rac.iitri.org/

Once the guidelines are in place, they can become part of the design review process. Derating should also be part of a design check off to verify compliance to the guideline prior to design review.

20.5   Design FMEA

The most powerful reliability tool in the design phase prior to building and testing any product is the FMEA tool. The design FMEA needs to be performed prior to signing off designs for procurement and build. The design FMEA activity takes place before any prototypes are ever built. If you've never performed FMEAs in your organization, this is one area where resistance is almost guaranteed. For some reason, designers have a hard time embracing the concept of a design FMEA. They will often point to the many design checks that are incorporated into the design process to eliminate errors as good enough, with "This is how we've always done it." Typically, design reviews to catch problems include a checklist based on common mistakes made, peer reviews, automated simulation programs, and automated design check programs to verify compliance to design guidelines. These processes are valuable and necessary to the design process. However, they have limitations in the types of problems they can identify. Because the FMEA is a concurrent effort, potential reliability and safety issues can be identified and fixed where it will have the least impact on the program.

In Chapter 7, the FMEA process is discussed in detail. We recommend that the material presented in Chapter 7 be used to develop an FMEA training program. Prior to a design FMEA, it is imperative that all participants have been trained in the process. Performing design FMEAs for the first time in an organization is difficult. If the participants are not trained in the process, then the FMEA meeting will be highly unproductive. Untrained participants have a tendency to steer the team off on tangents and are more likely to challenge the process. It is better to delay the FMEA design review by a week so that everyone is trained than to train as you go through the design FMEA. The training doesn't have to be long, but everyone needs to be familiar with the process.

FMEAs should be performed for all significant sections of the design. This includes the total system and subsystems including printed circuit board assemblies (PCBAs).

The best place to perform the design FMEA is after the design is completed but before the design goes through final design review. Design changes usually result from the FMEA, and so there is no point in doing the final design review until the FMEA items have been closed out. The benefits from a design FMEA include:

- Discovery of design errors
- Identification of system failures due to interconnects

- Identification of failure effects from grounding problems
- Identification of failure effects if voltages sequence at different times
- Analysis of impact when high-risk reliability components fail (i.e. What is the likely failure effect when a tantalum capacitor fails short?)
- Identification of safety, regulatory, or compliance issues
- Identification of failure effects due to software errors
- Test comprehensiveness

The output from the design FMEA is a list of design issues that require corrective action. The corrective action list is order ranked on the basis of the severity of each issue. After the team has completed the FMEA spreadsheet and the corrective action list has been generated, the next step is to decide what issues will be resolved and who will do it. There usually is confusion at this point over which issues should be resolved. Obviously, all safety, regulatory, and compliance issues need to be addressed. There unfortunately is no standard rule for deciding which of the non-safety-related issues should be resolved. Factors like available resources and time available to fix issues need to be considered. Some companies use the 80/20 rule where the top 20% (corrective action issues) represents 80% of the potential problems. Once it is agreed which issues will be fixed, the next challenge is tracking these issues to closure. Often, issues are identified as needing to be fixed but because there is no follow-up, they remain unresolved. A simple solution is to generate a single form for tracking FMEA issues to closure.

20.6 Installing a Failure Reporting Analysis and Corrective Action System

Failure reporting, analysis, and corrective action system (FRACAS) is a closed-loop feedback system used to collect and record data, analyze trends, and track problems to root cause and corrective action(s) for both hardware and software problems. FRACAS provides a cradle-to-grave solution for problem resolution. FRACAS is used to verify containment and resolution of failures. A good FRACAS identifies reliability problems when they surface and tracks the progress made in identifying root cause and corrective action. Finally, FRACAS is used to track problems to closure and without it, the impact to the bottom line can be significant and problem identification/resolution may be nothing more than guesswork.

FRACAS is installed in the design phase. The installation can be as simple as structuring the new product into the FRACAS database and ensuring that appropriate data entry fields are in place. FRACAS is first operated during prototype, where design bugs are entered after they are identified. The designers are responsible for entering this data; therefore they should be adequately trained and should be familiar with the FRACAS software and database. They should be able to easily access the database, and it should have sufficient capacity to manage the volume of activity anticipated. If the FRACAS is new, then sufficient time should be allowed to debug the software before beginning construction of the prototype. If the FRACAS is buggy, clumsy, or difficult to use when prototyping begins, it is likely that design problems will not be entered into the database. Instead, they will be recorded in notebooks, on scrap paper, and on personal computers, where they are likely to be misplaced or lost and never tracked to closure.

Implementation of the FRACAS will require the following:

1. Identification of the key product parameters that will be used to sort the information (i.e. date, manufacturer, part number, quantity, where used on, etc.). This is a much longer list but existing failure report forms can be used for this source.
2. Deciding if the FRACAS system will be manual (paper system) or if a computerized method will be chosen. This is not a trivial task, especially if it is computerized. FRACAS will take into consideration everything that is nonconforming or unacceptable from:
   (a) Engineering development data
   (b) FMEA recommendations
   (c) Highly Accelerated Life Test (HALT) findings
   (d) Highly Accelerated Stress Screens (HASSs)/Highly Accelerated Stress Audit (HASA) findings
   (e) Incoming material inspection nonconformance
   (f) In-process manufacturing failure reports
   (g) Field failure reports
   (h) Customer feedback
3. The identification of the personnel who will sit on and the one who will lead the failure reporting board (FRB). The quality manager often leads the FRB. The FRB lead must have the authority to drive all the issues to closure. The board will consist of personnel from manufacturing, purchasing, design engineering (sustaining engineering), marketing, product management, and perhaps others.

20.7 HALT Planning

Planning is a major part of HALT testing and can consume more time than the test itself. There are many issues that need to be worked out before HALT begins. First, there needs to be agreement regarding what will undergo HALT. After deciding the assemblies for HALT, there needs to be consensus on the number of assemblies that will be tested. There should be at least three to five assemblies for HALT, with an additional unit used for debugging only (a "gold" unit). HALT is a destructive test. After the test, the assemblies cannot be repaired and sold because a considerable amount of product life has been removed. If the assemblies are expensive, then debate is likely regarding the number of units that are destructively tested. Avoid testing only a single unit. This is impractical if you are doing the test at a test facility. The problem with having only one unit to test is that there will be significant downtime after each failure. The time it takes to troubleshoot and fix a failure can be significant. If there are multiple units, then testing can proceed with the next unit as you troubleshoot and fix the failed assembly. HALT planning flow is illustrated graphically in Figure 20.6.

The design team and its management must buy in. Management's commitment of resources to support the HALT effort communicates the commitment that reliability activities will be performed to improve product design. After it is agreed which assemblies will be HALT tested, the next step is to form HALT teams for each assembly. The teams are cross-functional and consist of members from software, test, manufacturing, design engineering, and reliability. The purpose of the cross-functional teams is to work

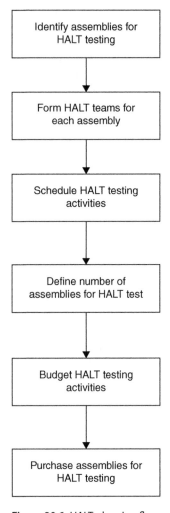

Figure 20.6 HALT planning flow.

out all issues related to supporting the HALT effort prior to the test. Use a checklist to ensure that all issues are addressed. An example of the HALT checklist is shown in Figure 20.7. If you plan to outsource HALT to a test facility, this step will help in managing test cost and test time.

20.8 HALT Test Development

In the HALT planning phase, the assemblies were identified for HALT testing along with the number of assemblies to be tested. Teams were then formed for each assembly to support the HALT activities. The HALT test development can begin after the planning is in place. The goal of test development is to have everything ready to support the HALT effort before testing begins (Figure 20.8). In HALT, development teams are formed for each assembly and will vary depending on the specific skills needed. The first

			HALT Planning Meetings held at _____AM/PM	
Latest rev.			= Need to discuss this week (*HALT Planner it to highlight AI# for discussion at next HALT Planning Meeting*)	
mm/dd/yyyy			= Action complete (*HALT leader is to darken the Done Date when action is completed*)	
Action owner	Done date	AI#	Activities	Notes
		1	HALT week date set	
		2	Lab is available for HALT week	Contact name, address & phone
		3	HALT Team identified	
		3.1	Designer	Name & phone
		3.2	Software	Name & phone
		3.3	Test	Name & phone
		3.4	Reliability	Name & phone
		3.5	Chamber technician	Name & phone
		3.6	Repair facilities	Contact name, address & phone
		4	Liquid nitrogen is available for HALT week	Order tank refill if needed
		5	Assemblies (DUTs) are available for HALT week	
		6	Extra interface unit(s) needed for HALT	Yes–no
		7	Extra interface unit(s) is available for HALT	
		8	Cables to connect from instruments to DUT are available	Are spare cables needed?
		9	Power suppllies are available for HALT	
		10	Power supply cables available for HALT	Are spare cables needed?
		11	Mechanical fixturing is available for HALT	
		12	Mechanical fixturing verified that it works with DUT	
		13	Test instrumentation for HALT test is identified	
		14	Test instrumentation is available for HALT	
		15	Make list of things to bring to HALT Lab	
		16	Make list of things to ship to HALT Lab	

Figure 20.7 HALT planning checklist.

activity of the team is to define the HALT stress tests that will be performed. Defining the HALT test starts with identifying which stresses will be exerted on the assembly to reveal reliability concerns.

After the stresses are identified for the each assembly (Table 20.2), upper and lower stress limits may need to be identified. There may not be an upper or lower limit that is known. Limits are defined when the component changes physical states due to a known stress level. An example of an upper temperature limit in an assembly is a temperature that causes a connector housing to melt. There's no reason to stress an assembly beyond a known physical limitation. The failures identified when the upper limit is exceeded do not relate to real field failures.

After the HALT test plan is defined, the remainder of the HALT test development activities can precede. The three areas of activity in the HALT planning are

- Mechanical fixturing
- Electrical test plan and execution
- Software test plan and execution

The mechanical test plan includes defining how the assembly will be mechanically fixtured in the HALT chamber. Mechanical fixturing should optimize the energy transfer from the vibration chamber into the DUT. Mechanical fixturing should not induce resonances into the assembly. The fixturing should be as light as possible and mechanically strong. There are several companies that make universal mechanical fixturing. A list of these companies can be found in Appendix A. After the mechanical fixturing is developed, it is a good idea to test the fixturing by attaching accelerometers to the DUT.

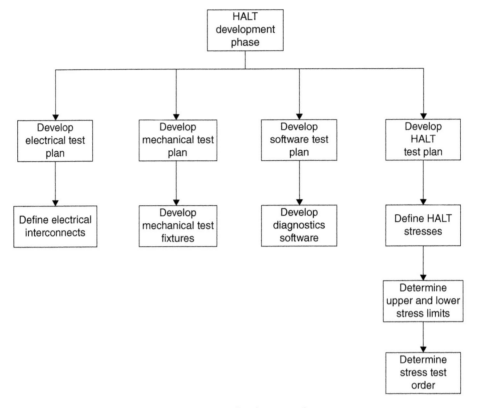

Figure 20.8 HALT development phase.

Place an accelerometer on the vibration table and several on the DUT to verify that the mechanical energy transfer is efficient. On the screw-type fasteners, use mechanical locking devices like split-lock washers to ensure the DUT is securely attached to the vibration chamber.

Developing the electrical test plan is usually more complicated. Ideally, it should be the same test plan that is being developed for the assembly in manufacturing. However, it is not unusual for the manufacturing test plan to be incomplete at the time of HALT testing. The test plan for manufacturing may also not transfer well to a HALT test. In-circuit bed of nails fixturing designed for manufacturing test will not perform well in a HALT chamber. The test development team may need to develop special fixturing for the HALT test. The electrical test plan must include how to power the DUT and what I/O signals will be connected to the assembly during test. The strategy for HALT testing is to have only the assembly under test in the HALT chamber and all external supplies, support logic, loading, and test I/Os external to the chamber. This may not always be easy.

Developing the software test plan is usually more straightforward. The software that is developed for testing the product in manufacturing can usually be used for the HALT test. It is important to identify the software that will be required for the HALT test and to make sure it is ready at the time of the test. The software that will be used for HALT needs to be checked out before HALT testing.

Before HALT begins, the following questions need to be answered:

1. What assemblies will be tested?
2. What assemblies can be tested?
3. What must be omitted?

For each particular assembly:

1. What is the quantity of each assembly for test?
2. What testing is required to verify proper operation?
3. What testing can be performed to verify proper operation in the HALT chamber?
4. What software is required to do the HALT testing?
5. What hardware is required to do the HALT testing?
6. Who will build the boards (use production process and tooling)?
7. Who will debug the board?
8. What mechanical test fixture(s) is required to do the HALT testing?
9. Are any special electrical test fixtures required to do the HALT testing?
10. What cabling and interconnect are required to do the HALT testing?
11. Are there any special cooling plate required for HALT testing?
12. Are there any special power requirements needed to do the HALT testing?
13. Are there any special test equipment required to do the HALT testing?

Logistics and scheduling issues:

1. What are the material cost for the reliability assemblies that will be tested?
2. What are the engineering development time and resources required?
3. What are the software engineering development time and resources required?
4. What are the mechanical fixturing development time and resources required?
5. What are the engineering test development time and resources required?
6. What are the manufacturing development time and resources required?

20.9   Risk Mitigation Meeting

By the end of the design phase, a significant portion of the product's development resources has been expended. A risk mitigation meeting is scheduled as the project nears completion of the design phase. The meeting should focus on the progress that has been made to mitigate the most severe risk issues. The progress made to mitigate the most severe risks before first prototype is a strong barometer of how the program is being managed. If the rate of new risk issues (risk mitigation slope) has not flattened out, then there is a good chance that the design is still in a state of flux. If the rate of closure on risk issues is not increasing, it can be a sign that there is a lack of commitment to resolve key issues. If the most significant risk issues are not addressed and mitigated until late in the program, then significant redesign and program setbacks are possible. In addition, if satisfactory progress has not been made on the most severe risk issues, alternative solutions must be initiated.

The risk mitigation meeting should focus on these issues. The functional groups meet periodically to review progress and strategies for closure on each risk issue. The purpose of the meeting is to determine if adequate progress has been made and whether the program should proceed into the design validation phase. The risk mitigation meeting reports to senior management on the progress made by the individual groups to mitigate risk since the last development phase. There is no need to expend significant capital for prototypes if the project is not likely to succeed.

Further Reading

FMEA

S. Bednarz, Douglas Marriot, Efficient Analysis for FMEA, 1998 Proceedings Annual Reliability and Maintainability Symposium (1998).

M. Kennedy, Failure Modes and Effects Analysis (FMEA) of Flip-Chip Devices Attached to Printed Wiring Boards (PWB), IEEE/CPMT International Manufacturing Technology Symposium, IEEE (1998).

M. Krasich, Use of Fault Tree Analysis for Evaluation of System Reliability Improvements in Design Phase, 2000 Proceedings Annual Reliability and Maintainability Symposium (2000).

K. Onodera, Effective Techniques of FMEA at Each Life-Cycle Stage, 1997 Proceedings Annual Reliability and Maintainability Symposium, IEEE (2000).

Prasad, S. (1991). Improving manufacturing reliability in IC package assembly using the FMEA technique. *IEEE Transactions of Components, Hybrids and Manufacturing Technology* 14 (3): 452–456.

SAE (2001). *Recommended Failure Modes and Affects Analysis (FMEA) Practices for Non-Automobile Applications*. SAE.

D. J. Russomanno, R. D. Bonnell, J. B. Bowles, Functional Reasoning in a Failure Modes and Effects Analysis (FMEA) Expert System, 1993 Proceedings Annual Reliability and Maintainability Symposium, IEEE (1993).

R. Whitcomb, M. Riox, Failure Modes and Effects Analysis (FMEA) System Development in a Semiconductor Manufacturing Environment, IEEE/SEMI Advanced Semiconductor Manufacturing Conference, IEEE (1994).

HALT

J. A. Anderson, M. N. Polkinghome, *Application of HALT and HASS Techniques in an Advanced Factory Environment*, 5th International Conference on Factory 2000 (April, 1997).

C. Ascarrunz, *HALT: Bridging the Gap Between Theory and Practice*, International test Conference 1994, IEEE (1994).

R. Confer, J. Canner, T. Trostle, S. Kurz, *Use of Highly Accelerated Life Test Halt to Determine Reliability of Multilayer Ceramic Capacitors*, IEEE (1991).

N. Doertenbach, High Accelerated Life Testing – Testing with a Different Purpose, IEST, 2000 Proceedings (February, 2000).

General Motors Worldwide Engineering Standards, Highly Accelerated Life Testing, General Motors (2002).

R. H. Gusciaoa, *The Use of Halt to Improve Computer Reliability for Point of Sale Equipment*, 1998 Proceedings Annual Reliability and Maintainability Symposium, IEEE (1998).

E. R. Hnatek, Let HALT Improve Your Product, *Evaluation Engineering*. (n.d.)

G. K. Hobbs, What HALT and HASS Can Do for Your Products, *Evaluation Engineering*. (n.d.)

Hobbs, G.K. (1997). What HALT and HASS can do for your products. In: *Hobbs Engineering, Evaluation Engineering*, 138. Qualmark Corporation.

Hobbs, G.K. (2000). *Accelerated Reliability Engineering*. Wiley.

P. E. Joseph Capitano, Explaining Accelerated Aging, *Evaluation Engineering*, p. 46 (May, 1998).

McLean, H.W. (2000). *HALT, HASS & HASA Explained: Accelerated Reliability Techniques*. American Society for Quality.

Minor, E.O. (n.d.). *Quality Maturity Earlier for the Boeing 777 Avionics*. The Boeing Company.

M. L. Morelli, *Effectiveness of HALT and HASS*, Hobbs Engineering Symposium, Otis Elevator Company (1996).

D. Rahe, *The HASS Development Process*, ITC International Test Conference, IEEE (1999).

D. Rahe, *The HASS Development Process*, 2000 Proceedings Annual Reliability and Maintainability Symposium, IEEE (2000).

M. Silverman, *HASS Development Method: Screen Development, Change Schedule, and Re-Prove Schedule*, 1998 Proceedings Annual Reliability and Maintainability Symposium, IEEE (1998).

Silverman, M.A. (n.d.). *HALT and HASS on the Voicememo II™*. Qualmark Corporation.

M. Silverman, *Summary of HALT and HASS Results at an Accelerated Reliability Test Center*, Qualmark Corporation, Santa Clara, CA, 1998 Proceedings Annual Reliability and Maintainability Symposium, IEEE (1998).

Silverman, M. (n.d.). *Summary of HALT and HASS Results at an Accelerated Reliability Test Center*. Santa Clara, CA: Qualmark Corporation.

Silverman, M. (n.d.). *Why HALT Cannot Produce a Meaningful MTBF Number and Why This Should Not be a Concern*. ARTC Division, Santa Clara, CA: Qualmark Corporation.

J. Strock, Product Testing in the Fast Lane, *Evaluation Engineering* (March, 2000).

W. Tustin, K. Gray, Don't Let the Cost of HALT Stop You, *Evaluation Engineering*, pp. 36–44. (n.d.)

21

Design Validation Phase

In the previous phase, the schematics, bill of materials, and outline drawings needed to design the product were developed. In addition, functional prototypes exist. Now, in the design validation phase, the functional prototypes are tested to verify that the design conforms to specification. This is the final opportunity to identify design, quality, reliability, manufacturing, test, and supplier issues before the design is released for production. Identifying all the design-related problems takes a cohesive effort between manufacturing engineering, test engineering, reliability, and design engineering to fully evaluate the design. At this point in time, all of these functional groups are working on the program in the design validation phase. Each has different concerns regarding the reliability of the product. Everyone is diligently working to resolve any remaining risk issues prior to production release. Manufacturing is validating special tooling and assembly processes for rampability. Test engineering is checking out test hardware, software, and test fixtures. Reliability is stressing the product to understand how it will fail. Engineering is testing prototypes to validate that the design meets the concept requirements with margin. The majority of the design problems (bugs) are identified in this phase.

This is engineering's last opportunity to identify and fix design-related problems before shipping the product to the customer. Once the product is released to production, the design team will be redirected toward the next platform or a derivative product. If design-related problems surface later in production, resolution of the problem usually is not the responsibility of the original design engineer. Instead, sustaining engineering will support this activity. This group probably will not have the technical experience and knowledge of the original design team. That is why it is so important to identify and fix design problems in the validation phase. The activities that take place in the validation phase are shown in Table 21.1.

21.1 Design Validation

At the end of the design phase, purchasing obtains material for prototype testing and evaluation. Boards are then built using the standard manufacturing process. Design engineers should not build the prototypes because this is an opportunity for manufacturing and test to spot problems early. Once built, engineering begins the process of validating the design's conformance to the requirements and specifications laid out in the concept phase. To fully validate the design takes time and patience. At this stage,

Improving Product Reliability and Software Quality: Strategies, Tools, Process and Implementation, Second Edition. Mark A. Levin, Ted T. Kalal and Jonathan Rodin.
© 2019 John Wiley & Sons Ltd. Published 2019 by John Wiley & Sons Ltd.

Table 21.1 Reliability activities in the design validation phase.

Participants	Validate design phase	
	Reliability activities	**Deliverables**
• Reliability engineering	1. Design and performance validation.	1. Product performance specifications are validated, and any limitations noted.
• Marketing	2. HALT working prototypes. Failures traced to root cause and corrected in design. Product undergoes a Final HALT to verify fixes.	2. HALT failures, stress levels, and root cause document in report. Corrective action plan to remove failure modes. Final HALT report verifies fix.
• Design engineering (electrical, mechanical, software, thermal, etc.)	3. HASS effectiveness is validated using a Proof Of Screen (POS) test.	3. POS verifies effectiveness of HASS protocol.
• Manufacturing engineer	4. Operate FRACAS. All failures during prototype are entered into FRACAS database and tracked to closure. accelerated reliability growth (ARG) and early life testing (ELT) planning.	4. FRACAS report.
• Test engineering	5. FMEA is performed (on any significant design changes only) and process FMEA.	5. Completed FMEA spreadsheet and closure on corrective action items.
• Field service/customer support	6. Closure on all risk issues. Review status and agree on each risk issue and mitigation plans. Risk mitigation meeting.	6. Risk mitigation meeting and agreement to proceed to next phase. Risk issues need to be mitigated before production phase.
• Purchasing/supply management		
• Safety and regulation		

Note: FMEA, failure modes and effects analysis; FRACAS, failure reporting, analysis, and corrective action system; HALT, Highly Accelerated Life Test; HASS, Highly Accelerated Stress Screens; POS, proof of screen.

unfortunately, many programs find themselves behind schedule and over budget. There is a natural tendency to shortcut the design validation process so that production can begin. Shipping without completing design validation will certainly result in significant retrofitting, engineering change order (ECO) activity, and higher failure rates. There is significant risk in releasing a product without knowing how it will perform under various customer environments and use conditions.

Design validation is the process of testing the design to learn how it will perform under various loads, inputs, environments, and use conditions. In essence, you are characterizing the performance capabilities and identifying the limitations in the design. Design validation testing also involves accelerated stress testing so that potential field failures can be identified. Once this information is known, the design can be refined to increase performance and enhance reliability. In addition, the product performance capabilities can be defined on the basis of what the design is capable of achieving. If certain aspects

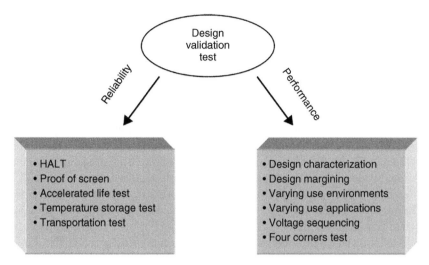

Figure 21.1 Reliability activities in the validation phase.

of the design specifications cannot be met, then reducing the product specification prior to release is possible. The design verification process improves both the reliability and performance of the design, as shown in Figure 21.1.

Design validation testing begins by testing the device's performance under ambient or nominal conditions. The device is then tested at the upper and lower specified temperature operating extremes to verify that it can operate safely and to specification. Then the product is tested at temperatures in excess of the specification to determine how much design margin there is. Design margin is important because there is a relationship between design margin and first-pass yield in manufacturing. Designs that have sufficient design margin also have high first-pass test yields. Conversely, if there is no design margin, the product is likely to have a higher failure rate in test. Design margin can compensate for component variability.

21.2 Using HALT to Precipitate Failures

Highly Accelerated Life Test (HALT) testing is performed on circuit boards, subassemblies, and at the system or product level. System level HALT can be difficult if the system is physically large in size. Three problems exist. One, most HALT chambers do not accommodate large systems. Two, it can be difficult to get sufficient vibrational energy into a large system to precipitate a failure. And three, temperature changes in the product in the chamber can be prohibitively long for efficient HALT. The only place where HALT testing is not performed is at the component level. Accelerated life testing, as defined in Chapter 15, is the best way to accelerate failures, identify failure modes, and determine reliability at the component level.

At some point in this phase, material will need to be purchased and assemblies built for HALT. For expensive assemblies, it is best to hold off buying the costly items for HALT until there is assurance that the design works. This isn't always possible if the items have long lead times, minimum purchase quantities, or high nonrecurring engineering (NRE) charges. If all the material is bought for HALT but the prototypes do not work, some of the material may well end up being scraped. The risk is greater on a new platform product than a derivative. It is also not recommended to perform HALT on a board that has had a massive amount of engineering fixes, jumper wires, and glued components. The problem with excessive amount of rework is that issues can surface in HALT due to the quality of the rework and not the quality of the design and manufacturing process. Circuit boards that have excessive amounts of rework should have the board artwork revised to incorporate the design changes. It is best to consult design engineering after the prototypes are built to get an early indication of the functionality of the prototype.

HALT should be performed on assemblies that are built using the same bill of material as the final product. Using material from a different supplier for prototype and HALT may identify problems with a prototype part or process that is not in the final product. In addition, some problems will go unidentified because problems with a component or manufacturing process cannot surface if it is not part of the prototype process. Some examples of parts that are different in prototype from the final product are as follows:

- *Machined parts.* They will respond differently under stress from a cast part.
- Hand soldering versus auto assembly.
- *Custom parts.* Sometimes a supplier provides a prototype using a different manufacturing process or tooling than what will be used in the final production version.
- Socketed parts versus nonsocketed parts.
- *Printed circuit board (PCB).* PCBs can be fabricated from a small quick-turn facility versus the standard fabricator.

There can be another problem with using different suppliers for the prototype. Suppose Supplier A provides a quick-turn delivery and is used for prototypes. Supplier B is used for production. If Supplier A finds a problem in the design and fixes it on his print, communicates it to the designer, but the designer fails to update his documents, then the problem may surface again in production, and there will be a scramble to figure out why. In PCB fabrication, manufacturers use different (often custom) software programs to check for layout problems in the board artwork. These problems affect the yield in PCB manufacturing, and the manufacturer will fix them. The PCB manufacturer considers this part of the fabrication service and routinely "cleans up" your design. However, the supplier may not notify you of the fixes. After the material is tested and found to be acceptable, it is transferred to a PCB manufacturer who will be used for volume production. The design issues were not fixed in the artwork, so problems surface in production that were not identified with the prototype. This kind of problem can be avoided. Do not allow suppliers to make any changes to the artwork or the design without submitting a timely engineering change request.

The assemblies for HALT testing should be built using the standard manufacturing process that will be used in production. Any special tooling required for production should be utilized in the assembly of the boards for test. Do not build the boards for HALT by hand in a prototype lab. Build test assemblies the same way you would production assemblies. HALT will reveal manufacturing as well as design-related problems. Therefore, it is best to mimic as close as possible the design and manufacturing process.

Before you HALT the product, some final planning is in order. There is a significant amount of preparation work required before HALT. The reliability engineer and the design engineer must ensure that everything is in place before HALT begins. The following is a list of items that have to be ready before the test:

1. The product that will be tested, five working units, and one more which is considered golden. The "golden unit" is not stress tested; it is used when there are testing issues or problems. The golden unit is used to identify if the problem is in the product under test or the test instrumentation. Inserting the golden unit will verify if the problem exists with the product. This will speed the troubleshooting process greatly. The golden unit is also used in troubleshooting, because the failed product can be compared to the golden unit to narrow the source of the failure. Hence, the information learned by using this unit is truly "golden."
2. *Test instrumentation.* This is probably the most important item on the list after the product itself because failures have to be discovered and corrected. Poor monitoring will miss some failures and render the HALT process less effective than it might otherwise have been.
3. The output specifications that will be monitored and the monitoring instrumentation.
4. Documentation, such as, schematics, assembly drawings, flowcharts, and so on.
5. A mechanical fixture to affix the product to the HALT table.
6. Input and output cabling.
7. Input and output liquid cooling hoses.
8. Special devices, that is, liquid cooling apparatus, chillers, air ducts, power sources, other support devices, and so on.
9. Test software if required.
10. The stress levels intended to be applied to the product (established and agreed to by the HALT team).
11. Scheduling the required time for testing.
12. The responsible design engineer is available to support the test for the entire HALT process.
13. A test engineer is available to debug test instrumentation problems and to assist in failure analysis for the entire HALT process.
14. The reliability engineer and a HALT chamber operator are available to support the test for the entire HALT process. (The reliability engineer documents the tests and writes the final HALT test report.)

After all the preparation work is complete, it is time to start HALT to precipitate failures. The tests can be as short as a few days or as long as several weeks. The length of the test is dependent on how many assemblies are available to test, the time it takes to fix failures, the frequency at which failures occur, and the frequency and length of downtime due to test setup and equipment problems.

It is recommended that you have six assemblies for the HALT test. One assembly is a "golden unit" and does not get stress tested. You can do HALT testing with less assemblies but the process will take longer and you may not identify fewer design issues. Use the "golden unit" to check out the system and as a troubleshooting aid when failures occur. The other five units are used for HALT testing. Place the first assembly into the chamber and orient it in a way to efficiently transfer the chamber air to the assembly. Then secure it to the chamber's vibration table. Use locking hardware to secure the assembly to the chamber. If locking hardware is not used, there is a risk that the assembly will loosen under vibration. Torque to specification all hardware with a calibrated torque driver. Connect power and I/Os to the system and secure them to the chamber (the cables should move with the vibrating table). After everything is set up, run a baseline ambient test to verify proper operation (a minimum of 10 minutes or as long as it will take to run the diagnostics two times). Next run a tickle vibration test (5 Grms) to verify that there are no loose electrical or mechanical connections. If everything passes, then you are ready to start the HALT testing.

There may be protection circuitry in the assembly that prevents the product from operating above a threshold point – e.g. temperatures above 85 °C. This type of circuitry may need to be disabled for HALT (unless it is for a safety issue). DC converters typically have thermal shutdown circuitry to prevent them from operating at temperatures above their specified maximum value. If the protection circuitry is needed or embedded in a component, then a strategy can be developed to locally control the temperature of that device so the rest of the product can be tested above the protection point. (You should wait to verify that the protection circuitry works before disabling it in the system.)

21.2.1 Starting the HALT Test

Once everything is in place for HALT, it is time to begin the test. HALT testing requires the device to be operational during the test. There is no reason to perform HALT if the device cannot be tested under operational conditions. Passive stress testing reveals little to no useful information about the design and is ineffective at precipitating failures. The HALT process flow is illustrated in Figure 21.2.

The HALT test begins with single stresses and is followed by combinational stress tests. We have suggested an order in which the testing proceeds, but there is nothing wrong with changing the order. It is recommended that testing begin with single stresses first before proceeding to combinational stress testing. A typical testing sequence for HALT is as follows:

1. Room ambient
2. Tickle vibration test
3. Temperature step stress test
4. Rapid thermal cycling stress test
5. Vibration step stress test
6. Combinational temperature and vibration test
7. Combinational search pattern test
8. Any additional stresses

Additional stresses are optional and based on the product and use environment) are line voltage and frequency margining, power sequencing, clock frequency, load variation, and so on.

The HALT test starts with placing the product into the HALT chamber and securing it mechanically to the chamber. The mechanical structure should be stiff, strong, and lightweight. The purpose of the fixturing is to firmly secure the product under stress without adversely affecting the test. Test fixtures that are heavy in mass complicate a rapid thermal cycling test and require longer dwell times for a product to stabilize. Next, connect the I/O (inputs and outputs) to the product (i.e. power sources, loads,

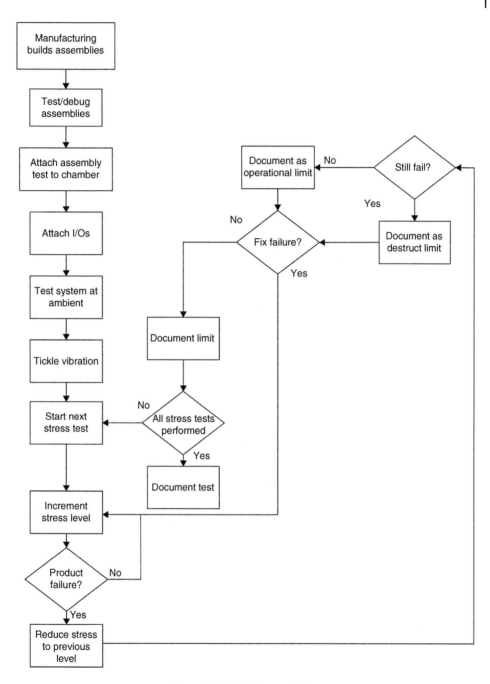

Figure 21.2 HALT process flow.

and instrumentation). Ensure that the I/Os are firmly attached to the product. Finally, attach the stress monitoring sensors to the product (i.e. accelerometers and thermocouples). The accelerometers can be attached using Super Glue® or any other type of Cyanoacrylate adhesive. The thermocouples are attached using either a thermally conductive adhesive or Kapton® adhesive tape.

21.2.2   Room Ambient Test

Once everything is secured to the chamber, the doors are closed and the test chamber is turned on. Set the chamber to room ambient and perform a nitrogen purge to evacuate any moisture residing in the chamber. The product is turned on and allowed to stabilize before performing the first diagnostic test to verify everything is operating properly. This step verifies that the test setup is functioning properly. The test time varies based on how long it takes to verify that everything is functioning properly. It is often a function of the time it takes the test software to run a complete functional test. The test software should achieve 100% test coverage. Being able to run complete functional testing is desirable but not always possible. When complete test coverage is not possible, there need to be assurances that there is at least sufficient test coverage to determine if the product is operating properly or a second HALT is performed when test software is complete.

21.2.3   Tickle Vibration Test

After the test setup, HALT chamber and product are confirmed to be operating properly, a test is run to ensure that all mechanical connections are secure. This is achieved using a tickle vibration test. The tickle vibration test applies low-level vibration between 2 and 5 Grms. The chamber temperature is set to maintain room ambient temperature and functional testing of the product is repeated to verify that the device is mechanically secure and that there are no loose connections (Figure 21.3).

21.2.4   Temperature Step Stress Test and Power Cycling

Before starting the temperature step stress testing, the upper and lower stress physical limits in temperature are defined (Table 21.2). These limits represent the point at which

Table 21.2   HALT Profile test limits and test times.

Profile for Standard HALT Instrument

Ambient temperature is considered to be 20 C.

Enter diagnostics run time and stress limits	Run Time (Minutes)
Calibration	10
Checkers/Diagnostics	5
Other diagnostics (describe)	0
Enter Lower temperature Limit (degrees C)	-40
Enter Upper Temperature Limit (degrees C)	120
Enter Upper Vibration Limit (Grms)	60
Total Diagnostic Run Time	**15**
Total Mandatory Test Time/DUT (hrs)	**10.4**

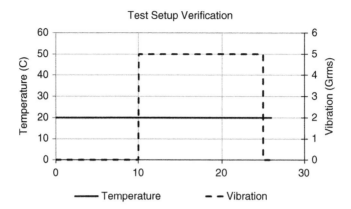

Figure 21.3 HALT test setup verification test.

there can be material changes physically (often referred to as a phase change) and results in a failure. The temperature limits may be based on the absolute maximum allowed limits where known catastrophic failure can be expected. The upper temperature limit may be based on internal thermal protection circuitry that causes shutdown and cannot be safely bypassed. An example of this would be a connector with a plastic body that melts or becomes soft when a temperature threshold is reached. Failure at this temperature is expected and represents limitations in the design and not a product failure.

The temperature step stress starts at ambient and proceeds to lower temperatures in typically 10 °C or 20 °C increments. Testing starts with the weakest stress and moves to stronger stresses as the testing continues. This way the subtle failures will not be lost with excessive stress testing. Once the temperature is reached, the product dwells long enough for the product to reach temperature stabilization, typically 10–15 minutes. The dwell time includes the time required to run functional tests to verify proper operation. Once testing has completed at the first temperature step stress, the product is power cycled to verify that it powers on, boots up, and comes into a known good state. Continue to the next temperature step stress and repeat the temperature dwell proceeded with running functional tests, checkers, diagnostics, built in self-tests or other program test protocol to verify proper operation. Once testing has completed, the product is power cycled like before to verify that it turns on in a known good state. This process continues until the lower "cold" temperature limit is reached. After reaching the lower temperature limit, the product is returned to room ambient temperature and functional testing is performed to verify that the product is operating correctly. The product now begins temperature step stress until the upper "high" temperature limit is reached.

The temperature step stress test continues until a failure is precipitated. Record the point where the failure occurs. Now reduce the stress to the previous stress level, to find out if the system recovers. If the system begins to work again, then the failure is identified as a soft failure. If the system does not recover, then the failure is identified as a hard failure. Document the failure and the stress level that caused it.

There may be the possibility to "band-aid" the failing element in order to continue increasing the stress. If the fix is simple, it may be possible to implement while the product is in the chamber. If troubleshooting is required, remove the product and place the

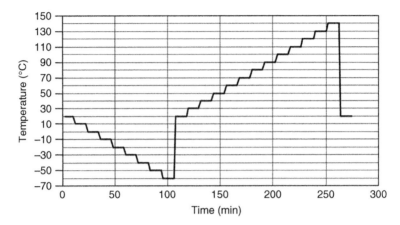

Figure 21.4 Temperature step stress.

next product in the chamber for testing. One advantage of having five units for test is that HALT testing can continue while the recently failed unit is being fixed.

After the high temperature has been reached, the product is returned to room temperature. At the end of this test, the upper and lower soft temperature limits (soft failure) and the upper and lower destruct limits (hard failure) have been identified. The first stress test, temperature step stress, is now complete. Figure 21.4 shows this test graphically.

21.2.5 Vibration Step Stress Test

The next stress test is a vibration step stress that is applied to the product. In this test, the product is maintained at ambient temperature while vibrational stresses are increased in 5–10 Grms increments. The test continues until the limit of the chamber's capability is reached, the upper vibrational limit is reached, or the product can no longer survive higher stress levels. With each step in vibration, the product is tested to verify proper operation. When stress levels exceed 20 Grms, it may be necessary to run a tickle vibration to detect failure. Many times vibration-caused failures do not reveal themselves to the test instrumentation at the higher vibration levels, but the failure becomes apparent at the lower levels. Document failures and troubleshoot to root cause. The vibrational step stress test is shown graphically in Figure 21.5.

21.2.6 Combinational Temperature and Vibration Test

After testing has been completed for single types of stresses, combinational stresses are applied. The first combinational stresses are temperature and vibration. In this test, the temperature starts at ambient and is stepped in 10–20 °C increments until just below the upper and lower destruct limits are reached. The product remains at each temperature stress, while vibrational stresses are induced on the product in 10–20 Grms increments. At each vibration stress level, the product is allowed to stabilize and functional testing is performed.

It is very important to record the stress levels at which the soft and hard failures occurred. Later, when you have made design corrections these stress levels should have

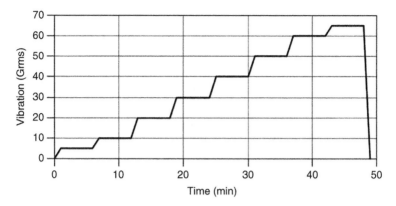

Figure 21.5 Vibration step stress.

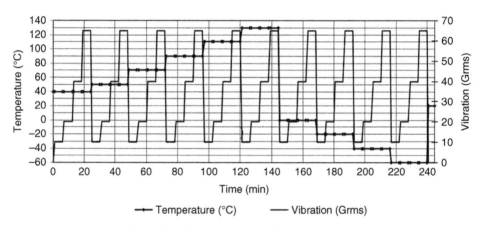

Figure 21.6 Temperature and vibration step stress.

increased, thus increasing the products operating margins. The combinational temperature and vibration test is shown graphically Figure 21.6.

21.2.7 Rapid Thermal Cycling Stress Test

The next stress test is rapid thermal cycling. The upper and lower operational limits (soft failure) were identified in the previous test. In the rapid thermal transition test, the product is rapidly transitioned to just below the upper and lower operational limits. In general, keeping the temperature limits to 5 °C below the operational limits is sufficient. The chamber temperature is made to change as rapidly as possible. Once the product reaches ramp temperature, it is allowed to dwell there typically for 10–15 minutes so that the product reaches that temperature before ramping to the next set point. If the dwell times are not long enough for the product to stabilize at the temperature, the product will see a lot less stress during temperature ramp. This test method uncovers the extreme thermal rate of change weaknesses. Running several rapid temperature excursions, between three and five cycles is sufficient. The rapid thermal cycling test is shown graphically in Figure 21.7.

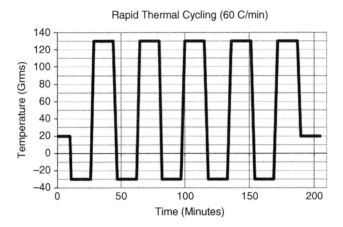

Figure 21.7 Rapid thermal cycling (60 °C min^{-1}).

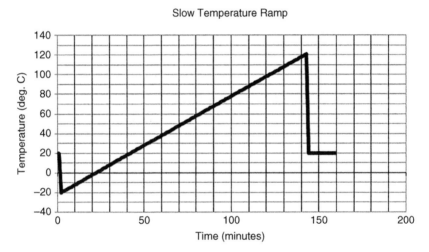

Figure 21.8 Slow temperature ramp.

21.2.8 Slow Temperature Ramp

Depending on the product, there may be additional stresses that are appropriate. The reliability tests that follow may be necessary depending on the product and the reliability requirements. The slow temperature ramp tests (Figure 21.8) can uncover instabilities that are temperature dependent. The temperature limits for this test are typically 10–20 °C below the high and low hard failure temperature limits determined in temperature step stress (basically the same temperature limits used for the combinational temperature and vibration step stress). The slow temperature ramp will uncover temperature related instabilities that can be extremely difficult to reproduce if you don't know the temperature window they occur in. The instability or anomaly may occur over a very narrow temperature range making it difficult to discover without running a so a slow temperature ramp. A failure reported by the user that occurs at a particular temperature will not have the temperature information included in the failure complaint. Unless you know to repeat the test at the temperature where the instability or anomaly occurs, you will not be able to duplicate the complaint. This type of failure often gets reported as a no fault found (NFF) or could not duplicate fault. The instability could be an oscillation, an increase in noise, a reduction in output gain, power supply instability, automatic gain control (AGC) instability, a loss of phase lock, or an increase in phase noise. These are just a few failure modes that can occur over a narrow temperature range. In addition, because the phenomena only occurs within a narrow temperature range a step stress may miss the window where this undesired behavior occurs. The slow temperature ramp provides you an opportunity to observe the instability or other unacceptable behavior as you sweep through the temperature changes.

21.2.9 Combinational Search Pattern Test

A relatively new HALT technique created by Dr. Greg Hobbs is the "search pattern technique." The idea is to slowly sweep temperature and rapidly sweep vibration simultaneously. Starting with the product at room temperature (or about 25 °C), the temperature is lowered to the lower stress limit, say −40 °C. At the same time, vibration is sweeping as fast as it can between 0 and 20 Grms. Typically, the vibration will go from the low level to the high level and back down again in less than 30 seconds (this is adjustable in some HALT chambers). Once the vibration stresses are started, the temperature is slowly swept from −40 to +140 °C (hypothetical values) and then back to room temperature. If the temperature rate of change is set to 2 °C per minute, the entire test will take 4.4 hours (refer to Figure 21.9).

The search pattern technique is valuable where the soft failure is very close to the hard failure. The temperature changes slowly while the product is being continuously monitored. This allows the test to be stopped before a hard failure is encountered. This opens opportunities for some failure investigation before the hard failure is found. Another advantage of slowly sweeping temperature over temperature step stress is that it will reveal any oscillations or instabilities that occur only at a specific temperature point or

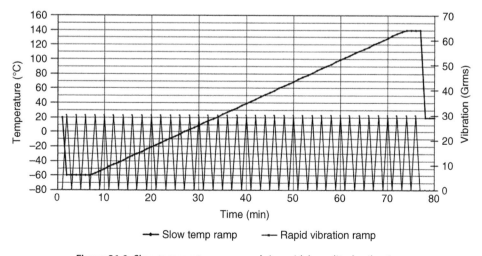

Figure 21.9 Slow temperature ramp and sinusoidal amplitude vibration.

narrow range. If you use step stresses, there is a possibility that you will pass over the point of instability.

21.2.10 Additional Nonenvironmental Stress Tests

Depending on the product, there may be additional stresses that are appropriate. Some additional stresses that can be applied are DC power supply voltage margining (first single supplies then different supply voltages in combination), AC line input voltage and frequency margining, timing margining, output loading, clock oscillator frequency variation, power cycling, and power sequencing.

21.2.11 HALT Validation Test

During HALT, failures will surface. Each HALT failure is documented either through a failure reporting, analysis, and corrective action system (FRACAS) system or in some other form to log failures (see Figure 21.10). Some of the failures will be fixed while the unit is in the chamber. Others may require a "band-aid" fix so that testing can continue. Often, components that have failed are removed that will require failure analysis to determine root cause. Getting to root cause for all failures is one of the requirements of HALT. After HALT is completed, there will be a list of failures identified along with the root cause and the stress required to precipitate the failure. Ideally, everything that fails is fixed through design changes. However, this is not always practical. Each design change has an associated economic and schedule impact that can be weighed against the improvement in design margin, reliability, and first-pass yield. The level of stress required to precipitate the change can also play a role in the decision to fix a particular

HALT FORM

Page___ of___

Assembly # __________ Rev: ____ S/N: ________

Date ___/___/___ Owner: __________

HALT Team Members __________

ACTION ITEM LIST & CORRECTIVE ACTION RECOMMENDATIONS

Item #	PROBLEM or ISSUE	CAUSE	EFFECT	T	G	V	Sf	RPN	RECOMMENDED CORRECTIVE ACTION	WHO	WHEN	A

HALT STRESS LEVELS

STRESS LEVELS - RECOMMENDED Other (Enter Value)	DESIGN LIMITS ( )	10% OVER DESIGN ( )	20% OVER DESIGN ( )	>30% OVER DESIGN ( )
WEIGHT FACTOR (T,G,V)	10	5	3	1
VIBRATION STRESS:				
TEMPERATURE STRESS:				
VOLTAGE STRESS				

Note any deviations from HALT Guideline: __________

Legend:
T = Temperature in °C; G = Vibration in Grms; V = Voltage margin level; Sf = Safety hazards; RPN = Risk priority number; A = Audit

Figure 21.10 HALT form to log failures.

failure. There should be a commitment to fix all failures through design change except when it can be shown to not make economic or business sense.

After all the agreed upon design changes have been implemented, a final validation HALT test is performed to verify the effectiveness of the design changes and to ensure no new failures were injected into the product. This test can be on a single device, although testing several is desirable. In addition, the validation HALT test doesn't need to be as rigorous as the original HALT; increasing the stress level increments will shorten the test time.

Finally, keeping track of all the testing that was performed in HALT can be difficult. There are many different types of tests performed and the sequence can vary. In addition, keeping track of which units were tested when, test equipment failures, and test anomalies can be challenging. Developing a form to track all this activity will be a life-saver later when it is time to write the HALT report. A sample form can be found in Figure 21.11.

21.3 Proof of Screen (POS)

During the HALT test, the product's soft and hard failure limits were identified. The HALT limits determine the appropriate Highly Accelerated Stress Screen (HASS) that will be used to weed out manufacturing defects in production. HASS is described in detail in Chapter 8. HASS applies accelerated stress levels to the product so that process-related defects are precipitated to fail. HASS replaces traditional burn-in or other forms of environmental stress screens (ESSs) because it is more efficient at removing process defects and is less damaging to the product life (cumulative stress from burn-in is less).

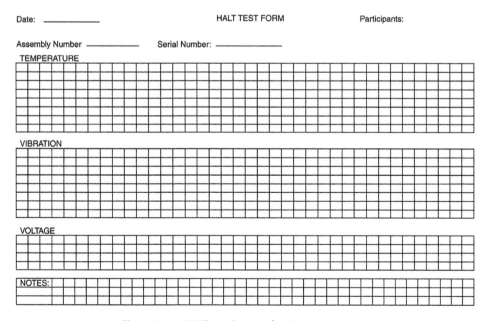

Figure 21.11 HALT graph paper for documenting test.

The HASS profile consists of two parts, the precipitation screen and the detection screen. The test begins with the precipitation screen. The precipitation screen is a stress level that is below the destruct limit and above the operating limit. The precipitation screen accelerates process defects to failure (refer to Figure 21.12). The HASS screening level applied to the product needs to be determined. A good stress level for temperature is between 80% and 50% of the destruct limits. The initial vibration stress level is at 50% of destruct limits. It is important to stay below the destruct limits; otherwise damage to a good product is likely. The purpose of the precipitation screen is to sufficiently damage defective products so it can be detected in test. Bad products are identified as the assembly passes through the detection screen.

During detection screening, the temperature stress is reduced to levels below the soft failure limit but above the product specification limit. The HASS profile is usually short, typically three to five cycles of precipitation and detection is adequate to detect failures before they occur during customer use.

Detection and precipitation screen is performed as one operation. You increase temperature past the soft failure range but below the damage level (this is the precipitation phase), then the stress is reduced below the soft failure level where a "good" assembly will recover (this is the detection range). If it doesn't recover, then a defect has been detected.

The HASS profile must not damage or severely degrade good products. Generally, if the right stress levels are applied, the defective assemblies will degrade at a significantly greater rate than good products so that they can be easily detected. The HASS profile must also be severe enough to precipitate a process defect to failure. A proof of screen (POS) is used to ensure that the HASS levels are not damaging good product by removing too much product life but effective enough to identify defective units.

The environmental stresses induced on the product by the HASS screen will remove some of the life expectancy of the product. This is unavoidable. The goal of the HASS screen is to provide a stress level high enough to precipitate manufacturing defects without removing an excessive amount of product life. How much product life is removed in HASS can be estimated through the POS.

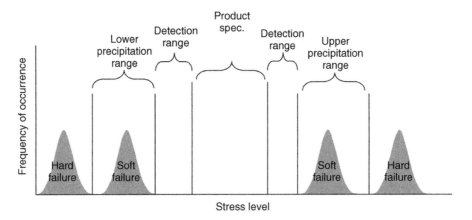

Figure 21.12 HASS stress levels.

The POS process repeats the HASS screen over and over again on a good product until it fails. Each HASS screen removes some product life. Applying the HASS stress repeatedly causes the product to continually degrade at an accelerated rate. Eventually, the product will fail because the stress continually degrades the product until it has reached wear out. If it takes 20 HASS cycles to render the product nonoperational, then it is reasonable to estimate that 5% of the product life is removed with each HASS screen. If the device failed after only four HASS screens, it can be assumed that 25% of the life of the product was removed each time. There is no minimum number of stress cycles desired before a product fails. Some companies want at least 20 cycles without a failure. The test should be run on a large enough sample size to ensure that normal manufacturing process variations are accounted for.

If, on the other hand, you run the HASS test for 100 cycles without a failure, the HASS stress levels may be set too low. Some practitioners recommend seeding products when doing POS to determine if the HASS screen is effective at detecting defective products. Seeding a board requires intentionally inserting a known manufacturing defect(s) into the product. The product undergoes HASS to determine if the defects can be found during the HASS screen. The problem with seeded defects is that it is difficult to insert seeded defects that are real representations of product defects (i.e. manufacturing process drift or supplier changes).

21.4   Highly Accelerated Stress Screen (HASS)

After HALT has completed and design changes implemented a re-HALT is performed to validated the effectiveness of the design fixes and ensure no new reliability issues were created in the respin/redesign. One of the outputs from HALT is identification of the product's soft and hard failure limits. The results from HALT were then used to develop a HASS profile for screening production/manufacturing material. Before the HASS profile can be released to production, it must go through a POS to verify the effectiveness of the HASS profile and ensure the HASS profile will not damage or degrade good product. The results is a HASS profile that is released to manufacturing for early production to help accelerate design maturity and ramp readiness. The result is a profile like Figure 21.13 for production HASS.

21.5   Operate FRACAS

A FRACAS was installed in the previous phase. The FRACAS database (often a purchased software program) is customized for the user's particular application. The customization of the database is part of the installation and checkout process for FRACAS.

With the product now in the design validation phase, prototypes are built and tested to validate the performance of the product. During development testing and design validation, failures will occur in the product. There is a natural tendency, especially early in the development phase, to treat these failures informally. They are often noted in a notebook, a piece of paper, or sometimes are fixed without any documenting at all. Treating any failure as irrelevant is shortsighted no matter how insignificant the failure may seem. These failures often resurface later in the program when it is more costly to rectify.

Variables				
	Temp. (C)	Vib. (Grms)	HALT Hard Fails	
			Temp	Vib
Start/End	20	0		
Min	-40	0	-50	
Max	80	30	100	60
Detect H	40	3		
Detect L	10	3		
Hot Soak	20 min.			
Cold Soak	15 min.			
Diagnostic run time	15 min.			
Thermal Ramp	60 C/min			
Transmissibility				40%
Total GRMS Minutes	1,016			

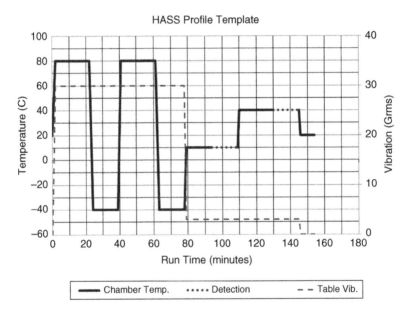

Figure 21.13 HASS profile.

FRACAS prevents this from occurring. FRACAS becomes operational once the first prototypes are built. Every failure that occurs from that time on is recorded in the FRACAS database. The failures are recorded as well as the activity to determine the root cause and corrective action. FRACAS will track the progress being made to resolve failures as well as indicate the rate at which new failures are occurring. This information will provide a good indication as to how fast the design is maturing or if it is still in a state of flux.

For FRACAS to be effective, everyone who identifies a failure must use it. Preventing designers from using their own system to document failures during prototype can be a difficult challenge. The problem can be resolved by providing training in the previous phase on how to use the FRACAS database. This, in conjunction with a commitment from the management that all designers will use the FRACAS database to record every failure, will ensure success.

21.6   Design FMEA

In the product design phase, a failure modes and effects analysis (FMEA) was performed on every subassembly, circuit board, and at the system level. There was a significant amount of effort and resources required to complete the task. The FMEA in the design validation phase is not intended to repeat the previous effort but to complement it. The design validation FMEA complements the previous effort by only evaluating significant design changes (ECOs) that resulted out of design validation, HALT, and other design-related failures. AN FMEA is not required for simple ECOs, such as a component value change. Significant ECO changes usually result in a new board or mechanical layout. The design FMEAs that are conducted in the design validation phase only address that which has changed. The FMEA is not repeated for the entire assembly; only significant changes made to the design need to be analyzed using with an FMEA. The time required to perform an FMEA for a design change will be significantly less than the time required for the original FMEA.

21.7   Closure of Risk Issues

At the end of the design validation phase, the product is complete and ready for market. The high-risk issues captured earlier should be closed before the end of the design validation phase. There should be no unresolved high-risk issues; it doesn't matter if the risk is a design, manufacturing, test, supplier, or a reliability issue. Any unresolved high-risk issues represent escapes in the risk mitigation process. Each high-risk issue has a contingency (backup) mitigation plan that should resolve the risk issue before entering the production phase. If a high-risk issue has not been resolved before entering the production phase, is required.

Escalation of unresolved high-risk issues is required because these issues often become costly problems once a product is on the market. The escalation process starts well before the completion of the design validation phase. Escalation begins by elevating the problem to senior management. Often, these problems are related to the way in which the risk is being managed, the type of resources used to solve the problem, or the skills of the people working to fix the problem. Senior management must determine why the problem is not being resolved and implement changes to fix it.

At the end of the design validation phase, the product is ready to be sold.

Further Reading

FMEA

S. Bednarz, D. Marriot, *Efficient Analysis for FMEA*, 1998 Proceedings Annual Reliability and Maintainability Symposium (1998).

M. Kennedy, *Failure Modes and Effects Analysis (FMEA) of Flip-Chip Devices Attached to Printed Wiring Boards (PWB)*, IEEE/CPMT International Manufacturing Technology Symposium, IEEE (1998).

M. Krasich, *Use of Fault Tree Analysis for Evaluation of System Reliability Improvements in Design Phase*, 2000 Proceedings Annual Reliability and Maintainability Symposium (2000).

K. Onodera, *Effective Techniques of FMEA at Each Life-Cycle Stage*, 1997 Proceedings Annual Reliability and Maintainability Symposium, IEEE (2000).

Prasad, S. (1991). Improving manufacturing reliability in IC package assembly using the FMEA technique. *IEEE Transactions of Components, Hybrids and Manufacturing Technology* 14 (3): 452–456.

D. J. Russomanno, R. D. Bonnell, J. B. Bowles, *Functional Reasoning in a Failure Modes and Effects Analysis (FMEA) Expert System*, 1993 Proceedings Annual Reliability and Maintainability Symposium, IEEE (1993).

SAE *Recommended Failure Modes and Affects Analysis (FMEA) Practices for Non-Automobile Applications*, SAE (Reaffirmed 2012) https://www.sae.org/standards/content/arp5580/.

R. Whitcomb, M. Riox, *Failure Modes and Effects Analysis (FMEA) System Development in a Semiconductor Manufacturing Environment*, IEEE/SEMI Advanced Semiconductor Manufacturing Conference, IEEE (1994).

Acceleration Methods

H. Caruso, A. Dasgupta, *A Fundamental Overview of Accelerated-Testing Analytic Models*, 1998 Proceedings Annual Reliability and Maintainability Symposium, IEEE (1998).

M. J. Cushing, *Another Perspective on the Temperature Dependence of Microelectronic-Device Reliability*, 1993 Proceedings Annual Reliability and Maintainability Symposium (1993).

J. Evans, M. J. Cushing, P. Lall, R. Bauernschub, *A Physics-of-Failure (POF) Approach to Addressing Device Reliability in Accelerated Testing of MCMS*, IEEE (1994).

Lall, P. (1996). Tutorial: temperature as an input to microelectronics-reliability models. *IEEE Transactions on Reliability* 45 (1): 3–9.

ESS

H. Caruso, *An Overview of Environmental Reliability Testing*, 1996 Proceedings Annual Reliability and Maintainability Symposium, IEEE (1996).

H. Caruso, A. Dasgupta, *A Fundamental Overview of Accelerated-Testing Analytic Models*, 1998 Proceedings Annual Reliability and Maintainability Symposium, pp. 389–393 IEEE (1998).

M. R. Cooper, *Statistical Methods for Stress Screen Development*, 1996 Electronic Components and Technology Conference, IEEE (1996).

G. A. Epstein, *Tailoring ESS Strategies for Effectiveness and Efficiency*, 1998 Proceedings Annual Reliability and Maintainability Symposium, 37–42, IEEE (1998).

S. M. Nassar, R. Barnett, *Applications and Results of Reliability and Quality Programs*, 2000 Proceedings Annual Reliability and Maintainability Symposium, IEEE (2000).

HALT

J. A Anderson, M. N. Polkinghome, *Application of HALT and HASS Techniques in an Advanced Factory Environment*, 5th International Conference on Factory 2000 (April, 1997).

C. Ascarrunz, *HALT: Bridging the Gap Between Theory and Practice*, International test Conference 1994, IEEE (1994).

R. Confer, J. Canner, T. Trostle, S. Kurz, *Use of Highly Accelerated Life Test Halt to Determine Reliability of Multilayer Ceramic Capacitors*, IEEE (1991).

N. Doertenbach, *High Accelerated Life Testing – Testing With a Different Purpose*, IEST, 2000 proceedings (February, 2000).

General Motors Worldwide Engineering Standards, *Highly Accelerated Life Testing*, GM (2002).

R. H. Gusciaoa, *The Use of Halt to Improve Computer Reliability for Point of Sale Equipment*, 1998 Proceedings Annual Reliability and Maintainability Symposium, IEEE (1998).

Hnatek, E.R. (1999). *Let HALT Improve Your Product*. Nelson Publishing.

Hobbs, G.K. (1997). What HALT and HASS can do for your products. In: *Hobbs Engineering, Evaluation Engineering*, 138. Qualmark Corporation.

Hobbs, G.K. (2000). *Accelerated Reliability Engineering*. Wiley.

Hobbs, G.K. (1977). *What HALT and HASS Can Do for Your Products*. Nelson Publishing Inc.

Joseph Capitano, P.E. (1998). Explaining accelerated aging,. *Evaluation Engineering* 46.

McLean, H. (1991). *Highly Accelerated Stressing of Products with Very Low Failure Rates*. Hewlett Packard Co.

McLean, H.W. (2000). *HALT, HASS & HASA Explained: Accelerated Reliability Techniques*. American Society for Quality.

Minor, E.O. *Quality Maturity Earlier for the Boeing 777 Avionics*. The Boeing Company.

M. L. Morelli, *Effectiveness of HALT and HASS*, Hobbs Engineering Symposium, Otis Elevator Company (1996).

D. Rahe, *The HASS Development Process*, ITC International Test Conference, IEEE (1999).

D. Rahe, *The HASS Development Process*, 2000 Proceedings Annual Reliability and Maintainability Symposium, IEEE (2000).

M. Silverman, *Summary of HALT and HASS Results at an Accelerated Reliability Test Center*, Qualmark Corporation, Santa Clara, CA, 1998 Proceedings Annual Reliability and Maintainability Symposium, IEEE (1998).

M. Silverman, *HASS Development Method: Screen Development, Change Schedule, and Re-Prove Schedule*, 1998 Proceedings Annual Reliability and Maintainability Symposium, IEEE (1998).

Silverman, M.A. (n.d.). *HALT and HASS on the Voicememo II™*. Qualmark Corporation.

Silverman, M. (1998). *Summary of HALT and HASS Results at an Accelerated Reliability Test Center*. Santa Clara, CA: Qualmark Corporation.

Silverman, M. (n.d.). *Why HALT Cannot Produce a Meaningful MTBF Number and Why this Should Not be a Concern*. Santa Clara, CA: Qualmark Corporation, ARTC Division.

J. Strock, Product Testing in the Fast Lane, *Evaluation Engineering* March, (2000).

Tustin, W. and Gray, K. Don't let the cost of HALT stop you. *Evaluation Engineering* 36–44.

22

Software Testing and Debugging

Obviously, software needs to be tested prior to releasing it. There are several stages of testing that are executed sequentially to make for the most effective defect detection. First developers test their code, then it gets handed off to software quality assurance (SQA) to continue testing. Developer testing consists of unit tests and integration tests. SQA testing focuses on system testing. Unit tests exercise individual components in isolation. Integration testing checks the interaction between modules in a controlled manner. System testing exercises the entire software package. Testing should be done in the following order:

Unit testing → Integration testing → System testing

where each step is successfully concluded before the subsequent testing step is begun. Testing in this order is significantly more productive than attempting to run these tests in a different sequence. When individual components are tested first (during unit test) then the subsequent testing can focus on finding integration or system defects with the assurance that individual modules are relatively robust. If all the software is tested at once, it becomes very difficult to locate the root cause of a defect. Lots of time is spent trying to determine where a defect is located. This causes significant loss to productivity.

Details about the different types of testing can be found in the following sections.

22.1 Unit Tests

Unit tests are run against the smallest testable unit of code, which are methods, functions, or procedures (depending on the language). The purpose of unit tests is to make sure that the lowest layer of code works properly before proceeding to deliver or test software at a higher level. Unit tests are run by the software developers. They should be run prior to delivering code into the code repository. They certainly should be run before handing software off to the SQA organization to test.

Generally, software developers write specific unit test code to exercise their low-level code. Typically, unit tests and product software development are an iterative process. A software developer will write product code, then execute their unit tests, find errors, fix their code, and repeat until all unit tests pass. In many cases, the unit tests are written prior to writing product code. There is a development methodology called *test-driven design* where all the tests are written before any of the code is developed as a means of

Improving Product Reliability and Software Quality: Strategies, Tools, Process and Implementation, Second Edition. Mark A. Levin, Ted T. Kalal and Jonathan Rodin.
© 2019 John Wiley & Sons Ltd. Published 2019 by John Wiley & Sons Ltd.

For C:	"#include <assert.h>"
For .Net in Visual Studio:	"using Microsoft.VisualStudio.TestTools.UnitTesting;"
For Java:	use "-ea" or "-enableassertions" on the Java command line

Figure 22.1 Assert functions can be used with an appropriate header.

forcing the developer to understand the requirements. Product code to be tested is often referred to as "code under test."

In some cases, the unit test code can be straightforward. The unit test code sets up some required environment, executes the code under test, and checks that the results are correct. Unit test frequently follow the arrange, act, assert (AAA) pattern. In this pattern, the unit test first arranges the environment to allow the code under test to be called. Then the unit test acts, by calling the code under test. Finally, the unit test uses an assertion to verify that the code under test worked as expected. An *assertion* (or an *assert*) is a function that verifies a condition or behavior against an expected result. Assert functions are generally defined in a library and can be used by including the appropriate header, as show in Figure 22.1.

Many modern IDEs (integrated development environments) have either built-in support or plug-ins available for creating, executing, and managing unit tests. There are also standalone unit test frameworks available that run outside of an IDE. Almost all modern software languages have either IDE support or available unit test frameworks to assist with unit test management.

There are many unit test frameworks available, in fact too many unit test frameworks to list them all, buts some of the more widely used frameworks include:

- Visual Studio Unit Testing Framework (formerly called MSTest) is a unit test framework for .Net languages that is integrated and ships with Visual Studio.
- JUnit is a free open source unit test framework designed for use with the Java programming language.
- CppTest is a free open source unit test framework for use with the C/C++ programming language.
- Boost Test Library is a free open source unit test framework for use with the C/C++ programming language.
- Google Test (aka GTest) is a free open source unit test framework for use with the C/C++ programming language.

In many cases, the code under test references system components outside of itself. In these cases, the unit test code must employ a technique called *mocking*, where those external references are provided (mocked up) by the unit test code. There are various open source and commercial mocking tools available to assist with this task. In some cases, the extent of external references is so great that they may not be mockable. In these cases, unit testing may not be applicable to the code under test, and other test techniques should be used.

Since unit tests are programs in and of themselves, it is easy to automate their execution. Automating the execution of unit tests increases the probability that these tests will be run completely and reliably. Automating the execution of unit tests guarantees that all of the unit tests are run during each test-repair-test cycle. Manually executing unit tests leaves open the possibility that some tests are inadvertently not executed after product code changes, thereby increasing the risk that a product code change has introduced an undetected error.

Unit test code, like product code, should also be reviewed for thoroughness, efficiency, and accuracy.

22.2 Integration Tests

Integration tests are a type of developer tests that check to see if different components operate together as expected. Integration tests are especially important when testing software that talks to hardware.

Integration tests are run after all of the unit tests have been executed and pass. Unit tests test code in isolation using mock objects or stubs to simulate external components. Integration tests explicitly test the interaction between two software components or a software component and a hardware component.

The integration test cases should attempt to test just the two integrating components in isolation. Just like unit tests, when possible, all variables, functionality, and components that are outside of the components under test should be stubbed or mocked out in order to isolate the test cases.

There are several different approaches to executing integration tests. These approaches are called *top down, bottom up, big bang,* and *sandwich.* Best practices in software design generally organize software into layers. With the top-down approach, integration testing starts with the highest layer and tests to the next layer down, repeating one layer at a time until all layers have been tested together. The bottom-up approach is the exact opposite, starting from the lowest software layer and iteratively testing upward. The sandwich approach starts at the top and bottom layers and meets in the middle. With big-bang integration testing, all components and layers are integrated and tested simultaneously.

It is strongly suggested that the big-bang approach be avoided if at all possible. Attempting to integrate and test all components at once can cripple productivity, because it is usually very difficult to determine where a defect or fault may be occurring. When only two components are being tested together, it is obvious that any defects are in one or the other or both of those components and nowhere else.

For hardware/software projects, it is recommended that integration start at the bottom layers of software, where the software interacts with the hardware. Since generally it is more time consuming and costly to fix hardware than it is to fix software, it is most cost-effective to test the software interaction with the hardware at the earliest possible time in the project in order to give the developers sufficient time to fix any discovered hardware defects.

If the software is sufficiently large and if the software team is large enough, it might be most efficient to use the sandwich approach to integration testing to allow multiple integration tests to proceed in parallel.

22.3   System Tests

System tests focus on overall software requirements and features. They test the software as a complete package. System testing is performed by the SQA engineers. It should be done after developers have successfully executed all unit tests and integration tests. It would be a waste of time if system tests were attempted prior to unit and integration test completion, since the software would probably be too buggy for SQA engineers to make significant progress.

The system test plans should have one or more test cases for each requirement, feature, or user story that has been specified. It is typical in system testing for each feature to have its own test plan with multiple test cases.

In a product with hardware and software, the system test must be executed on the actual hardware. If the hardware can be configured in multiple ways, then the overall system test plan should include running tests on each of the configurations. It is not necessary that every system test case be run on every hardware configuration. However, the overall system test plan must test every requirement and each configuration must have enough test cases executed to have confidence that the product works for each.

System tests clearly focus on testing the features of the integrated system. Test authors typically think about this in terms of testing of user visible functional requirements. For some products, a couple of implicit functional areas should be considered for inclusion in system tests. Those are performance and scalability testing. Performance testing verifies that the integrated product operates at the required speed. Depending on the product, this might include verifying functionality such as throughput, user interface response time, or transaction processing time.

Related to performance testing is scalability testing. Scalability testing verifies that the product can scale up to the largest configuration required. Scalability testing includes verifying functionality such as the support for the maximum number of users, the largest amount of data, the most interconnected nodes, maximum memory usage, peak performance, or the maximum number of monitored systems. Scalability testing is often used to determine the maximum supportable system size.

Scalability and performance testing are frequently combined to characterize the impact to performance caused by scaling the system up to its limits. A typical feature system test plan might look something like Figure 22.2.

22.4   Regression Tests

Regression means to return to a previous less-developed state. In software, regression specifically means a return from a working state to a nonworking state. Regressions occur when a code change introduces a new defect to working code. Often, a software defect that breaks working code is just referred to as a *regression*.

In any nontrivial software, many different components must interact properly for the software to work correctly. As noted in the unit test section, unit tests are not sufficient to ensure that the software system works correctly as an entirety. Thus, a developer might make changes to the code, achieve passing unit tests for those changes and yet still break something elsewhere in the system. Therefore, it is necessary that a full regression

Test Name	Obstruction response system test					
SQA Engineer(s):	D. Smith					
Approvers:	J. Jones					
Approval Date:	1/23/2018					

Plan Metrics:						
	Total test cases	5				
	Test cases executed	1				
	Test cases passing	1				
	Test cases to be run	4				

Test ID	Requirement / Feature	Test Name	Precondition(s)	Test Steps	Expected Result	Date Run	Status (Pass, Fail, NA)	Defects filed
1	Blocked forward	Right turn	Robot moving forward	Place obstruction in front of robot	Turn right 90°	2/14/18	Pass	
2	Blocked forward and right	Left turn	Robot moving forward	Place obstructions in front and to the right of the robot	Turn left to –90°			
3	Blocked front and sides,	U-turn	Robot moving forward	Place obstructions in front, to the right and left of the robot	Reverse direction, turn 180°			
4	Blocked all directions	Emergency stop	Robot moving forward	Place obstructions completely around the robot	Stop and sound buzzer.			
5	Sensor error	Sensor failure alert	Robot moving forward	Disconnect sensor wire	Stop and light LED.			

Figure 22.2 Sample test plan.

suite of tests be developed to regularly test the entire software system to ensure that no regressions are introduced.

Regression test suites should be automated. Manual execution of test suites is tedious, labor intensive, and error prone. Automated regression test suites can be run frequently. It is recommended that regression tests should be run every time there is a new software build. This allows for the earliest possible detection and repair of regressions.

Unit tests make a good foundation on which to start building a good regression test suite. It is recommended that unit test code should be gathered together and run as part of the regression test. Some integration tests and system tests should be automated and included in the regression test suite as well, since unit tests are not sufficient to discover all defects.

Tests added to the regression test suite should be chosen to provide adequate coverage of the entire software system. It might be possible, though unlikely, that every test developed could be automated and put into the regression test suite. There are several reasons why adding every possible test to the regression suite is generally not practical. First, not all tests can be automated. Second, regression test suites should run relatively quickly to give timely feedback to developers and to avoid interfering with subsequent builds. If a regression test were to take say a day or longer, then developers would not be able to quickly fix defects they may have introduced. This, in turn, leads to productivity loss when other developers are delayed due to a broken system preventing them from making progress.

One common approach is to have two different regression test suites. One suite would include a strategic subset of tests for a short regression suit that can be run frequently but still provides adequate overall coverage. This is generally referred to as a *smoke test*. The other regression test suite would be more complete and would run less frequently. The smoke test suite could run with each build, while the full regression test suite might run once a day, for instance overnight, or even once a week, perhaps during the weekend.

As the software is continuously developed over time, the number of tests in a test suite will continue to grow as well. This can lead to test suites that may take a long time to execute. For the purposes of keeping the test suite as efficient as possible, the individual tests in the suite should be examined periodically to eliminate unnecessary tests. For instance, tests that never fail may not be adding value and possibly be removed. Likewise, tests that always fail should be fixed or removed.

22.5 Security Tests

There are three main types of security testing: vulnerability scans, penetration testing, and static code analysis. Vulnerability scans are run internally on the system on which the product will be deployed. Penetration testing is run from outside of your product and its systems. There is often significant overlap between vulnerability testing and penetration testing. Static code analysis is run on the code base.

Vulnerability and security scanning takes place on the system itself to identify known vulnerabilities and risks. The term *system* as used here is defined as computers, operating system, and networks. Several industry groups research and publish lists of known security vulnerabilities. Those groups include the National Cybersecurity and Communications Integration Center's (NCCIC), the CERT Division of the Software Engineering

Institute (SEI) at Carnegie Mellon University and OWASP (The Open Web Application Security Project). Various security vendors develop vulnerability scanners that check for the known vulnerabilities published by these organizations. Vulnerability scans will find known issues of the systems on which the software will run. These tools will not find security vulnerabilities in the software itself. However, if the platform is not secure, neither is the product.

Penetration testing is a type of security testing where a simulated external attack is launched to determine if the software or system can be breached and exploited. Penetration testing (generally referred to as pen testing) is generally performed by an outside group who will attempt to hack into the product. Some of the vulnerabilities that pen testers use to hack into the product are identified by the vulnerability scanning tools already mentioned. So, it is recommended to run the vulnerability scans first and close up any vulnerabilities before running pen testing. Unlike vulnerability scans, which look for specific known security risks, pen testing can find some new vulnerabilities introduced by the code itself. Pen testers also often use various commercially or freely available tools to start their attacks.

Static code analyzers are tools that analyze the source code to look for specific types of code errors. In general, they act as automated code reviewers. SCA tools generally implement many rules that can be enabled to look for specific types of coding errors including known security code errors. SCA could be run on the source code to augment the code reviews. Effectively implementing an SCA tool can take a nontrivial amount of work, but can improve overall code quality in addition to finding security vulnerabilities.

There are many tools available to do security scans. Some of the common tools used for security testing are:

1. Vulnerability and security scanners and pen testers
   - Aircrack-ng, http://www.aircrack-ng.org
   - Burpsuite, https://portswigger.net/burp
   - OWASP Zed Attack Proxy (ZAP), https://www.owasp.org/index.php/OWASP_Zed_Attack_Proxy_Project
   - Nessus, Tenable, https://www.tenable.com/products/nessus/nessus-professional
   - Nexpose, Rapid 7, https://www.rapid7.com/products/nexpose
   - Nmap, https://nmap.org
2. Static code analyzers
   - Coverity, Synopsys, https://www.synopsys.com/content/dam/synopsys/sig-assets/datasheets/SAST-Coverity-datasheet.pdf
   - Flawfinder, open source, https://www.dwheeler.com/flawfinder
   - .NET Security Guard, open source, https://dotnet-security-guard.github.io
   - Parasoft, http://www.parasoft.com
   - SonarQube, https://www.sonarqube.org

22.6 Guidelines for Creating Test Cases

All types of tests (unit tests, integration tests, and system tests) should consist of a set of test cases. Test cases describe the inputs with expected outputs and behaviors. All test suites must have a set of test cases that test the following conditions:

- *Testing normal execution.* Test cases that provide legal input and verify that the component produces the correct results.
- *Testing boundary conditions.* Many defects occur when consumers of the system or component provides input that is right at the boundary conditions of acceptable values for the software. These test cases provide input right around the limits of legal input values. This should include test cases both legal and illegal near acceptable input conditions.
- *Testing error conditions.* Software needs to be tested for its ability to gracefully handle illegal usage or input. Expected results from error conditions could include displaying error messages or error recovery and error handling functions. Typically, a significant percentage of the code in a software system is designed to handle error conditions. If error conditions are not tested, then it is fairly certain that the software will not work as expected.

Additionally, the following types of test cases might be required as well:

- *Performance tests.* If software performance is important, then test cases should be developed. This may be for the software as a whole system or for subsystem components.
- *Usability tests.* A bad user interface can render an otherwise-perfect software system unusable. If the software presents any nontrivial user interface to the user, usability tests should be included. These are generally executed during the requirements and design stages for the development.
- *Longevity tests.* Certain kinds of software defects will not overtly manifest with short iterations of test runs. Looping test suites to run for long periods of time, such as non-stop for several days, will find defects that are not otherwise apparent. For example, memory leaks might be small enough that a few test executions will not be sufficient to cause them to be noticeable.
- *Exploratory testing.* A testing procedure where SQA engineers use the product in an unscripted way to find problems that may derive from a specific sequence of usage events or software state.

Adequate test coverage dictates that there should be many test cases for each of the categories of test cases described above. There is no magic number that describes the appropriate number of test cases. Analysis should be performed to determine how many of each type of test case is necessary to provide adequate test coverage. Review of code coverage metrics can provide input as to whether enough test cases exist. However, remember that code coverage by itself is not sufficient to determine if adequate testing has occurred.

22.7 Test Plans

Test cases for each type of testing should be documented in test plans. There should be multiple test plans, depending on the size of the software project. Each component should have its own unit test plan. Each set of components requiring integration should have an integration test plan. There should be one or more system test plans that document the system test cases. There could be a system test plan for each feature. Likewise, if performance or usability testing is performed, there should be test plans that document the test cases for those.

There are four main benefits derived from creating test plans:

1) Test plans can and should be reviewed. This provides a mechanism to make sure that the test cases are broad enough to provide adequate coverage of the code under test. It also acts as a means to check that the individual test cases are correct and will actually test the intended functionality.
2) Test plans allow for tracking of the test completion. As each test case is executed, its status should be updated. By creating and tracking test plans, it can be assured that all steps to test the software were indeed executed and passed.
3) Test plans provide traceability from requirements to test.
4) Tests can be repeated or reused exactly in a future test cycle to create an apples-to-apples comparison.

There is no required format for a test plan. It could be as simple as a list in a text file or a spreadsheet. However, the test plan is documented, it should have several attributes:

- The test plan should identify the code under test covered by the plan.
- For each test case, there should be a description of the test case and what it tests.
- For each test case, there should be a record of when the test case was executed, the results, and if any defects were discovered.

As each test case is executed, the corresponding record in the test plan should be updated with the date of execution and the results. Results should be documented regardless if the test case passes or fails. The results could be documented with a notation as simple as pass or fail. A performance test case might contain the value of the execution time of the test case.

Before the software is released to end users, the test plans should be reviewed to verify that all necessary test cases have been executed and have passed.

22.8 Defect Isolation Techniques

22.8.1 Simulation

Simulation can be used to validate architectural assumptions, model performance of a system, and test software when hardware is not available. This section is only going to focus on the last of those; testing software with simulated hardware. In the industry literature, testing software with simulated hardware is often called software-in-the-loop (SIL) and hardware-in-the-loop (HIL) testing. SIL and HIL techniques were pioneered by the aviation and automotive industries. SIL and HIL differ in that HIL more fully simulates the behavior of low-level circuits and is often aided by simulation running on custom hardware. Software that simulates only a subset of a hardware system, such as only the hardware interfaces, is often referred to as emulation. For the purposes of this section, I am not going to distinguish between simulation and emulation, and I will use the term *simulation* to refer to both.

Simulation is not a replacement for testing with actual hardware. Software obviously cannot be fully tested with simulated hardware. However, there are several major benefits of testing software with simulated hardware:

- There are software developer productivity improvements gained by testing with hardware simulation. Typically, hardware is never available soon enough, nor is there ever enough hardware available for all of the software developers to test. Thus, testing with simulations allows software developers to be more productive by removing a blocking dependency.
- Testing software with simulated hardware provides a mechanism to create repeatable deterministic tests in order to run frequent regression tests on the software. There may not be adequate hardware available to dedicate to a regression test environment. And early in the development cycle, what hardware that might be available may not be reliable enough to run with repeatable results.
- Testing software with simulated hardware helps isolate otherwise difficult-to-identify bugs. When integrating systems, especially software and hardware, it is often very difficult to determine where in the system the root cause of a given defect is. It could be in the software or in the hardware or in the interface between the two. This is especially true if either the hardware or the software are immature and have not been adequately tested before attempting integration. In such cases, the whole system might be so unstable as to cause huge productivity loss while trying to track down issues. When software can be tested without actual hardware before integration, the result is quicker and easier integration with faster time to fault identification.
- With real hardware, it is extremely difficult to test software handling of hardware error conditions that might be difficult to induce. Simulated hardware provides a mechanism to inject simulated faults to test the software's error handling code. This is even more useful when trying to test code that handles dangerous or hazardous conditions. How does the software handle a car crash or the right jet engine dying, or a runaway thermal event? It is very difficult, perhaps even impossible, and certainly very expensive for those types of conditions to be tested in live integrated systems. For this reason, the aviation and automotive industries have been early adopters and proponents of using simulated hardware to test software.

There are different levels of simulation and there are different methods to create the simulation. The type of simulation and means to create the simulation depend partially on the type of hardware to be simulated and the verification practices of the hardware development teams. Hardware simulation ranges from nearly complete simulated behavior to partial behavior to simple loopback/echo type of simulation.

It is common nowadays for hardware teams developing FPGAs and application-specific integrated circuits (ASICs) to create fairly complete simulation models in order to perform hardware design verification (DV) of their functionality. There are many tools available to create simulation or to generate C code as well as some form of HDL (hardware description language). Vendors such as Altera, National Instruments, Mathworks, Synopsys, Cadence, and Xilinx all provide tools to generate some form of simulation. Some hardware teams choose to verify their implementations with hand developed simulation rather than use tools to auto-generate the code. If your hardware team is already creating these kinds of simulations for DV, then they can be adapted to be used to test the software.

One limitation of testing the software with full simulations that have been created for hardware DV is that executing them can be very time-consuming. Software simulations of hardware might be more than 1000 times slower than the actual hardware. Operations that take milliseconds with real hardware will take seconds or longer to run in simulation. Simulation software obviously cannot run in real time. Actual runs of software use cases using full hardware simulation can take hours or days. Despite the length of time it takes to run these kinds of simulations, it is a useful software verification step to test the software against full simulation if it is available. It allows for a fairly complete testing of the software without having hardware. It may often be the case that hardware availability will lag due to long lead times.

Since fully executing hardware simulation is typically too slow to be practical to use in daily software testing and regression testing. For this more frequent use of testing with simulation, loopback, or echo simulation is recommended. Loopback simulation is where software has been developed to simulate a hardware interface where a call to that interface returns a preprogrammed value. This could be either a value specifically written to the interface or a hardcoded value. Unlike the fuller simulations already described, loopback testing does not allow the software to test with real hardware behavior, but it does provide the ability to test significant amounts of the software without having hardware available. Loopback simulations have fast-enough execution times to allow for frequent test runs. Loopback simulations also make it very easy for some kinds of hardware errors to be tested, such as hung hardware and timeouts, as well as incorrect results.

Generally, the software team creates the loopback simulation from published hardware interface specifications. It is incremental effort for the software verification team to create loopback simulation. It is typically only a couple of percentage points of the overall software verification effort to develop echo type simulators. The quality return on investment gained from creating and using loopback simulation is well worth the effort.

Once the simulators have been developed, they should be used in several ways. Simulators should be used while software is being developed to allow developers to test their code before delivery. Automated regression tests should be put in place to ensure that subsequent code deliveries do not cause any secondary breakage. Developer testing and regression testing should be run against a lightweight simulation, such as loopback simulator. The system integration test plan should include running the integrated software against full hardware simulation. The system integration testing should be run against the most complete hardware simulation available.

In all uses of testing with simulation, test cases should include simulating nominal hardware behavior as well as simulation of various error conditions. Use of testing software with simulated hardware will result in earlier detection of defects and better identification of defect location when doing integration with actual hardware.

22.9 Instrumentation and Logging

Debugging software is always a challenge. Debugging without adequate information as to the state of the software is much more difficult. Instrumenting code so that it generates logs of its state can provide much of the information needed to find the root cause of a defect. In some cases, logs might be the only way to debug a problem. This is often true when trying to debug a problem occurring on embedded systems or in the field.

It is generally better to have two versions of the software, one version instrumented with debug code and one version not instrumented. Having debug code execute all the time slows down software execution. Even in cases where performance is not an issue, there is still the management of debug files, which can easily fill up file systems. Code should include conditionally compiled logging statements. When the software is built, it should build both debug versions and non-debug versions. It might even be worthwhile having several levels of debug code where more or less details of internal software state are logged. If the product is going to be deployed in an environment where it will be actively monitored, such as a network, then it is appropriate to log error messages all the time, but not debug information.

At the most basic level, every method or function should log the following information:

- Timestamp of method or function entry
- Values of parameters passed into the method or function
- Timestamp of method of function exit
- Values of parameters and return codes at method or function exit
- Errors that occur during the method or function

Each log message should include a priority. The Syslog Protocol standard (RFC 5424) contains good definitions for log message priorities. They are:

0. *Emergency.* System is unusable.
1. *Alert.* Action must be taken immediately.
2. *Critical.* Critical conditions.
3. *Error.* Error conditions.
4. *Warning.* Warning conditions.
5. *Notice.* Normal but significant condition.
6. *Informational.* Informational messages.
7. *Debug.* Debug-level messages.

Additional information that could be logged includes the values of all the private properties or variables belonging to the function or method, as well as the values of any global variables that are used. Presumably, most log messages will be either "Informational" or "Debug." An (incomplete) snippet of instrumented code might look like Figure 22.3.

If instrumented properly, executing the debug version of the software will produce a timestamped trail of the software's operation. This will provide valuable information that will allow developers to determine when and where an issue or defect occurred. A snippet of log from properly instrumented debug code might look like Figure 22.4.

Log files should have a common format among all of the software components. This will make the overall log files easier to read and interpret. If every software component has a unique log format, it will take much longer to understand the flow of the software when analyzing logs to find the root cause of a problem.

Log file formats should be constructed with the intention to use automated tools to analyze the logs. In many cases there are log format standards, such as syslog, for which many analysis tools already exist. Leveraging those tools where possible can make life much easier for both the developer as well as the debugger. For instance, not only are there many open-source and off-the-shelf tools available for analyzing syslog files, but also many operating systems contain utilities to format, store, and forward syslog messages. In some cases, such as network management, it is assumed that all logs are in a common standard (in networking, it is syslog and SNMP), which allow centralized

```
// int ChangeOrientation(int angle)
// changes the robot's orientation by some angle
// angle = 0 to 360 degrees
// returns success or an error code
// see enums for return values

enum return_code { Success, Illegal_angle, Rotation_Failure };
enum priority { Emergency, Alert, Critical, Error, Warning, Notice, Informational, Debug };

int ChangeOrientation( int angle )
{
        int rc;
        double wheelAmount;

        // debug log rotate() entry
        #ifdef DEBUG
        log(Debug, get_system_time(), id->sin_addr, "enter ChangeOrientation(), turn angle = %d", angle);
        #endif /* DEBUG */

        // check that angle is legal, must be between 0 and 360
        // if the angle is 0 or 360, then do nothing
        if (angle > 0 && angle < 360) {
                // calculate how much to turn the wheels
                // if left turn (i.e. angle is > 180) then calculate negative turn value
                if ( angle <= 180)
                        wheelAmount = (angle / 360) * ((trackWidth * PI) / wheelCircumference);
                else {
                        angle -= 180;
                        wheelAmount = (angle / 360) * ((trackWidth * PI) / wheelCircumference);
                        wheelAmount *= -1;
                }

                // PointTurn() does a point turn where
                // the left wheel turns forward and the right wheel turns in reverse by wheelAmount
                        rc = PointTurn(wheelAmount);

                // handle failure of PointTurn()
                if (rc > 0) {
                        log(Error, get_system_time(), myIpAddr, "PointTurn() failed with error %d", rc);
                        return Rotation_Failure;
                }
        } else if (angle < 0 || > 360) {        // if angle is illegal, log it and return a failure
                log(Error, get_system_time(), id->sin_addr, "illegal angle specified %d", angle);
                return Illegal_angle;
        }

        #ifdef DEBUG
        log(Debug, get_system_time(), id->sin_addr, "exiting rotate() successfully");
        #endif /* DEBUG */
        return Success;
}
```

Figure 22.3 Sample log code.

management of all components. New standards are emerging for monitoring and tracking of equipment for various industries, e.g. there is an emerging standard (VDI 5600 for Factory 4.0) for equipment and sensors in automated factories. If your product is expected to be deployed in a heterogeneous environment, then it is suggested that you use whatever standard logging protocols and formats are appropriate for that environment if they have been defined.

```
<7> Mar 12 12:27:00.003 198.162.1.33 SensorCheck(), forward sensor: obstruction identified
<7> Mar 12 12:27:00.003 198.162.1.33 SensorCheck(), right sensor: clear
<7> Mar 12 12:27:00.003 198.162.1.33 SensorCheck(), calling ChangeOrientation(90)
<7> Mar 12 12:27:00.004 198.162.1.33 enter ChangeOrientation(), turn angle = 90
<7> Mar 12 12:27:00.004 198.162.1.33 enter PointTurn(), wheelAmount = 0.5
<7> Mar 12 12:27:00.006 198.162.1.33 PointTurn() exiting successfully
<7> Mar 12 12:27:00.006 198.162.1.33 ChangeOrientation() exiting successfully
<7> Mar 12 12:27:00.007 198.162.1.33 SensorCheck() exiting successfully
```

Figure 22.4 Example log file extract.

Further Reading

J. Engblom, G. Girard, B. Werner, Testing Embedded Software using Simulated Hardware, ERTS Conference 2006, 2006.

Gerhards, R. (2009). *RFC 5424 – The Syslog Protocol*. IETF Trust.

Júnior, J.C.V.S., Brito, A.V., and Nascimento, T.P. (2015). Verification of embedded system designs through hardware-software co-simulation. *International Journal of Information and Electronics Engineering* 5 (1).

King, P.J. and Copp, D.G. (2008). *Hardware in the loop for automotive vehicle control systems development and testing. Measurement and Control Journal* 39 (1).

Kohl, S. and Jegminat, D. (2005). *How to Do Hardware-In-The-Loop Simulation Right*. SAE International.

The CERT Division of the Software Engineering Institute (SEI), https://www.sei.cmu.edu/about/divisions/cert/index.cfm

The Open Web Application Security Project, https://www.owasp.org

The National Cybersecurity and Communications Integration Center's (NCCIC), https://www.us-cert.gov

23

Applying Software Quality Procedures

Once all the techniques are understood, they are put together in a process that enables the software organization to produce high-quality software. There are two generic types of processes that the organization should employ. The first is a software development process. The second is an organizational process improvement process. The software development process is the set of metrics and techniques used to deliver a specific software project with adequate quality. The organization process improvement process allows the group to improve software quality from release to release.

The overview of the process during a project has seven elements:

1. Develop requirements, including quality goals.
2. Size the work.
3. Create a defect run chart showing where the software is against the quality goals.
4. Use failure modes and effects analysis (FMEAs), reviews, inspections, and defensive programming techniques to avoid defect injection.
5. Employ the various testing techniques to discover defects.
6. Measure defects discovered during the project and keep the defect run chart up to date.
7. Use the defect run chart to determine when the software quality goals have been achieved.

The generic process described above is usable with both waterfall and Agile software life cycles. During waterfall life cycles, these activities map pretty cleanly into the various stages of the project. In an Agile process, each of these activities could be used during a sprint. However, in Agile, some of the activities might be used in between sprints for retrospectives, or applied to an entire epic. For instance, each sprint would have sizing, requirements, reviews, code inspections, and testing. Each sprint could have a defect run chart to track quality for that spring. However, the defect run chart might also be applied to an epic rather than a sprint. If each sprint will be released to a customer, it would be more appropriate to have a defect run chart per sprint. On the other hand, if an epic represents the actual customer release, then the defect run chart might be better applied to the epic. Likewise, the process improvement methods, such as root cause analysis (RCA) and FMEAs, might fit into the retrospective held between sprints. However, these mechanisms typically would take several days to perform well, so it is perfectly valid

Improving Product Reliability and Software Quality: Strategies, Tools, Process and Implementation, Second Edition. Mark A. Levin, Ted T. Kalal and Jonathan Rodin.
© 2019 John Wiley & Sons Ltd. Published 2019 by John Wiley & Sons Ltd.

to apply them to the epic and perform them outside of the sprint process as part of an overall process improvement process. Another option for the process improvement mechanisms using Agile is to create user stories that cover them and actually perform tasks such as RCA and FMEAs as deliverables during various sprints.

Long-term quality improvement processes generally do the following:

1. Set software quality goals primarily using escape defect rates.
2. Measure the escaped defect rate for each release.
3. In between releases (or as orthogonal activities during a release), use analytic techniques such as RCA and FMEAs to identify areas of weakness with the technology, training, and processes.
4. Address the weaknesses identified in the previous step.
5. Repeat this procedure.

23.1 Using Defect Model to Create Defect Run Chart

Start with the simplest type of defect model where only the total number of injected defects are estimated. To create this simple model, calculate the total number of defects expected to be injected in the project by multiplying defect density estimate with the lines of code (LOC) size estimate.

It is most accurate to use your development group's historically measured defect density data for this calculation. If you do not have historical data from your organization, you can use industry data as discussed in Section 9.3 on defect density; i.e. 10 defects per KLOC for low-level languages and 20 defects per KLOC for high-level languages.

Defects will not be injected at a steady rate during the course of the project. Nor will the defects be discovered at a steady rate. Defects will be injected during each activity in the project, i.e. requirements development, design, coding, and even while fixing other defects. Defects will be discovered during reviews, code inspections, and during each of the testing activities. Typically, most of the defects will be injected during coding, and most of the defects will be discovered and fixed during testing. A very sophisticated defect model might estimate the number of defects injected, discovered, and fixed over the course of time during the project, perhaps even taking into account the number of engineers developing, testing and fixing code at any given point in time. For organizations just starting out with defect run charts, it is sufficient to just create a simple burn up chart that plots the total number of defects expected and progress toward finding them as shown in Section 9.5 on defect run charts.

23.2 Using Defect Run Chart to Know When You Have Achieved the Quality Target

All defects found during the software development must be counted. This should include defects found during reviews, inspections, and developer testing in addition to those found during the traditional software quality assurance (SQA) type testing. Likewise, each defect that is fixed must be counted. Having accurate defect discovery and closure counts is critical to determining if the software is meeting its quality goals and is ready to release.

The defect discovered and fixed counts are used to update the defect run chart. The cadence of the defect run chart updates could vary, particularly depending on the type of software development lifecycle used. For a long waterfall project, it might be sufficient to update the defect run chart on a weekly basis. For an Agile process, where iterations are short, it might work better to update the defect run chart on a daily basis.

The defect run chart is both a burn-up and a burn-down chart. The burn-up part is the line that shows the number of discovered defects as it approaches the model number. The burn-down part of the chart shows the open defects as that count decreases toward the number of allowable escapes determined in the quality goal. Both of these are useful for taking management action to ensure that the software quality achieves its goals.

If the number of discovered defects is not approaching the estimated number of injected defects in a timely fashion, then this should be examined and mitigation strategies should be executed if appropriate. The same is true if the number of open defects is not decreasing to the target as expected by the schedule. Before deploying mitigation activities, the data should be examined to determine why the defect metrics are not converging as expected. There could be a number of factors (including simply that the software is better than expected) and a number of different mitigation tactics.

Possible reasons why the number of discovered and repaired defects is not conforming to the model as expected include the following:

- If the discovered defect counts are too low, it could be because some of the software quality methods have not been adequately executed. It should be determined if inspections, reviews, and various testing have been thoroughly carried out as expected. If these techniques have not been adequately performed, then remedial action should be taken to perform them as soon as possible.
- If the discovered and/or closed defect counts are too low, it might have been caused by the project having gotten off to a late start and is simply behind in overall execution. In this case, there are several options available to achieve the quality goals. The project schedule can be pushed out to make up for the delay. Or more staff can be added to get more work done. Or scope/content can be removed from the project to decrease the overall amount of work to be performed. A note on adding staff: This should be done very judiciously. Throwing bodies at a late project often will just be disruptive and make the project even later.
- Sometimes the discovered defect rate is higher than the model. This is a concern for the quality of a particular release if the closure rate is not keeping up. Even if the defect closure rate is keeping up with the incoming rate, it is still worth determining the root cause of the higher than expected discovery rate. In some cases, it might be nothing more than an opportunity to better calibrate the defect prediction model. However, in other cases, it might represent a particularly problematic piece of code that could benefit from additional attention.
- It is possible that the estimation used for the model is wrong. This would obviously require no mitigation at all.

23.3   Using Root Cause Analysis on Defects to Improve Organizational Quality Delivery

Before any process improvement processes are undertaken, it must be determined where to start improving. Pareto charts are a particularly useful means of identifying targets for improvement.

Ideally, when a defect is discovered, it should be categorized by what kind of defect and when it was discovered. The defect type categories include requirements, design, coding, testing, and, of course, escapes. The activity that discovered the defect could be different than the type of the discovered defect. For instance, a requirements defect could be discovered during testing or it could be discovered during requirements development. Multiple Pareto charts should be created to show where in the code defects are found (i.e. which components), where in the development process defects are injected, and where in the process defects are not being adequately discovered. These Pareto charts will show where to focus the improvement efforts.

The details on when defects are discovered and their type are both useful inputs into the quality process improvement process. There are numerous examples of how to use this categorization data:

- If it is determined that the number of defects is about right, but that many defects are found downstream from when they are introduced, then a target for improvement activities would be to find more of those defects when they are injected. This could have a major impact to project cost since the earlier defects are discovered, the less costly it is to fix them. For instance, if many requirements defects are discovered during system testing activities, it would be beneficial to focus on improving the requirements procedures to find those defects sooner.
- If it is found that a particular development activity generates more defects than the model expects, then there could be a deficiency in the defect prevention activities for that activity. For example, if it is determined that the software design process discovers many more defects than are expected during design, then the design process should be examined with the intent on improving it.
- If a particular software component or technology has a higher-than-expected defect rate, then investigation might focus on discovering the root causes. It could be that the code is old and unstable, or it could be that this area of the code is particularly complex.

Once it has been determined where to focus the improvement efforts, RCA can be used to find the reasons why the quality was not as expected.

23.4   Continuous Integration and Test

In previous sections of the book, there are recommendations to automate testing, to keeping up with the defect backlog, and to start testing early and often. One mechanism to accomplish this is continuous integration and test. With continuous integration and test, the build and test environments are automated such that each time there is a code delivery, the software is automatically built and the tests are automatically run.

This process uncovers defects (and build issues) at the fastest possible rate, allowing the software development organization to stay on top of the defect backlog and to help identify defect root causes in the most efficient way.

It is very difficult to determine which code introduced which defects when large amounts of new code are allowed to accumulate before being delivered and tested. Also, just the instability introduced by the amassed defects in that accumulated code decreases tester and debugger productivity. Another common problem is that a code change might break functionality in some other related area. If many code changes are delivered at the same time, this kind of ancillary breakage is extremely difficult to debug.

Continuous integration and test is a process where developers deliver their code as soon as it has been unit tested and code reviewed, and then each delivery initiates an automated build and test cycle. If a new defect is introduced in a particular delivery, it becomes obvious which code change caused the problem, because only a very small number of code changes are introduced at any given point in time.

The continuous integration and test process improves software development productivity in two ways. First, the debugging process is streamlined by improving the ability of developers to determine which code introduced a defect. Second, the overall defect density of the product, at any point in time, is kept lower, thus making testing of the entire product more efficient by providing a more stable platform to test.

Further Reading

Booch, G. (1994). *Object-Oriented Analysis and Design with Applications*, 2e. Addison Wesley Longman, Inc.

Humble, J. and Farley, D. (2011). *Continuous Delivery: Reliable Software Releases through Build, Test, and Deployment Automation*. Pearson Education, Inc.

Vandermark, M.A. (2003). *Defect Escape Analysis: Test Process Improvement*. IBM Corporation.

24

Production Phase

There are two major objectives in the production phase. The first addresses production ramp (Table 24.1). The objective here is to quickly achieve design maturity and ramp to the desired production levels. The second major objective is to have mechanisms in place (quality controls) to ensure the quality of the product before volume production is achieved. The activities that take place in the production phase are illustrated in Table 24.2.

24.1 Accelerating Design Maturity

When a product goes into production, there will inevitably be problems. These problems impede the ability to achieve volume production including the ability to quickly increase or decrease volume production. The problems affect manufacturing and test yields and are composed of both design and process issues. These problems are the escapes in the product development process that occurred due to inadequate execution of the design and reliability process. It is often a result of either not doing or failing to close in on issues identified in risk mitigation, Highly Accelerated Life Test (HALT), device verification test (DVT), failure modes and effects analysis (FMEA), and Highly Accelerated Stress Screens (HASSs). These reliability tools, when implemented correctly, enable a product to quickly achieve design maturity.

The design is considered mature when the inherent reliability of the design is achieved. Typically, early in production, design, manufacturing, test, and reliability problems surface, which are significant enough to require a design change to fix them. After all these fixes are implemented, the product begins to achieve the reliability goal. At this point, the design has reached maturity and the quality/reliability issues that surface are few and far between. The problem most companies have in reaching design maturity is that they take too long to identify these issues and even longer to identify root cause and corrective action.

By the time a product reaches volume production, the design must be in a mature state. When a design reaches maturity, the majority of engineering resources are no longer needed to support the product. These vital engineering resources can then be directed toward their primary goal, developing new products to increase market share, and expand the business. Achieving design maturity takes time, but it is accelerated by the reliability process.

Improving Product Reliability and Software Quality: Strategies, Tools, Process and Implementation, Second Edition. Mark A. Levin, Ted T. Kalal and Jonathan Rodin.
© 2019 John Wiley & Sons Ltd. Published 2019 by John Wiley & Sons Ltd.

Table 24.1 Reliability activities in the production ramp Phase 5.

Participants	Product ramp phase	
	Reliability activities	**Deliverables**
• Operations/ Manufacturing	1. All products have HASS until acceptable pass rate is achieved. ARG and ELT testing.	1. HASS pass-fail report. ARG and ELT report with run hours, number of units tested, all failures submitted to FRACAS.
• Reliability engineering	2. Operate FRACAS. All product and customer failures entered into FRACAS database and tracked to closure.	2. Periodic FRACAS and FRB meeting. Failures tracked to closure.
• Design engineering	3. Start reliability growth. Product operating time and failure events entered into reliability growth chart. Progress at removing failure modes is evaluated.	3. Reliability growth curves.
• Manufacturing engineer	4. FMEA is performed (on any significant design changes only) and process FMEA.	4. Completed FMEA spreadsheet and closure on corrective action items.
• Test engineering	5. SPC program initiated. Production yield is monitored and adjusted as needed.	5. Production quality data reported.
• Field service/ customer support		
• Purchasing/supply management		
• Safety and regulation		

Note: FMEA: failure modes and effects analysis; FRACAS: failure reporting, analysis, and corrective action system; HASS: Highly Accelerated Stress Screens; SPC: statistical process control; FRB: failure review board.

All too often, it takes between two to five years for a new product to reach design maturity. Initially, the product starts out with a high failure rate (low mean time between failures [MTBFs]). When the product reaches maturity, the failure rate flattens out to what the design MTBF is capable of achieving. At this point, design-related problems rarely surface; the product reliability stops improving in value (MTBF). At this point, the product has reached its achievable MTBF. Assuming that a lot of variability is not there in the manufacturing process, the first-pass yield in production will flatten out as well. Simply stated, when a product reaches design maturity, there is little more that can be done to further improve the quality and reliability of the product, without significantly changing the design.

The time it takes for a product to reach design maturity is important because this is when the product has reached its lowest cost structure. Therefore, it should be easy to understand why there is such a big push early in the production phase to reach the product rampability and yield targets. Product life cycles will continue to shorten and technology will drive product obsolescence at a greater rate. If the time it takes to reach

Table 24.2 Reliability activities in the production release Phase 6.

Participants	Product phase	
	Reliability activities	**Deliverables**
• Reliability engineering	1. Switch from HASS to HASA once production pass rate is achieved.	1. HASA pass-fail report.
• Manufacturing engineer	2. Operate FRACAS. All product and customer failures entered into FRACAS database and tracked to closure.	2. Periodic FRACAS and FRB meeting. Failures tracked to closure.
• Test engineering	3. Continue reliability growth. Product operating time and failure events entered into reliability growth chart. Progress at removing failure modes is evaluated.	3. Reliability growth curves.
• Field service/customer support	4. FMEA is performed (on any ECOs only) and process FMEA.	4. Completed FMEA spreadsheet and closure on corrective action items.
• Purchasing/supply management	5. SPC program continued. Production yield is monitored and adjusted as needed.	5. Production quality data reported.
• Safety and regulation		

Note: HASA: Highly Accelerated Stress Audit; ECO: engineering change order.

design maturity doesn't also reduce, the products developed can become obsolete before they ever reach design maturity. Fortunately, the reliability process we have presented is the most effective way to accelerate the maturity of a design.

All the hard work in the product development phase has resulted in a reliable and robust design. The product has been designed to be reliable, but this alone is no guarantee for success. There is still more work to be done. Poor quality control in manufacturing and test will make the product unreliable. However, achieving high quality in manufacturing is not difficult. Only a few tools are needed to ensure quality in the products produced. These tools fall into two categories, quality control and product improvement as illustrated in Figure 24.1.

24.1.1 Product Improvement Tools

As mentioned earlier, the first and foremost objectives during production ramp are to achieve volume production and design maturity as quickly as possible. Many problems can surface after product release that vary in severity. Some affect the customer, that is, "out-of-box" failures (also referred to as *dead on arrival* [*DOA*]), installation problems, and high failure rate. Other problems affect manufacturing, that is, low first-pass yield, significant rework and scrap, and high boneyard pile (faulty product waiting to be fixed).

The product improvement tools are designed to solve these problems by focusing on data collection and data trending to detect early problems in the product. Each issue is

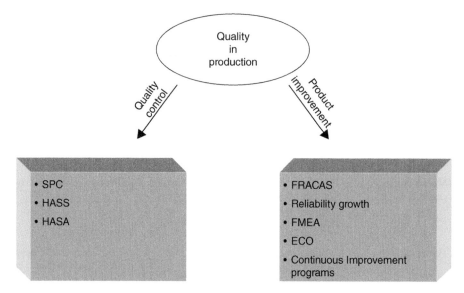

Figure 24.1 Achieving quality in the production phase.

then driven to root cause and corrective action. When issues arise that have significant impact but the problem resolution is not easy, a containment plan is put in place as a short-term fix. The containment solution is often costlier than the design fix. These same tools are then used to verify that the fix was effective.

24.1.1.1 FRACAS

Failure reporting, analysis, and corrective action system (FRACAS) began operating in the design validation phase. As product failures surfaced from prototype testing, design validation, and HALT, they were entered into the FRACAS database and their progress was tracked to root cause and corrective action. By the time the design is transferred to production, the problems that surfaced during design validation should be resolved, validated, and implemented. The FRACAS database provides the ability to make that determination.

Once the design enters production, the FRACAS system changes. During design validation, with the exception of customer alpha and beta testing, the product development team identified all the failures. Once the product is in production, the failure reporting data can come from many different places. Internally, failure data can come from assembly, test, receiving inspection, component engineering, supplier management, and reliability (the list goes on). In addition, failure reporting can also come from outside sources such as the customer, customer service and support, field service and repair, and marketing. The fact that failure reporting information comes from so many different sources can be a problem.

When failure reporting is coming from many different sources, there is a good likelihood that the data will be entered and stored in different places. FRACAS is an extremely effective tool in identifying failure patterns, so problems can be identified at the earliest possible point. If the data is stored in different databases, then it becomes difficult to impossible to gather the data needed to identify the trend. This is often one of the biggest problems with FRACAS. Different groups have different systems that are used for failure reporting. Getting everyone to use the same FRACAS system can be difficult to accomplish; often, groups are unwilling to change systems either because of the cost, time, and inability to perform their needs or the inability to transfer the existing database to the new format. All these issues are valid problems; the sooner they are the resolved, the better.

Another problem common to FRACAS systems is in the consistency of the data. As we have pointed out, failure reporting can come from many different sources. The quality of the reporting data entered into FRACAS is often a source of problems. Some of the common problems with recording failures into the FRACAS database are incomplete data, insufficient failure information, inconsistent failure reporting (the same failure is described differently so the magnitude of the failure may not be known), and failure data is not entered. These issues can be minimized and some can be eliminated entirely through a well-structured FRACAS system.

Data is entered into the FRACAS database on a continuous basis. This data is analyzed for failure trends and severity. However, just because there is a process to identify and track failures, it doesn't ensure that these failures will be resolved, let alone in a timely fashion. In order for a FRACAS system to be effective, it requires a Failure Review Board (FRB) to oversee problem resolution. The FRB board consists of a team of individuals who are responsible for the product and have the authority to resolve it. The FRB board consists of the following members:

- FRB leader (senior level manager)
- Design team representative (others may be pulled in)
- Reliability engineer
- Manufacturing representative.

This group represents the minimum team participation. The FRB team meets on a regular scheduled basis, often weekly. Attendance at the meeting is mandatory, so, if a team member cannot make it, their assigned backup fills in. The task of the FRB team is to review the failure reports, problem severity [FRACAS systems can assign design severity similar to the risk priority number (RPN) process used in FMEAs], and to prioritize problems. The FRB board then assigns appropriate team members who will be responsible for the top issues. These individuals resolve top issues to a root cause, determine corrective action, and implement fix. The FRACAS database is then used to validate if the fix was effective. An FRB member, who is assigned responsibility to resolve a problem, must have the authority to implement the needed changes. If manufacturing is assigned to fix a problem that requires an engineering design change, it is unlikely that they will be successful. Design issues should be assigned to the design team to fix; it is not for manufacturing to find a way to "band-aid" the problem. (As the top issues are resolved, the lower issues become the new top issues, and so on.)

24.1.1.2 Design Issue Tracking

A very simple alternative approach to FRACAS that tracks non-conformances to closure is design issue tracking. This is nothing more than an "Action Item" list with a stacked bar chart to graph the progress of the activity. As can be seen from Figure 24.2, the graph tracks failure bugs from documentation, to being assigned, to failure resolution and validation, and finally to closure. The height of the bars (*y*-axis) will stop increasing when

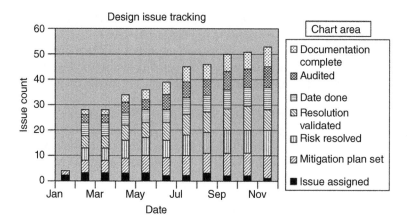

Figure 24.2 Design issue tracking chart.

new problems no longer surface. The graph tracks problem identification to closure and not the frequency at which a problem is occurring.

A design issue tracking system works well for small businesses and simple products. It is easy to set up and does not require custom-made or costly software to manage. This can be set up using any of the common computer software spreadsheet applications, such as Microsoft Excel. Design issue tracking is an alternative method to FRACAS. Whichever system you chose to use (FRACAS or design issue tracking), they need to work in conjunction with an FRB board to resolve problems.

24.2 Reliability Growth

FRACAS provides an effective process to identify problems quickly and monitor the progress made toward resolution. However, FRACAS does not indicate how well the product is performing against the reliability goals that were set in the concept phase. Early in the concept phase the reliability goal was set, that is, the product will have a 10 000-hours MTBF. After the product is designed and manufactured you need to have some level of assurance that the product will meet its reliability goals. To determine how well the product is performing against the reliability goal requires reliability growth. An example of reliability growth is shown in Figure 24.3. The graph shows the current cumulative and rolling average product reliability for the product. Product improvements that result from the FRACAS activity should be observable in the reliability growth chart. The improvement will be less noticeable in a cumulative MTBF graph where there is a lot of runtime accumulated compared to the short runtime of recent products with the latest improvements. This problem is easily solved by a 13-week rolling average reliability growth curve. The rolling average better illustrates short-term improvements made to the product.

Reliability growth works in conjunction with the FRACAS activity. Typically, when a new product is released, the initial reliability is lower because of the problems identified in FRACAS that are, as yet, unresolved. After each problem is resolved, the product

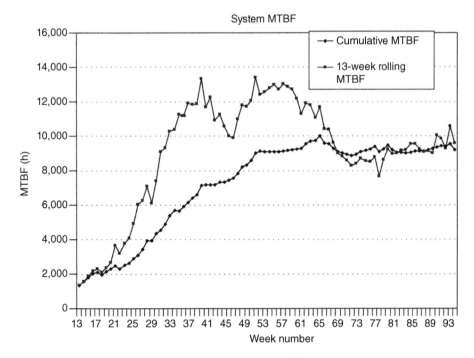

Figure 24.3 Reliability growth chart.

reliability increases because another failure mechanism has been designed out. This increasing product reliability (for new products that are produced) is tracked using reliability growth. But how do you know if the progress to improve product reliability is acceptable? Are the efforts being made to improve product reliability making a difference to product reliability? Are the reliability improvements being made quick enough to meet the business needs?

The reliability growth curve displays the effect that FRACAS is having on improving product reliability. This gives a snapshot of the current product reliability along with the rate at which it has improved. By tracking reliability growth, a decision can be made regarding the effectiveness of past reliability efforts. However, it is important to determine if the improvements in reliability are occurring too slowly to meet the business needs. Reliability growth can also be used to estimate what the future reliability of the product will be. Knowing this, a decision can be made regarding whether the product design is maturing at a rate fast enough to meet the business needs. An example of this is shown in Figure 24.4.

The future reliability growth can be estimated using a Duane curve. The Duane curve provides an estimate of the time required to achieve the reliability goal based on the current rate of reliability growth. The reliability growth rate is shown in Figure 24.5. Duane observed that the reliability improved on the basis of cumulative MTBF (θ_c) plotted against the total time, on a log–log scale. This plot could be represented as a straight line using

$$\text{Log } \theta_c = \log \theta_o + \alpha(\log T - \log T_o)$$

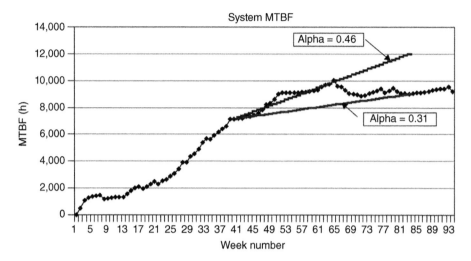

Figure 24.4 Reliability growth chart versus predicted.

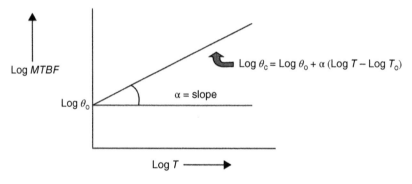

Figure 24.5 Duane curve.

where:

ϑ_c	=	$\vartheta_0 (T/T_0)^\alpha$
ϑ_c	=	Total cumulative time divided by the total number of failures
ϑ_0	=	Observed cumulative MTBF at time T_0
α	=	Growth rate, $0.1 \leq \alpha \leq 0.6$
T	=	Expected accumulated product hours, $T > T_0$
T_0	=	Actual accumulated product hours.

The rate of reliability growth is based on a value α, which is the slope of the reliability growth curve. If the value of α is closer to 0.1, then the reliability effort to improve product reliability is small and is having little effect. If the value of α is closer to 0.6, then the reliability effort is very ambitious and is having significant effect on improving product reliability. By comparing the desired reliability growth rate to the actual growth rate, a determination can be made regarding the need to make changes in the reliability activity.

The implementation of reliability growth requires assigning a lead person responsible for tracking reliability growth. Often, this is a reliability or quality engineer. However, the skill needed is not at the engineering level, so production personnel are well suited and can be trained for this task. The reliability growth report and the FRACAS report are the primary reports used by the FRB to evaluate progress.

24.2.1 Accelerated Reliability Growth (ARG)

The reliability activities through Phase 4 focused on improving design reliability and software quality. At the end of Phase 4, the product has gone through design verification, design validation, and alpha testing. The design is considered to be complete and beta testing under way. The focus now shifts toward production ramp readiness for volume manufacturing release at Phase 5 closure. However, in all likelihood there will be reliability escapes, software quality issues, manufacturing and supplier quality issues at the end of Phase 4 closure that surface after the product has been released for volume manufacturing. When these products start being returned in volume for issues relating to hardware reliability, software quality, workmanship, failed to work (referred to as *dead on arrival, or DOA*) or have a high infant mortality rate as reported by customers, engineering support teams (composing of hardware engineering, sustaining engineering, manufacturing engineering, and software engineering) are quickly redirected to work on identifying root cause and implement corrective action.

These activities are very disrupting to an organization, significantly impact future product development and cause future product release commitment dates to be missed. The organization can be tossed into a reactive firefighting effort. Waiting for the end user to notify you of a reliability or quality issue, makes this a reactive effort that can have significant financial ramifications if found after there is a large install base. These issues need to be discovered before Phase 5 closure when production is released for volume manufacturing. The question becomes, how can these issues be discovered before the product is released for volume manufacturing? In essence, can the customer experience and discovery of quality and reliability escapes be brought in house providing an opportunity to resolve or fix these issues at the lowest cost and least impact to product development? I have found that many of the hardware reliability and software quality issues can be discovered before Phase 5 closure when the operations group is focused on production ramp readiness. It is also easier to keep the product development team engaged on issues that surface in Phase 5, and will result in quicker time to root cause determination. Once Phase 5 closes, the design team typically disperses to work on the next product or project. It will be harder to pull those resources back in to investigate a reliability escape or quality issue. Two effective tools to help achieve this goal are called accelerated reliability growth (ARG) and accelerated early life test (ELT). Both of these tools help accelerate reliability growth and design maturity.

Accelerating design maturity is a proactive process that starts once Phase 4 closes with the first production units. The goal of an ARG program is to create the customer experience in house in a way that allows for the discovery of the design and quality escapes before high volume production ramp. This can be achieved through an accelerated stress with sufficient run time and sampling from production material to identify the reliability

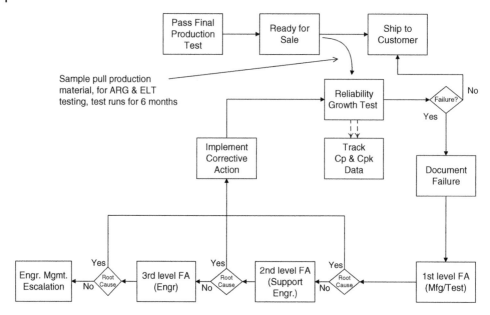

Figure 24.6 Phase 5 ARG process flow.

and quality escapes. Reliability growth testing is performed on production material that has gone through the manufacturing and test process and is ready for shipment to the end-user (Figure 24.6). Any quality or reliability issues found in this extended testing is treated like a field failure and is logged into the FRACAS system. The material used for ARG is material that has gone through manufacturing and passed, so any failures are like customer failures and should be investigated to determine root cause and have an associated corrective action.

The ARG testing should be an accelerated stress test, typically at elevated temperatures and include frequent power cycling.

24.2.2 Accelerated Early Life Testing (ELT)

The accelerated ELT is intended to identify unintended early life wear out mechanisms. These failures are systemic in nature so it can potentially affect the entire field population. The unintended early life wear out mechanism is an escape in the reliability process. It can be due to a component running hotter than anticipated, causing it to have a significantly shorter useful life. It can be a significant coefficient of thermal expansion (CTE) mismatch between a large component and the circuit board it is mounted to. The CTE mismatch can create significant strain energy as the product heats up causing it to develop solder cracks early in the product life. Once the solder crack is created, additional thermal cycles cause the stress intensity at the crack to be greater due to a stress concentration created at the solder fracture. The result is that the solder crack continues to propagate until there is a full separation of the solder connection. A component that is unintentionally used beyond its absolute maximum recommended limits is likely to suffer premature wear out. This may be due to a misunderstanding between the differences between absolute maximum limits and recommended operating limits. If you're lucky these types of design errors get detected during HALT when you step stress the part. However, if it doesn't get detected in HALT it is very likely to end up as an early life wear out failure.

The accelerated ELT is typically the same test used for ARG testing except that it runs for a significantly longer time. The sample size does not have to be large because the failure mechanism we are concerned about is wear out. Typically, ELT runs for a period of six months using the same test protocol as ARG.

24.3   Design and Process FMEA

Design FMEAs identify shortcomings in product design and safety and health hazards. When a design is changed because of market needs, added capability, new features, errors, field failures, and so on, a design FMEA can once again protect against design oversights. The entire design doesn't need to have a complete FMEA; usually, just that portion that has been changed. Having a record of the original design FMEA, will greatly speed this process to completion.

Can an FMEA be applied to more than just designs as a tool for improvement? Yes, an FMEA can be applied to manufacturing, assembly, test, receiving inspection, process equipment, and fabrication processes. In fact, the majority of the FMEA activity in this phase is process-related. The process FMEA should be used before implementing any significant changes in the manufacturing process. This ensures that the changes do not impact product quality and reliability. As with design FMEAs, having a record of the original process FMEA will greatly speed this process. This way, any mistakes can be detected and corrected before HASS. This will save time and money.

The FMEA process was described in detail in Chapter 7. The FMEA in this situation is a structured method to study a process that seeks to anticipate and minimize unwanted performance or unexpected failures.

The process is the same for both a design and process FMEA. The major difference is that the participants required will be different. The other major difference in a process FMEA is that it asks the question "what can go wrong with the process?" A process FMEA can also be an effective tool to determine the critical process to monitor for quality control. An FMEA, in this situation investigates the critical processes in the manufacturing process that effect product quality. Next, it determines what process controls need to be in place to prevent rejects and defective products from reaching the customer. The results can then be part of the quality controls used to ensure product quality.

24.3.1   Quality Control Tools

The reliability activities throughout product development ensured that the product was designed for reliability and has ample design and process margin. When a product is designed with sufficient process margin, then controlling quality in manufacturing is all that is needed for success. The quality control tools are designed to achieve this task. There is always a risk of escapes in the manufacturing process where nonconforming products are released for sale. The rate of escapes is related to the first-pass yield of the product. Products with poor yields tend to have higher escape rates than products with high first-pass yields. This is one of the major reasons it is important to have sufficient design and process margin in order to ensure product quality.

In the traditional quality program approach, identify critical processes and continuously monitor and control these processes to an acceptable variability. The focus is on detection and correction of process defects in order to ensure the maximum level of quality. These activities react to process variations that are caught downstream of the manufacturing process. In order for these techniques to be effective, they need to be pushed as far upstream as possible, so that manufacturing escapes are minimized. Ideally, this needs to be done right after any process, where variability is critical. In an assembly process, it can require automated inspection equipment (i.e. X-ray and optical) to detect variation in a critical process such as solder paste deposit, component placement, and solder joint quality. Early detection is the key. Do not rely solely on in-circuit and functional testing to determine product quality. Not only does this minimize the rework cost, and the amount of product affected, but it also minimizes the escape rate of nonconforming products. The technique used to monitor process control in manufacturing is called *statistical process control* (SPC).

There are alternative approaches to quality control. Most notable, is the technique of quality function deployment (QFD). QFD is the process of identifying all factors that might affect the ability of the product to satisfy customer needs and requirements. In essence, identify factors that may affect customer satisfaction. The QFD approach has the advantage of soliciting "voice of the customer" inputs, so issues such as feel and appearance can be considered as well.

24.3.1.1 SPC

The majority of high-technology products manufactured today are produced to a defined quality standard. The quality standards are achieved through the use of quality control processes. The quality control process incorporates SPCs that define upper and lower process control limits. When the process goes out of limits, the production line is stopped, forcing high visibility on the problem and the appropriate people are notified.

SPC charts use statistical monitoring to control the manufacturing process and maintain tolerances. The benefits are lower production cost, higher yields, less material scrap, reduced warranty cost, and rampability. The process parameters that are chosen to monitor are important. The difficulty comes in being able to identify the critical process parameters that effect product quality. These critical parameters must be controllable.

A generic control chart is shown in Figure 24.7. The control chart has an upper and a lower control limit that must be maintained by the process. The process cannot vary outside these two limits. For this example, a bar is cut to a critical length of 100 cm. A sampling method is selected along with a sampling plan. After a predetermined number of products have been made, a randomly selected sample lot (typically 30) is pulled to verify that the process is in control. The sample lot is measured for cut length and an average length is recoded and plotted on the control chart. The Range chart is the average of the process variations (plus and minus values) around the desired cut length. The range chart indicates that the process is in control as long as the averages of the variations are well within the acceptable control limit. Over time, the process may drift out of control. The control chart ensures that unacceptable process drift is identified, and that the correction actions bring the process back into control.

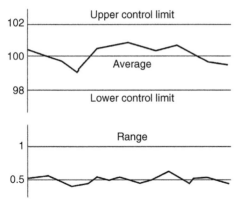

The SPC chart depicts a process that can vary around 100 cm, ±2 cm (i.e., a bar cut to 100 cm in length).

From the charts the process is in control and does not vary outside the control limits.

Note that the range is an absolute number, so it is never a negative number.

Figure 24.7 Typical SPC chart.

The time or number of units processed before the process reaches nonconformance is the sample rate of the sampling process. The number of samples and the sampling rate can vary. For stable processes, fewer samples and less frequent sample periods are required for process control. There are techniques to optimize the sampling size and frequency so that inspection cost is minimized. Another benefit to this process is that a minimal skill level is required to run the process control chart. This means that those closest to the process can be responsible for ensuring that the process remains in control and possibly make adjustments when needed. Only when the production personnel can no longer control the process engineers are needed to diagnose and correct the situation.

Control charts can be used to control important parameters that establish the quality of the end product. A few examples are: drill hole depth, output voltage, power out, light level, adhesion, weight, display accuracy in monitors, and so on. The list is determined by what the product quality specifies.

There are many variations of control charts but they *all* must be continuously monitored. The time period between monitoring points varies as a function of the rate of change or process drift before unacceptable errors are encountered. Each SPC data collection system is empirically determined, and each monitoring step in the whole SPC system may well have different monitoring periods. It is best to automate this task. With today's process tools, there is a wide variety of computerized and hand help SPC data collection and charting devices. They can take the drudgery out of the work of data collection, calculation, and charting the results. Applying SPC to the manufacturing process affords a straightforward method of controlling the process for an already well-designed product. For process control, where much greater control and control accuracy is required, the Six Sigma method can be applied.

Implementing SPC into the manufacturing process will require the following:

1. Selecting the key personnel who will learn the SPC skills. They will be the in-house trainers who will work to develop the SPC skills internally. (The quality control manager is a good candidate.)
2. Identifying the key parameters that need to be in control. Which of these should be done first, second, and so on? (Review and Pareto failure reports to reveal nonconformance that is not under control.)

3. Determining the hardware and software that is needed to implement the SPC process with ease and acceptable cost. (There are many companies that produce SPC tools; using the Internet will reveal many.)
4. Defining a "kickoff" day in which everyone involved in the new SPC process will participate.

An example of an SPC process is a wave-soldering machine used in circuit board manufacturing. The product subassembly was an electronic printed circuit board with plated-through holes for the component leads. After the components are "stuffed" onto the board, the leads are trimmed to length and the board is placed on a conveyer system. Then it is slowly passed through a wave-soldering process where the leads are passed over a bath of molten solder. The bath has a device that creates a wave or hump in the bath, so the component leads are immersed for the correct amount of time. Passing through this process the solder cools and the leads remain solidly soldered to the board. The process was not under SPC at the time. One day, the test group found a very high solder defect rate and determined that it was due to the wave-soldering process.

A review of the process, the machine, and the personnel revealed nothing that could be attributed to the cause of the solder defects. However, the problem persisted. Then, the quality manager decided to implement SPC on the wave-soldering process. Soon, he was able to correlate the solder defects to the time of the day. He was still unaware of the cause of the problem but his control charts revealed that the solder temperature took a sudden drop at a certain time of the day, and this could be tracked to the circuit boards that exhibited the defects. He sat on the production line and just observed. Then he saw the new wave-soldering person toss a few ingots of solder into the bath on his way to lunch while the machine was in process. This caused the solder to be low in temperature for a short time. It was later determined that this was the cause of the solder defects. Here, the SPC temperature revealed that the solder bath was driven low enough to cause the defects but not low enough to cause much concern at the wave machine.

24.3.1.2 Six Sigma

Six Sigma is a process-control method that has gained a lot of acceptance in the past quarter century. It assumes that everything has a normal distribution and that the production process error-detection system can actually measure differences down to one part per million. This accuracy is rarely required. The training required to be an expert in Six Sigma is not trivial. This special skill creates a problem. The talent and knowledge that is already in a company must step aside for the Six Sigma experts, who often can only offer statistical support. Unless the experts are the ones doing the actual corrective actions, there will always be a disconnect between the Six Sigma statisticians and the actual knowledge base in the company.

Establishing the Six Sigma capability within your existing staff will take some months to develop. It is recommended that the simpler SPC process be implemented immediately to get that control mechanism in place, and up and running. After the SPC system is established the time may be right to send key personnel for Six Sigma training. You may find that it may well not be needed for your company, especially in the early stages of reliability improvement.

24.3.1.3 HASS and HASA

HASS, as described earlier, is a process that is applied to detect unacceptable changes in the manufacturing process or materials that are going into the product. HASS is a more efficient process than the traditional product burn-in commonly used to reduce infant mortality failures. Highly Accelerated Stress Audit (HASA) is simply a HASS process that is implemented on a sampling or audit basis.

Traditional product burn-in processes are generally ineffective and costly. environmental stress screens (ESSs) is a more effective method to reduce product infant mortality and takes significantly less time to run. However, HASS is, usually, more effective at identifying manufacturing-related defective products than ESS and is less damaging to the product.

All of these "burn-in" techniques take time to run, they require capital cost to set up, and reduce first-pass yield. Convincing management on the need to stress test final products to reduce infant mortality failures is often difficult. The need to ramp product and not impede product ramp is strong. So often, a compromise is required, so both business and quality objectives are achieved. If the product development successfully followed the reliability process, the design should be reliable and have sufficient design margin. If this turns out to be the case, then HASA will turn out to be the most economical end effective method to ensure product quality is maintained.

If the reliability process during product development was cut short, skipped, or omitted, it is likely some form of 100% ESS or HASS testing will be required to weed out defective products. This testing will continue to be required until design changes bring the product failure rate to an acceptable level. The payback from a well-implemented reliability program is seen early in manufacturing, because costly burn-in testing quickly changes to an auditing (HASA) process. There is a significant cost savings from being able to quickly switch to HASA, and not screening every product built.

These tools are used to effect control, monitoring, and the identification of an unacceptable process before nonconformance becomes part of the end product. In addition, there are tools that afford continuous improvement.

Further Reading

FMEA

S. Bednarz, D. Marriot, *Efficient Analysis for FMEA*, 1998 Proceedings Annual Reliability and Maintainability Symposium (1998).

M. Kennedy, *Failure Modes and Effects Analysis (FMEA) of Flip-Chip Devices Attached to Printed Wiring Boards (PWB)*, IEEE/CPMT International Manufacturing Technology Symposium, IEEE (1998).

M. Krasich, *Use of Fault Tree Analysis for Evaluation of System Reliability Improvements in Design Phase*, 2000 Proceedings Annual Reliability and Maintainability Symposium (2000).

K. Onodera, Effective Techniques of FMEA at Each Life-Cycle Stage, 1997 Proceedings Annual Reliability and Maintainability Symposium, IEEE (2000).

Prasad, S. (1991). Improving manufacturing reliability in IC package assembly using the FMEA technique. *IEEE Transactions of Components, Hybrids and Manufacturing Technology* 14 (3): 452–456.

D. J. Russomanno, R. D. Bonnell, J. B. Bowles, *Functional Reasoning in a Failure Modes and Effects Analysis (FMEA) Expert System*, 1993 Proceedings Annual Reliability and Maintainability Symposium, IEEE (1993).

SAE (2001). *Recommended Failure Modes and Affects Analysis (FMEA) Practices for Non-Automobile Applications*. SAE.

R. Whitcomb, M. Riox, *Failure Modes and Effects Analysis (FMEA) System Development in a Semiconductor Manufacturing Environment*, IEEE/SEMI Advanced Semiconductor Manufacturing Conference, IEEE (1994).

Quality

Gupta, P. (1992). Process quality improvement – a systematic approach. *Surface Mount Technology*.

Johnson, C. (1997). Before you apply SPC, identify your problems. *Contract Manufacturing*.

Kelly, G. (1992). SPC: another view. *Surface Mount Technology*.

Lee, S.B., Katz, A., and Hillman, C. (1998). Getting the quality and reliability terminology straight. *IEEE Transactions on Components, Packaging, and Manufacturing* 21 (3): 521–523.

Mangin, C.-H. (1996). *The DPMO: Measuring Process Performance for World-Class Quality*. SMT.

S. M. Nassar, R. Barnett, *IBM Personal Systems Group Applications and Results of Reliability and Quality Programs*, 2000 Proceedings Annual Reliability and Maintainability Symposium (2000).

Oh, H.L. (1995). A Changing Paradigm in Quality. *IEEE Transactions on Reliability* 44 (2): 265–270.

Pearson, T.A. and Stein, P.G. (1992). On-line SPC for assembly. *Circuits Assembly*.

Ward, D.K. (1999). A formula for quality: DFM + PQM = single digit PPM. *Advanced Packaging*.

Reliability Growth

H. Crow, P. H. Franklin, N. B. Robbins, *Principles of Successful Reliability Growth Applications,* 1994 Proceedings Annual Reliability and Maintainability Symposium, IEEE (1994).

J. Donovan, E. Murphy, *Improvements in Reliability-Growth Modeling*, 2001 Proceedings Annual Reliability and Maintainability Symposium, IEEE (2001).

L. Edward Demko, *On reliability Growth Testing*, 1995 Proceedings Annual Reliability and Maintainability Symposium, IEEE (1995).

G. J. Gibson, L. H. Crow, *Reliability Fix Effectiveness Factor Estimation,* 1989 Proceedings Annual Reliability and Maintainability Symposium, IEEE (1989).

D. K. Smith, *Planning Large Systems Reliability Growth Tests*, 1984 Proceedings Annual Reliability and Maintainability Symposium, IEEE (1984).

J. C. Wronka, Tracking of Reliability Growth in Early Development, 1988 Proceedings Annual Reliability and Maintainability Symposium, IEEE (1988).

Burn-In

T. Bardsley, J. Lisowski, S. Wislon, S. VanAernam, *MCM Burn-In Experience*, MCM '94 Proceedings (1994).

D. R. Conti, J. Van Horn, *Wafer Level Burn-In*, Electronic Components and Technology Conference, IEEE (2000).

J. Forster, *Single Chip Test and Burn-In*, Electronic Components and Technology Conference, IEEE (2000).

T. Furuyama, N. Kushiyama, H. Noji, M. Kataoka, T. Yoshida, S. Doi, H. Ezawa, T. Watanabe, Wafer Burn-In (WBI) Technology for RAM's, IEDM 93–639, IEEE (1993).

R. Garcia, IC Burn-In & Defect detection Study, (September 19, 1997).

Hawkins, C.F., Segura, J., Soden, J., and Dellin, T. (1999). Test and reliability: partners in IC manufacturing, part 2. *IEEE Design & Test of Computers*, IEEE.

T. R. Henry, T. Soo, *Burn-In Elimination of a High Volume Microprocessor Using IDDQ*, International Test Conference, IEEE (1996).

Jordan, J., Pecht, M., and Fink, J. (1997). How burn-in can reduce quality and reliability. *The International Journal of Microcircuits and Electronic Packaging* 20 (1): 36–40.

Kuo, W. and Kim, T. (1999). An overview of manufacturing yield and reliability Modeling for semiconductor products. *Proceedings of the IEEE* 87 (8): 1329–1344.

W. Needham, C. Prunty, E. H. Yeoh, *High Volume Microprocessor Test Escapes an Analysis of Defects our Tests are Missing*, International Test Conference, pp. 25–34.

Pecht, M. and Lall, P. (1992). A physics-of-failure approach to IC burn-in. *Advances in Electronic Packaging*, ASME 917–923.

A. W. Righter, C. F. Hawkins, J. M. Soden, P. Maxwell, *CMOS IC Reliability Indicators and Burn-In Economics*, International Test Conference, IEEE (1988).

T. Sdudo, *An Overview of MCM/KGD Development Activities in Japan*, Electronic Components and Technology Conference, IEEE (2000).

Thompson, P. and Vanoverloop, D.R. (1995). Mechanical and electrical evaluation of a bumped-substrate die-level burn-in carrier. *Transactions On Components, Packaging and Manufacturing Technology, Part B* 18 (2): 264–168, IEEE.

HASS

Lecklider, T. (2001). How to avoid stress screening. *Evaluation Engineering* 36–44.

D. Rahe, *The HASS Development Process*, 2000 Proceedings Annual Reliability and Maintainability Symposium, IEEE (2000).

Rahe, D. *HASS from Concept to Completion*. Qualmark Corporation.

M. Silverman, *HASS Development Method: Screen Development, Change Schedule, and Re-Prove Schedule*, 2000 Proceedings Annual Reliability and Maintainability Symposium, IEEE (2000).

25

End-of-Life Phase

25.1   Managing Obsolescence

Eventually, all products reach the end of life when they can no longer serve the needs of the customer both in overall performance and capability. There are many reasons that products suffer in performance. Sometimes the product eventually wears out from use; but more than likely it is because other products can outperform the older units.

Computers that were state of the art two to three years ago are being replaced with computers with significant improvements, allowing the user greater productivity. In this case, the older computer may be performing to the original specifications, but if the market needs higher capabilities, the older unit is rendered obsolete. Perfectly good audio devices and cell phones are continuously being replaced with newer devices that, in one way or another, outperform their predecessors. This is performance obsolescence through new improvements. The time between new product release and obsolescence is getting shorter and shorter. The time to make profits for a given product is getting is narrower and narrower. So early market entry and high reliability are very important for profit capture.

Some products are expected to have a long product use life (e.g. automobiles, CAT scanners, consumer electronics). Reliability concerns can surface if the product starts to wear out before its end of life. When this happens, expect more frequent component failures. There may even come a time when the product cannot be repaired because replacement parts are no longer available. Planning for this eventuality can be a lifesaver.

If a product has a long product life, that is, customer demand is sufficient to keep the product on the market for five years or longer, then there is a good chance that some of the parts required to build the system may no longer be available. These components may be obsolete because next-generation technology replaced them, the supplier is no longer in business, or demand dropped below a manufacturer's minimum requirement. Usually, there will be an advance notice of a parts discontinuance or change; this should be part of any purchasing agreement. Once a part obsolescence is noted, a plan for mitigation is required.

Some strategies for dealing with obsolescence are as follows:

- Seek an alternative supplier (but often similar suppliers obsolete similar component products for similar reasons).

Improving Product Reliability and Software Quality: Strategies, Tools, Process and Implementation, Second Edition. Mark A. Levin, Ted T. Kalal and Jonathan Rodin.

- Look for suppliers who specialize in obtaining and storing these components.
- On occasion, a producer will learn that a component has become obsolete and a last buy will not satisfy their needs. (These suppliers who specialize in this market may well be used as an early warning ally to help avoid this problem.)
- Determine how many of these soon to be unobtainable parts will be needed to support product life for a last time purchase to meet this need.
- Once a part is identified as soon to be obsolete, it should be flagged in as "DISCONTINUED" and "DO NOT USE IN NEW DESIGNS."
- As part of the new product development process, ensure that there is a bill of materials (BOMs) check step that reviews all BOMs for part obsolescence. This will avoid designing in parts whose obsolescence may not be realized until production.

When replacement parts are found, it is good practice to perform a Highly Accelerated Life Test (HALT) on the product to validate performance to specification.

25.2 Product Termination

At some point in time, a business is compelled to stop manufacturing a product for sale. In addition, a decision is required to stop supporting a product. The two decisions do not need to occur at the same time. When a product is being phased out, the following activities should take place for proper closure:

1. Remove items no longer needed from the warehouse:
   (a) Parts and assemblies
   - Scrapped
   - Sold at discount
   - Reworked for new products
   (b) Literature
   - Schematics
   - Manuals
   - Bills of material
2. Transform the old manufacturing processes out of production:
   (a) Manufacturing lines and/or cells
   (b) Production line inventories
   (c) Test fixtures and equipment
   (d) Processes
3. End sustaining engineering support.

25.3 Project Assessment

Finally, a review should be done to study the lessons learned throughout the product life cycle. What plans could have been put in place to mitigate problems that occurred that might well happen on the next product? Should additional checks be added to the process? Do the planned life cycles match what really happened with the product? Should life cycles be reviewed, and how often? There are many items and subsequent actions that can be added to the process to develop smoother product life cycles, and going through an end-of-life review will help ensure better outcomes.

Further Reading

R. Solomon, P. A. Sandborn, M. G. Pecht, Electronic Part Life Cycle Concepts and Obsolescence Forecasting, IEEE (2000).

26

Field Service

You do your job well and still there are some failures, both in-house and in the field. Knowing this, you must make this eventuality as painless as possible. Doing so will reduce cost and customer dissatisfaction. Taking repair and maintenance into consideration is part of the product design requirements. By design, you can provide a product that is easily maintained and quickly returned to service.

26.1 Design for Ease of Access

When subassemblies or component parts fail, easy access makes the repair faster and more reliable. Designing for easy access usually adds little to no cost to the product; that is, as long as this aspect of the product design is kept in mind during the design phase.

Typically, a product will have several subassemblies that make up the whole. Designing the system so that these smaller units can be removed and replaced quickly and easily is an accessibility plus. Designs should allow the service technician the ability to get to any assembly that is likely to fail without the need for loosening or removing other assemblies. When this is not the case, the handling of the other subassemblies, twisting cables, removing pulleys, belts, brackets, and so on can lead to unanticipated failures in the future or can extend the service time.

When a failed assembly can be removed and replaced without removing other parts, the reliability after repair will be higher. This is because in removing other assemblies to service a part, there is a risk of damaging unaffected assemblies and the added complexity associated with reassembly. We all have the experience where something is serviced for one problem and soon a new problem surfaces that was probably caused by improper reassembly or adjustments.

26.2 Identify High Replacement Assemblies (FRUs)

Accept that your product will have some failures. Identify those assemblies that are most likely to need service more often than others. Design the system so that these sections can be easily removed, serviced, and replaced, without having to access other parts of

Improving Product Reliability and Software Quality: Strategies, Tools, Process and Implementation, Second Edition. Mark A. Levin, Ted T. Kalal and Jonathan Rodin.
© 2019 John Wiley & Sons Ltd. Published 2019 by John Wiley & Sons Ltd.

the system. Doing so will make the system service event less painful to you and your customer.

When considering replacement items, be sure to identify those that will need periodic servicing and or replacement, for example, fans, filters, belts, drive wheels, fluids, circuit boards, batteries, and so on. Knowing what will need replacement and making this step easy is part of the design effort. Where applicable, identify those items that should be replaced on schedule.

Power supplies that have cooling fans have two failure specifications. The mean time between failures (MTBF) of the supply may be several hundred thousand hours while the fan in the supply may have only a 20 000- to 80 000-hour life expectancy. When the fan fails, the power supply will sense fan rotation stoppage and shut down to prevent failures from overheating. Select power supplies with long fan life, not just high MTBFs. See that these fans can be easily replaced – even in the field. Fans will be less expensive to stock for repairs than power supplies. In cases where you return these power supplies to their manufacturer for service, make fan replacement part of the repair contract to ensure that newly repaired power supplies do not fail soon because the fan is about to expire.

Often, several fans are used in larger systems. In a group of the same fans, each having the same life expectancy, one fan in the group will fail first. It is best to replace the whole group, all at the same time. Lubricant loss is the main cause of fan failure. All of the fans in the group will have a similar environment, so the oil loss from each fan will be very similar. The fan that failed first is an indicator that the others will soon follow. Replace them all at once. After you learn the fan failure rate, you can initiate a preemptive replacement schedule. They can be replaced during scheduled maintenance, thus avoiding a failure during use.

Fans, motors, pumps, filters, seals, and so on are components that will need to be replaced periodically, much like the fan belt in an automobile. Some subassemblies are attached to these wearout items. Often, a fan is made part of the whole unit, for example, a power supply. This means that the whole subassembly will need to be replaced when a fan fails. Avoid this by making the fans a subassembly in itself. Subassemblies with fans usually have many wires to disconnect and later reattach when the replacement unit is obtained. Having the fans in a self-contained unit allows for a simpler wiring removal and attachment because only a few wires must be disconnected and reattached, even for complex fan systems. Field replacement units (FRUs) need to be identified as part of the design and passed on to those who provide these items to the field. Remember that the failure modes and effects analysis (FMEA) process can help identify FRUs.

Preemptive service planning can make the total service cost lower. Use failure reports (failure reporting, analysis and corrective action system [FRACAS]) to identify those areas that are more predictable with regard to wearout, and install service schedules intended for scheduled down times. Apply usage rates and inventory these items to ensure that there are no outages.

In cases where there is a need to replace filters, consider adding a filter flow-sensing component to the design. This can help your customer to avoid downtime during operation. Again, ensure that the filter can be easily replaced. Add low-fluid-level indicators to avoid catastrophic failures that could be avoided by a simple oil change or a fluid topping-off step.

26.3   Wearout Replacement

Some wear items need special accommodations. Incandescent lamps need sockets as do some relays, contactors, circuit breakers, and so on. Sockets can be added to these components to facilitate fast service. However, adding connectors (sockets) of any sort lowers reliability. Be prudent when adding connectors.

26.4   Preemptive Servicing

When servicing your own product, consider replacing some parts or components that have shown predictable wear out times. An electrolytic capacitor will lose the fluid in the capacitor much like a fan loses oil from evaporation. The hotter the environment, the sooner this will happen. Check from FRACAS, if this is an issue and replace these parts as part of the repair process. Consider using components that can either last longer in this environment (move from 85 to 105 °C electrolytic capacitors because they last approximately four times longer at elevated temperatures). Rotating devices like fans can be affected by how they are mounted (ball-bearing devices are more reliable with a vertical shaft position than sleeve-bearing devices; however, they are much more susceptible to shock). The key is to use FRACAS to identify wearout items that have less-than-expected product life.

Some sophisticated systems provide monitoring of critical components, that is, X-ray tubes in CAT scanners, the number of contact insertion/removal cycles in large circuit board arrays, the number of times recording tapes/disks are rerecorded, and so on. Some of these top-end systems actually connect to the factory service group via the internet, to preemptively order replacement parts for servicing, again during scheduled maintenance. These systems ensure that the service personnel and the replacement hardware arrive at the same time for efficient servicing. This maintains a very high availability.

When examining current field replacement items, it may become apparent that some wear items need replacement too frequently. Here, you can make the change to a more reliable component before the design is finalized. Fixing this problem early in the product development cycle is less costly than doing it after the product is already in the field.

26.5   Servicing Tools

When a system needs an adjustment, see that the design takes into consideration where the adjusting tool will have to go to make the adjustment. Make it easy to insert this tool to make the adjustment. Be aware of the environment in which the system will be. Will there be adequate lighting, or will the covers of the system block light needed for proper servicing? Are some adjustments read on a poorly illuminated scale? Make this area more visible in your design.

The selection of service tools is important, too. Design to accommodate tools that are inexpensive and readily available. If your product sales are worldwide, consider tool availability in foreign countries. Even though slotted and Phillips screwdrivers are common worldwide, avoid them because they can leave small shards caused by scraping the screwhead during tightening. These metal icicles can get into places and cause other failures and damage. A slotted-hexhead machine screw is a good alternative. Torx screws are even better and last even longer as replacement screwdriver bits on the manufacturing floor, but they are not as available worldwide. In some cases, a design requires a special tool. When special tools are needed for your product, ensure that it is readily available through your facilities. Remember, however, special tools are expensive to design, manufacture, and stock. Ordinary tools cost much less.

Larger systems may require the removal of very heavy subassemblies. Human strength sometimes cannot do the job. Larger systems often have several heavy subassemblies, such as power units, air conditioners, large circuit card assemblies, and so on. Make sure that the design teams interface when there is a need for special tools (again consult with the field service personnel). Work to design one tool for all these tasks, instead of one special tool for each major subassembly. Even consider the more ordinary tool need. Select screw sizes that need only one size screwdriver. This too cuts down the cost and quantity of service tools.

The field service group will have a prescribed set of tools that they carry to service the equipment in the field. Get the list, make copies, and pass it on to every designer in the company. Make it clear that these are already the tools they can consider for field use. New tools can be added only upon acceptance by the field service personnel. Make as part of the design verification process a day when the designers get to remove, adjust, recalibrate, and replace their designs. Have everyone perform this service step. The design group will do a much better tool selection job, even beforehand, because they will be thinking of the tool set when they design the product. New tools add cost, make the toolbox heavier, and are often unnecessary. However, sometimes special tools are necessary.

When you find that the design team has a need for a special tool, ask why. Review all the tools that are part of the service kits and work with the service personnel to see if the existing set will do the job. If all that fails, a special tool may be necessary.

True, service personnel are proud of their tools, and to many, the more tools, the better. But after they have to carry them long distances, they tend to rethink this idea. Take the time to audit the toolboxes of the service personnel. See if there are tools in their kits that are not on the prescribed list. Learn why any have been added. You may learn that the assembly where this added tool is used cannot be repaired or removed without it, or it is a better tool than the one the designers specified. Service technicians will inevitably add some tools to their service kit. Looking over what they have added, however, can be very instructive.

26.6 Service Loops

Some assemblies will need interconnecting cables. Add some length to these cables where possible to allow for easy removal and replacement. This is called *a service loop*. Locate the connectors for these internal assemblies so that they can be unplugged before the subassembly is removed and plugged back in after the subassembly is securely attached. This added safety precaution is a valuable service feature. Design so that one person can do the job. Sending two repairmen to a job is more expensive. Where customers have in-house service personnel, this can help lower their costs, too.

Design cabinetry so that the service personnel can do the work and use meters and other tools without clumsy handling. Adding a power outlet to the cabinet for test equipment is a plus.

When parts are easily removed for service, the time to return the system back to operation is lowered. This too can be part of the design. Some systems require calibration after servicing. Design subassemblies so that this is taken into consideration. Perhaps, compartmentalization of the subassemblies can make it so most servicing requires no new adjustments or calibration procedures. This speeds the servicing and makes the system quickly available for use.

26.7   Availability or Repair Time Turnaround

Availability is an important part of a design. Design for quick service periods. Availability can be measured and is a metric that can be used with your reliability efforts. The average time between failures is known as the MTBF. The time to return a system to the user is the mean time to repair (MTTR). They mathematically relate in a term known as *availability*, expressed as a percentage of uptime:

$$Availability = MTBF/(MTBF + MTTR)$$

$$\times 100 \text{ (expressed as a percentage of uptime)}$$

The equation shows that the availability is greatest when the MTTR is the lowest. A substantial MTBF can be greatly impacted by a large MTTR. (An automobile that has few failures but requires replacement parts from another country is undesirable.)

26.8   Avoid System Failure Through Redundancy

Where high reliability is required, there may be a need for redundancy. This is where the designers use extra components in tandem, such that when one fails the others can still handle the load. This is commonly done with power supplies. Having five, 500 000-hour MTBF power supplies operating as a group where all are needed means that the combined MTBF is 100 000 hours. Adding one more power supply in tandem (six in all), can extend the real reliability of the supply group to several million hours depending on the time planned to inspect for a failed unit and replacing the failed unit before another unit fails. This is referred to as an $N + 1$ *redundancy*. This can be done with switch and connector contacts, fluid lines, and hard drives. The list is endless.

When the reliability of a component is very high and the probability of failure is extremely low, the need for redundancy or local spare parts can be eliminated. Some high-end automobiles have run-flat tires; there is no spare for replacement on the road. These tires can run flat at highway speeds for up to 100 miles so the driver can get to a service facility where replacement is done.

26.9   Random versus Wearout Failures

It is important to point out that the failures in the field are driven by *random failures* and by *wearout-type failures*. Wearout can be, to a large degree, accommodated in the service planning by design. The accumulation of data in the FRACAS system can help identify the amount of replacement items needed.

Further Reading

Steinberg, D.S. (2000). *Vibration Analysis for Electronic Equipment*, 3e, 9. Wiley.

Appendix A

The information in this appendix may contain inaccuracies or typographical errors. Every effort was made to ensure the accuracy of this information at the time of final edit. However, business and web information changes over time and without notice. In addition, the information here does not constitute an endorsement by the authors.

A.1 Reliability Consultants

Company	General information
DfR Solutions 9000 Virginia Manor Road Suite 290 Beltsville, MD 20705 https://www.dfrsolutions.com	Offers innovative reliability software, comprehensive reliability services, and industry expertise to empower customers to maximize and accelerate product design and development while saving time, managing resources, and improving customer satisfaction. DfR Solutions supports clients across electronic technology markets as well as throughout the electronic component and material supply chain.
	DfR Solutions will work with you throughout the lifecycle of the product and lend a guiding hand regarding quality, reliability, and durability (QRD) issues. The services it offers include technology insertion, design review, supply chain, testing/product quality, field/customer returns, and more.
Accendo Reliability (Reliability Engineering Professional Development) https://accendoreliability.com	Providing consulting services for reliability engineering and reliability management. The website is a great resource for a vast array of reliability information including training courses, webinars, consultants, reliability articles, ebooks, and a calendar linking to an extensive list of reliability activities taking place now and in the future.

Improving Product Reliability and Software Quality: Strategies, Tools, Process and Implementation, Second Edition. Mark A. Levin, Ted T. Kalal and Jonathan Rodin.
© 2019 John Wiley & Sons Ltd. Published 2019 by John Wiley & Sons Ltd.

Company	General information
Hobbs Engineering A Division of ESPEC North America Inc. 10390 East 48th Avenue Denver, CO 80238 Phone: (303) 655-3051 https://hobbsengr.com	Hobbs Engineering Corporation specializes in teaching and consulting accelerated reliability techniques such as HALT and HASS, which were invented by Dr. Hobbs. The corporation offers some 20 courses in classical and accelerated reliability methods.
Ops A La Carte LLC 990 Richard Ave #101 Santa Clara, CA 95050 408-654-0499 ext. 204 http://www.opsalacarte.com	Ops A La Carte is a professional reliability engineering firm with a team of reliability consultants offering a vast array of reliability services for every phase of the product life cycle. It offers a wide range of reliability services including: reliability goal setting and assessment, benchmarking, risk assessment, gap analysis, FMEA, consulting, HALT and HASS testing, training, webinars, and software reliability.
Accelerated Reliability Solutions 1115 Nottingham St. Lafayette, Colorado 80026 512-554-3111 http://www .acceleratedreliabilitysolutions.com/ services.html	Consulting, training, and test strategies for HALT and HASS.
Apex Ridge Reliability E-mail: abahret@apexridge.com http://apexridge.com 978-879-8617	Offers a full suite of reliability services including: reliability program plans, reliability testing, system reliability, reliability tools, reliability training, and organizational evaluation.
Carl S. Carlson E-mail: Carl.Carlson@effectivefmeas .com www.effectivefmeas.com	FMEA consultant. Author of the book *Effective FMEAs: Achieving Safe, Reliable, and Economical Products and Processes using Failure Mode and Effects Analysis*.
Dr. Jean-Paul Clech EPSI, Inc. P. O. Box 1522 Montclair, NJ 07042 973-746-3796 E-mail: JPClech@aol.com http://www.jpclech.com	Dr. Jean-Paul Clech consults through EPSI, Inc., a reliability engineering firm serving the electronics industry providing cost effective solutions to build in the reliability of electronic packages and circuit board assemblies. Specialty services include SMT/BGA/Flip-Chip/CSP solder joint reliability assessment, and package thermal stress analysis.
Equipment Reliability Institute (ERI) (805) 456-4274 E-mail: info@equipment-reliability .com https://equipment-reliability.com/ consulting-services	Offering consulting services on numerous reliability subjects with a focus on vibration, shock, HALT, HASS. A list of their on-site classes can be found at: https:// equipment-reliability.com/training/onsite-courses

Company	General information
FMS Reliability Fred Schenkelberg, CQE, CRE 15466 Los Gatos Blvd Los Gatos, CA 95032 E-mail: fms@fmsreliability.com 408-710-8248 http://www.fmsreliability.com	Providing consulting services for reliability engineering, reliability management, and reliability testing. Offers certified reliability engineering (CRE) preparation classes and material. The website (Accendo Reliability) is a great resource for a vast array of reliability information including training courses, webinars, consultants, reliability articles, ebooks and a calendar linking to an extensive list of reliability activities taking place now and in the future.
HBM PRENSCIA - ReliaSoft Arizona: 520-886-0410 Michigan: 248-350-8300 Europe, Middle East, and Africa +48 22 436 67 70 Asia and Asia Pacific +65 6272 7422 India +91 44 4208 7785 E-mail: ReliaSoft@ReliaSoft.com http://www.ReliaSoft.com	Offering global professional reliability services for reliability engineering and asset management for a vast array of products, including microelectronics, appliances, advanced weapons systems, and off-shore oil well drilling equipment.
Item Software (USA) Inc. 2875 Michelle Drive Suite 300 Irvine, CA 92606 U.S.A. US (240) 297 4442 e-mail: sales@itemsoft.com http://www.itemsoft.com 10:107:135:13A1412:133:13A1412: 132:13A1412:131:132:13	Providing consulting services for reliability, safety, and risk assessment products.
Patrick O'Connor Engineering Management, Quality, Reliability, Safety: Consultancy and Training 62 Whitney Drive, Stevenage, Hertfordshire SG1 4BJ, UK Tel: +44(0)1438 313048 Fax: +44(0)1438 223443 E-mail: pat@pat-oconnor.co.uk	Pat O'Connor provides consulting and training in the practical aspects of quality, reliability, and safety engineering, emphasizing the effective use of design analysis methods, testing, and management. His teaching is based on his books: *Practical Reliability Engineering, Test Engineering,* and *The Practice of Engineering Management.*
Reliability Analytics 24 Van Vorst st. Utica, NY. 13501-5620 315-765-0001 Email: seymour .morris@reliabilityanalytics.com http://reliabilityanalytics.com	Specializes in developing custom analytical models for evaluating system reliability, reliability prediction, product support, product maintenance, warranty, and RMA.

Company	General information
Reliability Center, Inc. P.O. Box 1421 501 Westover Ave. Hopewell, VA 23860 804-458-0645 804-452-2119 (Fax) info@reliability.com www.reliability.com	Reliability Center, Inc. specializes in helping businesses, industry, government, and healthcare organizations improve reliability in all aspects of their operations. The firm provides consulting services, training programs, and software products to clients using their exclusive Opportunity Analysis/Basic FMEA LEAP™System and Root Cause Analysis PROACT® System.
Reliass Cams Hall Fareham, Hampshire PO16 8AB United Kingdom +44 1329 227 448 +44 1329 227 449 fax http://www.reliability-safety-software .com	Reliass offers a wide range of consultancy services and can provide world-renowned experts in the field of reliability, availability, safety, and logistics consultancy services worldwide.

Note: ESS: Environmental stress screening; FMEA: failure modes and effects analysis; HALT: Highly Accelerated Life Test; HASS: Highly Accelerated Stress Screens; HAST: Highly Accelerated Stress Test; MTBF: mean time between failures.

A.2   Graduate Reliability Engineering Programs and Reliability Certification Programs

Name of University	Graduate Reliability Programs			On line
	MS	PhD	Cert.	
Arizona State University Master of Engineering in Quality, Reliability, and Statistical Engineering 699 S Mill Ave Tempe AZ 85281 480-965-2100 E-mail: asuonline@asu.edu https://asuonline.asu.edu/online-degree-programs/graduate	Yes	No		Yes

Name of University	Graduate Reliability Programs			On line
	MS	PhD	Cert.	
Beihang University	Yes			
Safety and Reliability Engineering for Aeronautics and Astronautics Systems Training Programs for MA Students				
No.37 XuanYuan Road				
Beijing, China				
010-82317114				
E-mail: cst@buaa.edu.cn				
http://rse.buaa.edu.cn/plus/view.php?aid=295				
California State University Dominguez Hills	No	No	Yes	Yes
1000 E. Victoria Street				
Carson, CA 90747				
(310) 243-3696				
E-mail: msqa@csudh.edu				
https://www.csudh.edu/qa-ms/certificates/reliability-engineering-schedule				
CALCE Center for Advanced Life Cycle Engineering	Yes	Yes	Yes	Yes
MS in Reliability Engineering				
PhD in Reliability Engineering				
University of Maryland				
1103 Engineering Lab Building				
University of Maryland				
College Park, MD 20742				
Phone: 301-405-5323				
E-mail: calce-education@umd.edu				
https://calce.umd.edu/graduate-program				
Clemson	No	No	Yes	Yes
1000 E. Victoria Street				
Carson, CA 90747				
(310) 243-3880				
E-mail: msqa@csudh.edu				
https://www.csudh.edu/qa-ms/certificates/reliability-engineering				
Drexel University	No	No	Yes	Yes
Graduate Certificate in Systems Reliability Engineering				
3025 Market Street				
Philadelphia, PA 19104				
(877) 215-0009				
E-mail: DUonline@drexel.edu				
https://online.drexel.edu/online-degrees/engineering-degrees/cert-systems-reliability/index.aspx				

Name of University	Graduate Reliability Programs			On line
	MS	PhD	Cert.	
Federation University of Australia Master of Maintenance and Reliability Engineering +61 3 5327 9018 E-mail: international@federation.edu.au https://study.federation.edu.au/#/course/GMR9	Yes	No		Yes
Heriot-Watt University Safety, Risk and Reliability Engineering, MSc/Diploma +44 (0)131 451 4665 E-mail: g.h.walker@hw.ac.uk https://www.hw.ac.uk/study/uk/postgraduate/safety-risk-reliability-engineering.htm	Yes	No	No	Yes
Norwegian University of Science and Technology (MSc in) Reliability, Availability, Maintainability and Safety +47 73595000 https://www.ntnu.edu/studies/msrams	Yes	No	No	No
Ohio State University 122 Hitchcock Hall 2070 Neil Avenue Columbus, OH 43210 614-292-7153 E-mail: eng-profed-REC@osu.edu https://professionals.engineering.osu.edu/reliability-engineering-certification	No	No	Yes	No
Rutgers, The State University of New Jersey Master of Business and Science Degree with Quality and Reliability Engineering concentration 118 Frelinghuysen Rd., SERC Building, Rm. 221 Piscataway, NJ 08854 (848) 445-5117 E-mail: hopham@rci.rutgers.edu https://mbs.rutgers.edu/program/quality-reliability-engineering	Yes	No		No
Subir Chowdhury School of Quality and Reliability Indian Institute of Technology Kharagpur, Kharagpur - 721302 + 91-3222-282230 E-mail: np-off@hijli.iitkgp.ernet.in http://www.sqr.iitkgp.ac.in	Yes		Yes	

Name of University	Graduate Reliability Programs			On line
	MS	PhD	Cert.	
University of Arizona	Yes	No		No
Master of Business and Science Degree with Quality and Reliability Engineering concentration				
Aerospace and Mechanical Engineering Dept.				
Bldg 16, room 200B				
Tucson, AZ. 85721-8191				
(520) 621-6551				
E-mail: graduateadvisor@sie.arizona.edu				
E-mail: jianliu@sie.arizona.edu				
http://www.sie.arizona.edu/ms-quality-and-reliability-option				
UCLA	Yes	No	Yes	Yes
Master of Reliability Engineering				
Engineering VI				
404 Westwood Plaza				
Los Angeles, CA 90095				
(310)794-5141				
E-mail: info@risksciences.ucla.edu				
https://www.risksciences.ucla.edu				
University of Kansas			Yes	No
Edwards Campus				
Professional and Continuing Education				
Regents Center 125				
12600 Quivira Rd.				
Overland Park, KS 66213				
University of Manchester	Yes	No		No
MSc Reliability Engineering and Asset Management				
Oxford Rd.				
Manchester				
M13 9PL UK				
+44(0)161 306 9219				
E-mail: pg-mace@manchester.ac.uk				
http://www.manchester.ac.uk/study/masters/courses/list/09822/msc-reliability-engineering-and-asset-management				
University of Maryland	Yes	Yes	Yes	Yes
MS in Reliability Engineering				
PhD in Reliability Engineering				
Building 89, Room 1103				
College Park, MD 20742				

Name of University	Graduate Reliability Programs			On line
	MS	PhD	Cert.	
301-405-5323 301-314-9269 (fax) E-mail: webmaster@calce.umd.edu Electronic Products and Systems Center http://www.calce.umd.edu				
University of Tennessee 506 East Stadium Hall Knoxville, Tennessee 37996-0750, USA Phone: (865) 974-9625 Fax: (865) 974-4995 E-mail: mrc@utk.edu Master of Science in Reliability and Maintainability Engineering http://www.engr.utk.edu/mrc	Yes	No	Yes	Yes
Vanderbilt University Master of risk, reliability, and resilience engineering Loews Vanderbilt Plaza Hotel Office Complex 2100 West End Ave., Suite 1100 Nashville, TN 37203 (615) 322-6397 E-mail: vuse.rrr@vanderbilt.edu https://news.vanderbilt.edu/2018/01/31/vanderbilt-school-of-engineering-offers-new-master-of-risk-reliability-and-resilience-engineering	Yes	No		No

A.3 Reliability Professional Organizations and Societies

American Society for Quality
600 North Plankinton Avenue
Milwaukee, WI 53203 USA
800-248-1946
414-272-1734 fax
help@asq.org email
http://www.asq.org

IEEE Corporate Office
3 Park Avenue, 17th Floor
New York, New York
10016-5997 USA
Tel: +1 212 419 7900
Fax: +1 212 752 4929

Society of Automotive Engineers
SAE World Headquarters
400 Commonwealth Drive
Warrendale, PA 15096-0001 USA
1-877-606-7323 USA
724/776-4841 outside USA

Society of Reliability Engineers
250 Durham Hall
Virginia Tech
Blacksburg, VA 24061-0118
http://www.sre.org

NASA preferred reliability practices and guidelines for design and test http://msfcsma3
.msfc.nasa.gov/tech/practice/prctindx.html

NASA preferred reliability practices with links to other NASA reliability and maintainability sites
http://www.hq.nasa.gov/office/codeq/overvw23.htm

Reliability Analysis Center (RAC)
http://rac.iitri.org
Society of Reliability Engineers (SRE)
This resource provides education, social contact, and insight to foster understanding of reliability, maintainability, and life testing.
http://www.sre.org

IEEE Reliability Home
http://www.ewh.ieee.org/soc/rs

IMAPS – International Microelectronics And Packaging Society http://www.imaps
.org

Emerald Library Sign-on
http://www.emerald-library.com/cgi-bin/EMRlogin

The Annual R & M Symposium (RAMS)
http://www.rams.org

A.4   Reliability Training Classes

Company	Available training	Webinars
Accelerated Reliability Solutions 1115 Nottingham St. Lafayette, Colorado 80026 512-554-3111 http://www .acceleratedreliabilitysolutions.com/ services.html	Offering an extensive list of reliability training teaching reliability principles and software tools. Classes include: FMEA, fault tree, reliability prediction, Weibull analysis, accelerated life modeling, reliability growth models, FRACAS, reliability-centered maintenance, probabilistic event and risk analysis, risk-based inspection analysis, and reliability block diagrams	

Company	Available training	Webinars
Accendo Reliability Fred Schenkelberg CQE, CRE 15466 Los Gatos Blvd Los Gatos, CA 95032 (408) 710-8248 https://accendoreliability.com	A resource for a vast array of reliability information, including training courses, webinars, consultants, reliability articles, ebooks, and a calendar linking to an extensive list of reliability activities taking place now and in the future. https://accendoreliability.com/courses	X
Apex Ridge Reliability E-mail: abahret@apexridge.com http://apexridge.com 978-879-8617	Hobbs Engineering Corporation specializes in teaching and consulting accelerated reliability techniques such as HALT and HASS, which were invented by Dr. Hobbs. The corporation offers some 20 courses in reliability and accelerated reliability methods, a wide variety of training seminars and webinars.	X
ASQ Reliability and Risk Division	Offers a regular series of webinars on reliability, maintainability, and risk. Can access past webinars through vimeo. http://www.asqrd.org/webinars	X
DfR Solutions 9000 Virginia Manor Road Suite 290 Beltsville, MD 20705 https://www.dfrsolutions.com	Offers over 17 different courses in reliability including in-house training and free webinars on reliability design, failure analysis, manufacturing, reliability testing, reliability modeling, and simulation and system reliability. Previous webinars are available from its website.	X
Equipment Reliability Institute (ERI) (805) 456-4274 info@equipment-reliability.com https://equipment-reliability.com/consulting-services	Offering a various array of on-site classes on reliability testing (HALT, HASS, ELT) and test equipment. The full list of classes can be found at: https://equipment-reliability.com/training/onsite-courses	
HALT and HASS Consulting NZ Ltd. 218A Annex Road Middleton, Christchurch 8024 Canterbury New Zealand Landline (NZ): 03 390 1255 https://www.halthass.co.nz	A very large list of reliability classes and training seminar series. Classes are available as open courses, live webinars, and on-demand and distance learning.	X
HBM PRENSCIA - ReliaSoft Arizona: 520-886-0410 Michigan: 248-350-8300 Europe, Middle East, and Africa +48 22 436 67 70 Asia and Asia Pacific +65 6272 7422 India +91 44 4208 7785 E-mail: ReliaSoft@ReliaSoft.com http://www.ReliaSoft.com	A resource for webinars, conference proceedings, articles, best practices, metrics, and guidelines.	X

Company	Available training	Webinars
Ops A La Carte LLC 990 Richard Ave #101 Santa Clara, CA 95050 408-654-0499 ext. 204 http://www.opsalacarte.com	Ops A La Carte is a professional reliability engineering firm with a team of reliability consultants offering a vast array of reliability services for every phase of the product life cycle. They offer an extensive list of reliability courses and webinars that are outlined on their website, www.opsalacarte.com.	
Relyence 540 Pellis Road Greensburg, PA 15601 USA Tel 724.832.1900 Email info@Relyence.com www.relyence.com	Offering webinar and on-site training classes for: FMEA, FRACAS, fault tree, reliability prediction, and reliability block diagrams (RBD)	X
Technology Training, Inc. 866-TTi-4Edu (866-884-4338). Training@ttiedu.com https://ttiedu.com	Provide training in engineering applications not found at conventional educational institutions. They offer a very extensive list of courses and training topics. The training is available on demand, distance learning, and live open courses. A complete list of all their courses and training topics can be found at: https://ttiedu.com/course_list	X

A.5   Environmental Testing Services

Environmental test facility	HALT/ HASS	Environmental testing	Electrical testing	Inspection and F/A
Accelerated Reliability Solutions 1115 Nottingham St. Lafayette, Colorado 80026 512-554-3111 http://www .acceleratedreliabilitysolutions.com/ services.html	Yes	Yes		
Anecto Ltd. Mervue Ind Estate Galway, Ireland +353(0) 91 7574 04 https://www.anecto.com/tests	Yes	Yes	Yes	Yes
Apex Ridge Reliability E-mail: abahret@apexridge.com http://apexridge.com 978-879-8617	Yes	Yes		Yes

Environmental test facility	HALT/ HASS	Environmental testing	Electrical testing	Inspection and F/A
Cascade Engineering Services, Inc. 6640 185th Ave. NE Redmond, WA 98052 (425) 895-8617 www.cascade-eng.com	Yes	Yes	Yes	Yes
Contech Research, Inc. 750 Narragansett Park Drive Rumford, Rhode Island 02916 (401) 865-6440 http://www.contechresearch.com		Yes	Yes	Yes
CSZ 12011 Mosteller Rd. Cincinnati, OH. 45241-1528 USA (513) 772-8810 http://www.cszindustrial.com	Yes			
DELSERRO ENGINEERING SOLUTIONS 3900 Broadway Road Easton, PA 18040 (610) 253-6637 E-Mail: info@desolutions.com https://www.desolutions.com/testing-services/reliability-tests/halt	Yes	Yes		
HALT & HASS Consulting NZ Ltd. 218A Annex Road Middleton, Christchurch 8024 Canterbury New Zealand Landline (NZ): 03 390 1255 https://www.halthass.co.nz	Yes	Yes	Yes	Yes
Intertek Global Branches 800 967 5352 http://www.intertek.com/ performance-testing/halt-and-hass	Yes	Yes		
NTS Numerous locations (844) 332-1885 https://www.nts.com/services/testing		Yes	Yes	Yes

Environmental test facility	HALT/ HASS	Environmental testing	Electrical testing	Inspection and F/A
Ops A La Carte LLC 990 Richard Ave #101 Santa Clara, CA 95050 408-654-0499 ext. 204 http://www.opsalacarte.com/Pages/ reliability/reliability_prot_halt.htm	Yes	Yes		
Reliant Labs, Inc. 925 Thompson Place Sunnyvale, CA 94085 (408) 737-7500 http://reliantlabs.com	Yes	Yes		
Sonoscan Numerous locations http://www.sonoscan.com				Yes
System Effectiveness Associates, Inc. 20 Vernon Street Norwood, MA 02062 Phone: (781) 762-9252 Fax: (781) 769-9422 Email: info@sea-co.com http://www.sea-co.com/index.cfm	Yes			
Westpak, Inc. 10326 Roselle Street San Diego, CA 92121 858-244-9193 https://www.westpak.com	Yes	Yes		

A.6   HALT Test Chambers

Manufacturers of HALT and HASS equipment

Angelantoni Test Technologies Srl (ACS)
Loc. Cimacolle, 464 - 06056 Massa Martana (PG)
c.f./p. iva/r.i.: 03216310544 – Italy
+39-075-89551
http://www.acstestchambers.com/Product/Prodotto?id_fam=11&id_prod=0

CSZ
12011 Mosteller Rd.
Cincinnati, OH. 45241-1528 USA
(513) 772-8810
(800) 989-7373
Email: indsales@genthermcsz.com
http://www.cszindustrial.com/Products/Vibration-Test-Systems/HALT-HASS-Chambers.aspx

Manufacturers of HALT and HASS equipment

ESPEC Qualmark North America Inc.
Colorado Office
10390 East 48th Ave.
Denver, CO 80238
(303) 254-8800
https://www.qualmark.com/products

Controltechnica (CTS)
Lotzenäcker 21
72379 Hechingen – Germany
Tlf: +49 747 198 50-0
+34 916 613 004
e-mail: vertrieb@cts-umweltsimulation.de
www.cts-umweltsimulation.de
https://www.cts-clima.com/en/vibration/halt-hass-chambers.html

Thermotron
291 Kollen Park Drive
Holland, Michigan 49423, USA
(616) 649-2373
(800) 409-3449
http://thermotron.com/equipment/halt-hass-chamber.html

A.7 Reliability Websites

Barringer and Associates, Inc. Links to Other Reliability Sites
http://www.barringer1.com/links.htm

Physics of Failure Homepage
http://amsaa-web.arl.mil/rad/pofpage.htm (check link)

Adams Six Sigma
http://www.adamssixsigma.com

Accelerated Reliability Solutions
1115 Nottingham St.
Lafayette, Colorado 80026
512-554-3111
http://www.acceleratedreliabilitysolutions.com/services.html

A.8 Reliability Software

Reliability software	Description
DfR Solutions 9000 Virginia Manor Road Suite 290 Beltsville, MD 20705 https://www.dfrsolutions.com	Sherlock is an automated reliability design analysis software program. The software can be licensed or you can use their consulting services to model degradation based on physics of failure models. Used to predict failure for components on a circuit board that see thermal and mechanical stresses. Can be used as a DFM/DFR tool to mitigate risk.
EPSI, Inc. (Electronics Packaging Solutions International Inc.) Dr. Jean-Paul Clech P. O. Box 1522 Montclair, NJ 07042 973-746-3796 E-mail: JPClech@aol.com http://www.jpclech.com	SRS software will estimates solder joint fatigue life for a vast array of components (LCCC, BGAs, CSP, Gull wing, J-lead, S-shaped). The reliability model is based on life test results from over 60 experiments. http://www.jpclech.com/SRS_Software.html
Relyence 540 Pellis Road Greensburg, PA 15601 USA Tel 724.832.1900 Email info@Relyence.com www.relyence.com	Offering software for: FMEA, FRACAS, fault tree, reliability prediction, and RBD
Item Software (USA) Inc. 2875 Michelle Drive Suite 300 Irvine, CA 92606 U.S.A. US (240) 297 4442 e-mail: sales@itemsoft.com http://www.itemsoft.com	Providing reliability, safety, and risk assessment software products.
HBM PRENSCIA - ReliaSoft Arizona: 520-886-0410 Michigan: 248-350-8300 Europe, Middle East, and Africa +48 22 436 67 70 Asia and Asia Pacific +65 6272 7422 India +91 44 4208 7785 e-mail: ReliaSoft@ReliaSoft.com http://www.ReliaSoft.com	Offering a suite of software packages for FMEA, fault tree, reliability prediction, Weibull analysis, accelerated life modeling, reliability growth models, FRACAS, reliability centered maintenance, probabilistic event and risk analysis, risk-based inspection analysis, and RBD

Reliability software	Description
Reliability Analytics Corporation 24 Van Vorst Street Utica, NY. 13501-5620 (315) 765-0001 seymour.morris@Reliabilityanalytics.com https://reliabilityanalytics.com	Offers a free reliability toolkit with about 30 free reliability calculators for all types of common reliability calculations. Link to tool kit is at: http://reliabilityanalyticstoolkit.appspot.com
Relyence 540 Pellis Road Greensburg, PA 15601 USA Tel 724.832.1900 Email info@Relyence.com www.relyence.com	Offering a suite of software products (including training classes) for: FMEA, FRACAS, fault tree, reliability prediction and RBD
Reliass Cams Hall Fareham, Hampshire PO16 8AB United Kingdom +44 1329 227 448 +44 1329 227 449 fax http://www.reliability-safety-software .com	Reliass offers RAMS software tools and solutions including: reliability software, fault tree software, FRACAS software, integrated logistics support (ILS), logistic support analysis record (LSAR), FMEA, FMECA, Six Sigma, and much more

A.9   Reliability Seminars and Conferences

Applied Reliability and Durability Conference (ARDC)
http://www.ardconference.com/index.html
http://www.arsymposium.org
https://www.linkedin.com/company/international-applied-reliability-symposium
International Conference on Reliability Engineering (ICRE)
E-mail: icre_conf@outlook.com
Tel: +861-32-7777777-0
http://www.icre.org
International Society of Science and Applied Technologies (ISSAT)
2101 State Route 27
Box 281
Edison, NJ 08818 U.S.A.
Email: cs@issatconferences.org
https://issatconferences.org/index.html
International Reliability Physics Symposium:
Contact:
Eric Snyder
Sandia Technologies, Inc.
Albuquerque, NM 87109

505-872-0011
505-872-0022 fax
Eric– Snyder@irps.org(email)

http://www.irps.org

International Conference on System Reliability and Safety (ICSRS)

Tel: +861-32-7777777-0
E-mail: icre_conf@outlook.com

http://www.icsrs.org

International Symposium on the Physics and Failure Analysis of Integrated Circuits:

IPFA Secretariat
Kent Ridge Post Office

P.O. Box 1129
Singapore 911105

65-743-2523
65-746-1095 fax
ipfa@pacific.net.sg (e-mail)

http://www.ewh.ieee.org/reg/10/ipfa

International Symposium on the Testing and Failure Analysis:

Contact:
AMS International
Materials Park, OH.
44073-0002
440-338-5151
440-338-4634 fax
shapowa@asminternational.org (e-mail)

http://www.asminternational.org

RAMS:
Contact:
Dr. John English, General Chair

University of Arkansas

Department of Industrial

Engineering

4207 Bell Engineering Center

Fayetteville, AK. 72701

479-575-6029
chair@rams.org (e-mail)

http://www.rams.org

Prognostics and Health Management (PHM) Society

http://www.phmsociety.org

Society for Maintenance and Reliability Professionals (SMRP)

3200 Windy Hill Rd., SE, Suite 600W

Atlanta, GA 30339

info@smrp.org

http://smrp.org/Conference

A.10   Reliability Journals

Engineering Failure Analysis	https://www.journals.elsevier.com/engineering-failure-analysis
IEEE Transactions on Device and Materials Reliability	https://ieeexplore.ieee.org/xpl/RecentIssue.jsp?punumber=7298
IEEE Transactions on Reliability	https://accendoreliability.com/journals/ieee-transactions-reliability
International Journal of Prognostics and Health Management	http://www.phmsociety.org/journal
International Journal of Reliability and Safety	http://www.inderscience.com/jhome.php?jcode=ijrs
International Journal of Reliability, Quality and Safety Engineering	https://www.worldscientific.com/worldscinet/ijrqse
Journal of Failure Analysis and Prevention	https://www.springer.com/materials/surfaces+interfaces/journal/11668
Journal of Quality Technology	http://asq.org/pub/jqt
Microelectronics and Reliability	https://www.journals.elsevier.com/microelectronics-reliability
Quality and Reliability Engineering International	https://onlinelibrary.wiley.com/journal/10991638
Reliability Engineering and System Safety	https://www.journals.elsevier.com/reliability-engineering-and-system-safety

Appendix B

B.1   MTBF, FIT, and PPM Conversions

One of the most often used numbers in reliability is the mean time between failures (MTBF) number. MTBF represents the average time one can expect a device to operate without failing. There is no assurance that the consumer will realize this failure-free time period because the MTBF is a statistical average. In fact, if a consumer experiences a failure, the likelihood of an additional failure is the same before the failure occurs as it is after the failure is repaired (Table B.1).

For example:

If the MTBF = 8,760 hours

Then on average a unit will fail every 8,760 hours or once a year.

1 year = 356 days × 24 h d⁻¹ = 8,760

Viewed another way:

If there are 10,000 of these systems in the field, then the manufacturer can expect 10,000 failures every year (for a repairable system), and if the product is a nonrepairable system, there will be about 6,700 failures. This is covered in greater detail in the next section.

The failures in time (FIT) rate is defined as the failures in time per billion hours.

It is easy to convert between MTBF, FIT, and parts per million (PPM) rates.

B.2   Mean Time Between Failure (MTBF)

There is a lot of confusion about the term MTBF. When people hear that a device has an MTBF of 10,000 hours, they often think that this means that this device will not have a failure for at least 10,000 hours. This is not the case. What this means is that for a group or fleet of systems with an MTBF of 10,000 hours, the *average rate of failure* will be 10,000 hours. Some of the units in this larger group will actually have a failure rate of the stated MTBF rate, while some will fail sooner, and some later. It is understood that with a population of units the average or mean failure rate will be the stated MTBF rate.

Reliability defined: The probability that a product will operate satisfactorily for a required amount of time under stated conditions to perform the function for which it was designed.

Improving Product Reliability and Software Quality: Strategies, Tools, Process and Implementation,
Second Edition. Mark A. Levin, Ted T. Kalal and Jonathan Rodin.

Table B.1 Conversion tables for FIT to MTBF and PPM.

FIT	MTBF	PPM
1	1,000,000,000	9
2	500,000,000	18
3	333,333,333	26
4	250,000,000	35
5	200,000,000	44
6	166,666,667	53
7	142,857,143	61
8	125,000,000	70
9	111,111,111	79
10	100,000,000	88
20	50,000,000	175
30	33,333,333	263
40	25,000,000	350
50	20,000,000	438
60	16,666,667	526
70	14,285,714	613
80	12,500,000	701
90	11,111,111	788
100	10,000,000	876
200	5,000,000	1,752
300	3,333,333	2,628
400	2,500,000	3,504
500	2,000,000	4,380
600	1,666,667	5,256
700	1,428,571	6,132
800	1,250,000	7,008
900	1,111,111	7,884
1,000	1,000,000	8,760
2,000	500,000	17,520
3,000	333,333	26,280
4,000	250,000	35,040
5,000	200,000	43,800
6,000	166,667	52,560
7,000	142,857	61,320
8,000	125,000	70,080
9,000	111,111	78,840
10,000	100,000	87,600
20,000	50,000	175,200
30,000	33,333	262,800

Table B.1 (Continued)

FIT	MTBF	PPM
40,000	25,000	350,400
50,000	20,000	438,000
60,000	16,667	525,600
70,000	14,286	613,200
80,000	12,500	700,800
90,000	11,111	788,400
100,000	10,000	876,000
200,000	5,000	1,752,000
300,000	3,333	2,628,000
400,000	2,500	3,504,000
500,000	2,000	4,380,000
600,000	1,667	5,256,000
700,000	1,429	6,132,000
800,000	1,250	7,008,000
900,000	1,111	7,884,000
1,000,000	1,000	8,760,000
1,100,000	909	9,636,000
1,200,000	833	10,512,000
1,300,000	769	11,388,000
1,400,000	714	12,264,000
1,500,000	667	13,140,000
1,600,000	625	14,016,000
1,700,000	588	14,892,000
1,800,000	556	15,768,000
1,900,000	526	16,644,000
2,000,000	500	17,520,000
3,000,000	333	26,280,000
4,000,000	250	35,040,000
5,000,000	200	43,800,000
6,000,000	167	52,560,000
7,000,000	143	61,320,000
8,000,000	125	70,080,000
9,000,000	111	78,840,000
10,000,000	100	87,600,000
20,000,000	50.0	175,200,000
30,000,000	33.3	262,800,000
40,000,000	25.0	350,400,000
50,000,000	20.0	438,000,000
60,000,000	16.7	525,600,000

Table B.1 (Continued)

FIT	MTBF	PPM
70,000,000	14.3	613,200,000
80,000,000	12.5	700,800,000
90,000,000	11.1	788,400,000
100,000,000	10.0	876,000,000
200,000,000	5.0	1,752,000,000
300,000,000	3.3	2,628,000,000
400,000,000	2.5	3,504,000,000
500,000,000	2.0	4,380,000,000
600,000,000	1.7	5,256,000,000
700,000,000	1.4	6,132,000,000
800,000,000	1.3	7,008,000,000
900,000,000	1.1	7,884,000,000
1,000,000,000	1.0	8,760,000,000
1	1,000,000,000	8,760,000,000
2	500,000,000	4,380,000,000
3	333,333,333	2,920,000,000
4	250,000,000	2,190,000,000
5	200,000,000	1,752,000,000
6	166,666,667	1,460,000,000
7	142,857,143	1,251,428,571
8	125,000,000	1,095,000,000
9	111,111,111	973,333,333
10	100,000,000	876,000,000
20	50,000,000	438,000,000
30	33,333,333	292,000,000
40	25,000,000	219,000,000
50	20,000,000	175,200,000
60	16,666,667	146,000,000
70	14,285,714	125,142,857
80	12,500,000	109,500,000
90	11,111,111	97,333,333
100	10,000,000	87,600,000
200	5,000,000	43,800,000
300	3,333,333	29,200,000
400	2,500,000	21,900,000
500	2,000,000	17,520,000
600	1,666,667	14,600,000
700	1,428,571	12,514,286
800	1,250,000	10,950,000

Table B.1 (Continued)

FIT	MTBF	PPM
900	1,111,111	9,733,333
1,000	1,000,000	8,760,000
2,000	500,000	4,380,000
3,000	333,333	2,920,000
4,000	250,000	2,190,000
5,000	200,000	1,752,000
6,000	166,667	1,460,000
7,000	142,857	1,251,429
8,000	125,000	1,095,000
9,000	111,111	973,333
10,000	100,000	876,000
20,000	50,000	438,000
30,000	33,333	292,000
40,000	25,000	219,000
50,000	20,000	175,200
60,000	16,667	146,000
70,000	14,286	125,143
80,000	12,500	109,500
90,000	11,111	97,333
100,000	10,000	87,600
200,000	5,000	43,800
300,000	3,333	29,200
400,000	2,500	21,900
500,000	2,000	17,520
600,000	1,667	14,600
700,000	1,429	12,514
800,000	1,250	10,950
900,000	1,111	9,733
1,000,000	1,000	8,760
1,100,000	909	7,964
1,200,000	833	7,300
1,300,000	769	6,738
1,400,000	714	6,257
1,500,000	667	5,840
1,600,000	625	5,475
1,700,000	588	5,153
1,800,000	556	4,867
1,900,000	526	4,611
2,000,000	500	4,380

Table B.1 (Continued)

FIT	MTBF	PPM
3,000,000	333	2,920
4,000,000	250	2,190
5,000,000	200	1,752
6,000,000	167	1,460
7,000,000	143	1,251
8,000,000	125	1,095
9,000,000	111	973
10,000,000	100	876
20,000,000	50.0	438
30,000,000	33.3	292
40,000,000	25.0	219
50,000,000	20.0	175
60,000,000	16.7	146
70,000,000	14.3	125
80,000,000	12.5	110
90,000,000	11.1	97
100,000,000	10.0	88
200,000,000	5.0	44
300,000,000	3.3	29
400,000,000	2.5	22
500,000,000	2.0	18
600,000,000	1.7	15
700,000,000	1.4	13
800,000,000	1.3	11
900,000,000	1.1	10
1,000,000,000	1.0	9
1	8,760,000,000	0.1
2	4,380,000,000	0.2
3	2,920,000,000	0.3
4	2,190,000,000	0.5
5	1,752,000,000	0.6
6	1,460,000,000	0.7
7	1,251,428,571	0.8
8	1,095,000,000	0.9
9	973,333,333	1.0
10	876,000,000	1.1
20	438,000,000	2.3
30	292,000,000	3.4
40	219,000,000	4.6

Table B.1 (Continued)

FIT	MTBF	PPM
50	175,200,000	5.7
60	146,000,000	6.8
70	125,142,857	8.0
80	109,500,000	9.1
90	97,333,333	10.3
100	87,600,000	11.4
200	43,800,000	22.8
300	29,200,000	34.2
400	21,900,000	45.7
500	17,520,000	57.1
600	14,600,000	68.5
700	12,514,286	79.9
800	10,950,000	91.3
900	9,733,333	103
1,000	8,760,000	114
2,000	4,380,000	228
3,000	2,920,000	342
4,000	2,190,000	457
5,000	1,752,000	571
6,000	1,460,000	685
7,000	1,251,429	799
8,000	1,095,000	913
9,000	973,333	1,027
10,000	876,000	1,142
20,000	438,000	2,283
30,000	292,000	3,425
40,000	219,000	4,566
50,000	175,200	5,708
60,000	146,000	6,849
70,000	125,143	7,991
80,000	109,500	9,132
90,000	97,333	10,274
100,000	87,600	11,416
200,000	43,800	22,831
300,000	29,200	34,247
400,000	21,900	45,662
500,000	17,520	57,078
600,000	14,600	68,493
700,000	12,514	79,909

Table B.1 (Continued)

FIT	MTBF	PPM
800,000	10,950	91,324
900,000	9,733	102,740
1,000,000	8,760	114,155
1,100,000	7,964	125,571
1,200,000	7,300	136,986
1,300,000	6,738	148,402
1,400,000	6,257	159,817
1,500,000	5,840	171,233
1,600,000	5,475	182,648
1,700,000	5,153	194,064
1,800,000	4,867	205,479
1,900,000	4,611	216,895
2,000,000	4,380	228,311
3,000,000	2,920	342,466
4,000,000	2,190	456,621
5,000,000	1,752	570,776
6,000,000	1,460	684,932
7,000,000	1,251	799,087
8,000,000	1,095	913,242
9,000,000	973	1,027,397
10,000,000	876	1,141,553
20,000,000	438	2,283,105
30,000,000	292	3,424,658
40,000,000	219	4,566,210
50,000,000	175	5,707,763
60,000,000	146	6,849,315
70,000,000	125	
80,000,000	110	9,132,420
90,000,000	97	10,273,973
100,000,000	88	11,415,525
200,000,000	44	22,831,050
300,000,000	29	34,246,575
400,000,000	22	45,662,100
500,000,000	18	57,077,626
600,000,000	15	68,493,151
700,000,000	13	79,908,676
800,000,000	11	91,324,201
900,000,000	10	102,739,726
1,000,000,000	9	114,155,251

Taking an example of 100 units that have a 1,000-hour MTBF; let's find out more about how many failures there will be, how many units will fail before the stated 1,000 failure rate, how many after, and how many units will have more than one failure.

The rate of failure is exponential. Here the expression is:

$$R(t) = N\varepsilon^{-\lambda t} \tag{B.1}$$

the number still surviving without a failure.

N is the number of units shipped; we will use 100.
$\varepsilon = 2.718$ (or the natural logarithm).
λ is the constant failure rate (in failures per million hours).
$t = 1,000$ hours (for this example).

FIT is sometimes used in place of λ, but it is smaller by three orders of magnitude, or one failure per billion hours of operation. It is read as "failures in a thousand million." Therefore 1,000 λ is one FIT.

$$\lambda = 1/\text{MTBF} \tag{B.2}$$

λ and MTBF are inversely related MTBF is mean time between failures. So Eq. (B.1) becomes:

$$R(t) = N\varepsilon^{-(t)/(\text{MTBF})} \tag{B.3}$$

Example : Let $t = 1,000\,\text{h}$

$\text{MTBF} = 1,000\,\text{h}$

$\qquad N = 100$ new VCRs or TV sets, or any other type of system

$\qquad R(t) = 100 \times \varepsilon^{-(1,000\text{ h})/(1,000\text{ h between failures})}$

$\qquad\qquad = 100 \times 2.71^{-1,000/1,000}$

$\qquad\qquad = 100 \times 2.71^{-1}$

$\qquad\qquad = 100 \ \times 0.37$

$\qquad R(t) = 37$ Units "STILL WORKING WITHOUT A FAILURE"

This also means that 63 *units had failures*. But in 1,000 hours, shouldn't all 100 units have had a failure? No; but there still were 100 failures!

At first, it seems impossible that there were 100 failures and 37 units were still working; but the answer is that of the 63 units that had failures, some had more than one failure. Some had two or three or even more failures. That's where the total of 100 failures comes from. The only way this could happen is when one unit fails, it is quickly repaired and placed back into service. Even after one failure, as soon as it is repaired there are 100 units that are operating that all have an MTBF of 1,000 hours. Even after 25 or 50 or 63 failures, as soon as that last failure was repaired there were always 100 units operating; all with an MTBF of 1,000 hours. This is considered the number of failures in "repairable" systems.

B.3 Estimating Field Failures

Suppose a product has a 1,000-hour system MTBF. Then λ will be 1/1,000 or 0.001. This means that every 1,000 hours a system will have a failure. With 100 systems then there will be 100 failures in those 1,000 hours. Remember that these failures will show up in only 63 units; the other 37 units will exhibit no failures during this time period.

So how many of the 63 units had 1, 2, 3, ...or n failures?

The number of units having more than one failure can be determined using:

$$P(n) = [(\lambda^n \times t^n)/n!] \times \varepsilon^{-\lambda t} \tag{B.4}$$

where:

$P(n)$ is the percent of units exhibiting n failures.
t is the time duration,
n is the number of failures in a single system, (e.g. 1, 2, 3,...n).

Let's learn how many units will have 1, then 2, then 3, and so on, failures per unit in the group of 63 units that will exhibit these 100 failures.

But first a short refresher in factorials:

Note: 0! is defined as equaling 1; and 0! is read as "zero factorial." See Table B.2 for a list of common factorials.

For zero failures: (This is the group of 37 units that had no failures in 1,000 hours.)

$$P(0) = [(0.001^0 \times 1,000^0)/1] \times 2.71^{-(0.001 \times 1,000)}$$
$$= [(1 \times 1)]/1 \times 2.71^{-1}$$
$$= 1 \times 0.37$$
$$P(0) = 0.37 \text{ or } 37\%$$

So with 100 units there will be 37 units exhibiting zero failures in one MTBF time period.

How many units will have one failure in 1,000 hours?

Substitute 1 for n:

$$P(1) = [(0.001^1 \times 1,000^1)/1] \times 2.71^{-1}$$
$$P(1) = (1/1) \times 0.37$$
$$P(1) = 0.37, \text{ or } 37\% \text{ will exhibit one failure.}$$

So with 100 units, there will be 37 units exhibiting one failure in one MTBF time period (1,000 hours).

How many units will have two failures in 1,000 hours?

$$P(2) = [(0.001^2 \times 1,000^2)/2] \times 0.37$$
$$P(2) = (1/2) \times 0.37$$
$$P(2) = 18\%$$

Table B.2 Factorials.

n	n!	n factorial	The math
0	0!	Zero factorial	Defined as 0
1	1!	One factorial	$1 \times 1 = 1$
2	2!	Two factorial	$1 \times 2 = 2$
3	3!	Three factorial	$1 \times 2 \times 3 = 6$
4	4!	Four factorial	$1 \times 2 \times 3 \times 4 = 24$
5	5!	Five factorial	$1 \times 2 \times 3 \times 4 \times 5 = 120$

So with 100 units there will be 18 units exhibiting two failures in one MTBF time period (1,000 hours).

$P(3) = 6$ units exhibiting 3 failures in one MTBF.

$P(4) = 1$ units exhibiting 4 failures in one MTBF.

$P(5) =$ may be 1 unit exhibiting 5 failures in one MTBF

(numbers are rounded).

A more simple way of finding the percentage of failures encountered in a given time period is:

$$P(f) = \lambda t \tag{B.5}$$

Find how many will fail in *one-hundredth* of an MTBF time period.

$P(f) = 0.001 \times 1{,}000/100 \, h$

$P(f) = 0.001 \times 10$

$P(f) = 0.01$ or 1%

Using 100 units, this means that 1 unit exhibits the very first failure in 10 hours. So the time to first failure is 10 hours!!!

Which one it will be in the 100 units is a mystery, however ....

Interestingly enough, 1 unit will last for 5,000 hours before it finally has its first failure.

Note: These failures have been considered where the failure rate was exponential. There are other failure rates that are Weibull, lognormal, and more.

B.3.1 Comparing Repairable to Nonrepairable Systems

If the system is comprised of nonrepairable systems, the number of failures in one MTBF period is lower.

In repairable systems, when a unit fails it is quickly repaired and placed back into service. If the unit cannot be repaired, then when 1 unit fails in 100 systems, there will be 99 units operating after the first failure. Then 98, 97, and so on until they all eventually fail. In 100 systems that are nonrepairable, there will be 67 units that will fail in one MTBF time period. If the MTBF were 10,000 hours, then two MTBFs would be 20,000 hours. The comparison between repairable systems and nonrepairable systems is shown in Table B.3.

Table B.3 Repairable versus nonrepairable systems still operating (in MTBF time units).

	MTBF time periods					
	1 MTBF		2 MTBFs		3 MTBFs	
	# Fails	# Still operating	# Fails	# Still operating	# Fails	# Still operating
Repairable systems	100	100	100	100	100	100
Nonrepairable systems	63	37	86	14	95	5

Index

Improving Product Reliability and Software Quality: Strategies, Tools, Process and Implementation,
Second Edition. Mark A. Levin, Ted T. Kalal and Jonathan Rodin.
© 2019 John Wiley & Sons Ltd. Published 2019 by John Wiley & Sons Ltd.